Walter Gander • Jiří Hřel

Solving Problems in Scientific Computing Using Maple and MATLAB®

Third, Expanded and Revised Edition 1997

With 161 Figures and 12 Tables

Springer

Walter Gander
Institute of Scientific Computing
ETH Zürich
CH-8092 Zürich, Switzerland
e-mail: gander@inf.ethz.ch

Jiří Hřebíček
Department of Information Technology
Faculty of Informatics
Masaryk University of Brno
Botanická 68a
CZ-60200 Brno, Czech Republic
e-mail: hrebicek@informatics.muni.cz

Library of Congress Cataloging-in-Publication Data

Gander, Walter.
    Solving problems in scientific computing using Maple and MATLAB /
Walter Gander, Jiří Hřebíček. -- 3rd, expanded and rev. ed.
        p.   cm.
    Includes bibliographical references and index.
    ISBN 3-540-61793-0 (pbk.)
    1. Maple (Computer file) 2. MATLAB. 3. Science--Data processing.
I. Hřebíček, Jiří, 1947-   . II. Title.
Q183.9.G36 1997
530'.0285'53--dc21   97-20204
                                                                       CIP

The cover picture shows a plane fitted by least squares to given points (see Chapter 6)

Mathematics Subject Classification (1991): 00A35, 08-04, 65Y99, 68Q40 68N15

ISBN 3-540-61793-0 Springer-Verlag Berlin Heidelberg New York

ISBN 3-540-58746-2 2nd edition Springer-Verlag Berlin Heidelberg New York

Typesetting: Camera-ready copy from the authors
SPIN 10544315   41/3143-5 4 3 2 1 0 – Printed on acid-free paper

Preface

Modern computing tools like MAPLE (a symbolic computation package) and MATLAB® (a numeric and symbolic computation and visualization program) make it possible to use the techniques of scientific computing to solve realistic nontrivial problems in a classroom setting. These problems have been traditionally avoided, since the amount of work required to obtain a solution exceeded the classroom time available and the capabilities of the students. Therefore, simplified and linearized models are often used. This situation has changed, and students can be taught with real-life problems which can be solved by the powerful software tools available. This book is a collection of interesting problems which illustrate some solution techniques in Scientific Computing. The solution technique for each problem is discussed and demonstrated through the use of either MAPLE or MATLAB. The problems are presented in a way such that a reader can easily extend the techniques to even more difficult problems.

This book is intended for students of engineering and scientific computing. It is not an introduction to MAPLE and MATLAB. Instead, it teaches problem solving techniques through the use of examples, which are difficult real-life problems. Please review the MAPLE and MATLAB documentation for questions on how to use the software.

All figures in the book were created either by using graphic commands of MAPLE and MATLAB or by direct use of *xfig* on a SUN workstation. Occasionally changes were made by Dr. S. Bartoň in the postscript files to improve the visual quality of the figures. These changes include different font sizes, line types, line thicknesses, as well as additional comments and labels.

This book was written as a collaboration between three institutes:

- the Department of Theoretical Physics and Astrophysics of Masaryk University, Brno, Czech Republic,

- the Institute of Physics of the University of Agriculture and Forestry, Brno, Czech Republic, and

- the Institute of Scientific Computing ETH, Zürich, Switzerland.

The authors are indebted to the Swiss National Science Foundation which stimulated this collaboration through a grant from the "Oststaaten-Soforthilfeprogramm". An additional grant from the ETH "Sonderprogramm Ostkontakte" and support from the Computer Science Department of ETH Zürich made it possible for Dr. S. Bartoň to spend a year in Zürich. He was the communication link between the two groups of authors and without him, the book would not have been produced on time. We would also like to thank Dr. L. Badoux, Austauschdienst ETH, and Prof. C.A. Zehnder, chairman of the Computer Science Department, for their interest and support.

Making our Swiss- and Czech-English understandable and correct was a major problem in producing this book. This was accomplished through an internal refereeing and proofreading process which greatly improved the quality of all articles. We had great help from Dr. Kevin Gates, Martha Gonnet, Michael Oettli, Prof. S. Leon, Prof. T. Casavant and Prof. B. Gragg during this process. We thank them all for their efforts to improve our language.

Dr. U. von Matt wrote the LaTeX style file to generate the layout of our book in order to meet the requirements of the publisher. We are all very thankful for his excellent work.

D. Gruntz, our MAPLE expert, gave valuable comments to all the authors and greatly improved the quality of the programs. We wish to thank him for his assistance.

The programs were written using MAPLE V Release 2 and MATLAB 4.1. For MAPLE output we used the ASCII interface instead of the nicer XMAPLE environment. This way it was easier to incorporate MAPLE output in the book. The programs are available in machine readable form. We are thankful to The MathWorks for helping us to distribute the software.

Zürich, September 13, 1993 — Walter Gander, Jiří Hřebíček

Preface to the second edition

The first edition of this book has been very well received by the community, and this has made it necessary to write a second edition within one year. We added the new chapters 20 and 21 and we expanded chapters 15 and 17. Some typographical errors were corrected, and we also rephrased some text. By doing so we hope to have improved our English language.

All programs were adapted to the newest versions of the software i.e. to MAPLE V Release 3 and to MATLAB Version v4. In order to simplify the production of the book we again chose the *pretty print* output mode for the MAPLE output.

We dedicate the second edition to our late colleague František Klvaňa. We all mourn for our friend, a lovely, friendly, modest person and a great scientist.

Druhé vydání je věnováno památce našeho zesnulého kolegy Františka Klvani. Všichni vzpomínáme na našeho drahého přítele, milého a skromného člověka a velkého vědce.

Zürich, October 7, 1994   Walter Gander, Jiří Hřebíček

Preface to the third edition

In the present edition the book has been enlarged by six new chapters (Chapters 22–27). Some of the previous chapters were revised: a new way to solve a system of differential equations was added to Chapter 1. Chapter 17 on free metal compression was completely rewritten. With the new approach, the compression of more general bodies can be simulated.

The index has been considerably enlarged and split into three parts, two of them containing all MAPLE and MATLAB commands used in this book. We are indebted to Rolf Strebel for this work.

All chapters have been adapted to the newest versions of MAPLE (Version 5 Release 4) and MATLAB 5. The calculations for MAPLE were done on Unix workstations by Standa Bartoň and Rolf Strebel, who also produced the worksheets. Notice that the order of the terms in sums and products and the order of the elements in sets is unspecified and may change from session to session. When the MAPLE commands are re-executed, one may get results in a different representation than those printed in the book. For example, the solution of a set of equations may depend on different free parameters. Commands which depend on the order of previous results (like accesses to sets and expression sequences) may have to be adjusted accordingly. Since we have re-executed the MAPLE examples with Release 4, some statements have changed compared to the previous editions of this book.

All MATLAB computations were performed on a PC, equipped with an Intel Pentium Pro Processor running under Windows NT 4.0 at 200 Mhz using MATLAB 5.0. We are indebted to Leonhard Jaschke for taking care of these test runs. MATLAB 5 offers new M-files for the integration of differential equations. While in the older versions one had to specify an interval for the independent variable, there are now new possibilities to stop the integration process. We have made use of this new feature and simplified our codes.

A criticism by some reviewers that the ASCII output of MAPLE does not look nice has been taken into consideration. We have transformed all the formulas using the MAPLE `latex`-command into LATEX. We thank Erwin Achermann who checked and adapted the layout.

The two systems MAPLE and MATLAB seem to come closer to each other. There is the Symbolic Math Toolbox for MATLAB which can be used to call MAPLE from a MATLAB program. Also, there are plans that in the near future a similar mechanism will be available on the other side. We have not made use of the Symbolic Math Toolbox, mainly because we do want to use both systems equivalently and complementarily.

The MATLAB and MAPLE programs (and worksheets) are available via anonymous ftp from `ftp.inf.ethz.ch`[1]

We dedicate this edition to one of our co-authors—the one with the highest seniority—Professor Heinz Schilt, the expert in Switzerland for computing and constructing sun dials with typical Swiss precision.

Zürich, March 18, 1997 Walter Gander, Jiří Hřebíček

---

[1]URL: `ftp://ftp.inf.ethz.ch/pub/software/SolvingProblems/ed3/`

List of Authors

Stanislav Bartoň   Institute of Physics and Electronics
Mendel Univ. of Agriculture and Forestry Brno
Zemědělská 1
CZ-613 00 Brno, Czech Republic
barton@vszbr.cz

Jaroslav Buchar   Institute of Physics and Electronics
Mendel Univ. of Agriculture and Forestry Brno
Zemědělská 1
CZ-613 00 Brno, Czech Republic

Ivan Daler   Air Traffic Control Research Department
Smetanova 19
CZ-602 00 Brno, Czech Republic

Walter Gander   Institute of Scientific Computing
ETH Zürich
CH-8092 Zürich, Switzerland
gander@inf.ethz.ch

Walter Gautschi   Department of Computer Sciences
Purdue University
West Lafayette, IN 47907-1398, USA
wxg@cs.purdue.edu

Gaston Gonnet   Institute of Scientific Computing
ETH Zürich
CH-8092 Zürich, Switzerland
gonnet@inf.ethz.ch

Dominik Gruntz   Oberon Microsystems, Inc.
Technopark Zürich
Technoparkstr. 1
CH-8005 Zürich, Switzerland
gruntz@oberon.ch

Jürgen Halin Institute of Energy Technology
ETH Zürich
CH-8092 Zürich, Switzerland
halin@iet.ethz.ch

Jiří Hřebíček Faculty of Informatics
Masaryk University Brno
Botanická 68a
CZ-602 00 Brno, Czech Republic
hrebicek@informatics.muni.cz

Leonhard Jaschke Institute of Scientific Computing
ETH Zürich
CH-8092 Zürich, Switzerland
jaschke@inf.ethz.ch

František Klvaňa †

Urs von Matt ISE Integrated Systems Engineering AG
Technopark Zürich
Technoparkstr. 1
CH-8005 Zürich, Switzerland
vonmatt@ise.ch

Michael H. Oettli Department of Computer Science
University of Minnesota
200 Union Street SE
Minneapolis, MN 55455, USA
na.moettli@na-net.ornl.gov

Tomáš Pitner Faculty of Informatics
Masaryk University Brno
Botanická 68a
CZ-602 00 Brno, Czech Republic
tomp@informatics.muni.cz

Heinz Schilt Dorfstrasse 24
CH-3506 Grosshöchstetten, Switzerland

Rolf Strebel Institute of Scientific Computing
ETH Zürich
CH-8092 Zürich, Switzerland
strebel@inf.ethz.ch

Jörg Waldvogel Seminar of Applied Mathematics
ETH Zürich
CH-8092 Zürich, Switzerland
waldvoge@math.ethz.ch

Contents

Chapter 1. The Tractrix and Similar Curves

W. Gander, S. Bartoň, and J. Hřebíček

1.1 Introduction

In this section we will use MATLAB to solve two similar systems of differential equations. First we generalize the classical tractrix problem to compute the orbit of a toy pulled by a child, and then we compute the orbit of a dog which attacks a jogger. We also show how the motions may be visualized with MATLAB.

1.2 The Classical Tractrix

In the 17th century Gottfried Wilhelm Leibniz discussed the following problem, see [2, 1]. *Given a watch attached to a chain, what is the orbit in the plane described by the watch as the endpoint of the chain is pulled along a straight line?*

Let a be the length of the chain. The problem is easily solved if we assume that the point-like watch is initially on the x-axis at the point $(a, 0)$, and that starting at the origin we pull in the direction of the positive y-axis, [2], (cf. Figure 1.1).

From Figure 1.1 we immediately obtain the following differential equation for the unknown function $y(x)$:

$$y' = -\frac{\sqrt{a^2 - x^2}}{x}. \tag{1.1}$$

To solve Equation (1.1) we only need to integrate:

```
> assume(a>=0);
> y:= -int(sqrt(a^2-x^2)/x,x)+c;
```

$$y := -\sqrt{a^{\tilde{}2} - x^2} + a^{\tilde{}} \operatorname{arctanh}(\frac{a^{\tilde{}}}{\sqrt{a^{\tilde{}2} - x^2}}) + c \tag{1.2}$$

MAPLE does not include the constant when performing indefinite integration. So we added an integration constant c. We can determine its value by using $\lim_{x \to a} y(x) = 0$:

```
> c:= solve(limit(y,x=a),c);
```

$$c := \frac{1}{2} I a^{\tilde{}} \pi$$

FIGURE 1.1. *Classical Tractrix.*

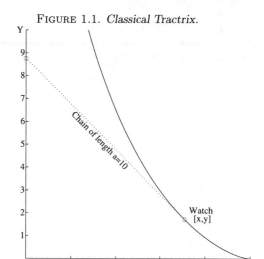

With MAPLE V Release 3 we used to obtain directly a real expression from `int`. We did not not need to compute a complex integration constant because for the real expression the constant turned out to be zero.

```
> yold:= -sqrt(a^2-x^2)+a*arctanh(sqrt(a^2-x^2)/a);
```

$$yold := -\sqrt{a^{\tilde{}2} - x^2} + a^{\tilde{}} \operatorname{arctanh}(\frac{\sqrt{a^{\tilde{}2} - x^2}}{a^{\tilde{}}}) \qquad (1.3)$$

To show that both results are equivalent, we have to subtract them, convert the result into a exponential-logarithmic representation, simplify it and expand it

```
> e:= expand(simplify(convert((y-yold)/a,expln)));
```

$$e := -\frac{1}{2} \ln(-a^{\tilde{}} + \sqrt{a^{\tilde{}2} - x^2}) + \frac{1}{2} I \pi + \frac{1}{2} \ln(-\sqrt{a^{\tilde{}2} - x^2} + a^{\tilde{}})$$

It is not possible to simplify this expression any further in MAPLE. But we should notice that the argument of the first logarithm is the negative argument of the second one. Therefore if we substitute $a - \sqrt{a^2 - x^2}$ by a positive unknown d and similarly substitute $\sqrt{a^2 - x^2} - a$ by $-d$, MAPLE is indeed able to simplify this expression:

```
> assume(d>0);
> simplify(subs({(a^2-x^2)^(1/2)-a=-d,-(a^2-x^2)^(1/2)+a=d},e));
```

$$0$$

Let us now assume that the object to be pulled is initially on the y-axis at the point $(0, a)$ and that we start pulling again at the origin, but this time in the direction of the positive x-axis.

Consider the point $(x, y(x))$ on the orbit of the object. The endpoint of the chain on the x-axis is at the point $(x - y(x)/y'(x), 0)$, that is where the tangent intersects the x-axis (this is the same point which would be obtained for one step of Newton's iteration!). Therefore, the condition that the chain has the constant length a, leads to the differential equation

$$\frac{y(x)^2}{y'(x)^2} + y(x)^2 = a^2, \tag{1.4}$$

which can no longer be solved directly by quadrature. Therefore we need to call the differential equation solver `dsolve`,

```
> restart;
> assume(a>0);
> eq := (y(x)/diff(y(x), x))^2 + y(x)^2 = a^2;
```

$$eq := \frac{y(x)^2}{(\frac{\partial}{\partial x} y(x))^2} + y(x)^2 = a^{\sim 2}$$

```
> p:=dsolve(eq,y(x));
```

$$p := \sqrt{-y(x)^2 + a^{\sim 2}} - a^\sim \operatorname{arctanh}(\frac{a^\sim}{\sqrt{-y(x)^2 + a^{\sim 2}}}) + x = \_C1,$$

$$-\sqrt{-y(x)^2 + a^{\sim 2}} + a^\sim \operatorname{arctanh}(\frac{a^\sim}{\sqrt{-y(x)^2 + a^{\sim 2}}}) + x = \_C1$$

and we obtain two solutions. From the initial condition $y(0) = a$ and from physics it follows for $a > 0$ that $y(x) > 0$ and $y'(x) < 0$. Thus the first solution is the correct answer to our problem:

```
> p[1];
```

$$\sqrt{-y(x)^2 + a^{\sim 2}} - a^\sim \operatorname{arctanh}(\frac{a^\sim}{\sqrt{-y(x)^2 + a^{\sim 2}}}) + x = \_C1$$

We obtain the solution $y(x)$ in implicit form. To find the integration constant `_C1` we have to use the constraint $\lim_{x \to 0} y(x) = a$

```
> _C1:= limit(lhs(subs({x=0,y(x)=y},p[1])),y=a);
```

$$\_C1 := \frac{1}{2} I a^\sim \pi$$

Therefore the solution satisfies the equation

$$\sqrt{-y(x)^2 + a^{\sim 2}} - a^\sim \operatorname{arctanh}(\frac{a^\sim}{\sqrt{-y(x)^2 + a^{\sim 2}}}) + x = \frac{1}{2} I a^\sim \pi$$

Again with MAPLE V Release 3 we obtained the real expression

$$\sqrt{a^2 - y(x)^2} - a \operatorname{arctanh}\left(\frac{\sqrt{a^2 - y(x)^2}}{a}\right) + x = 0$$

We could, of course, have obtained these equations also by interchanging the variables x and y in Equation (1.2) and (1.3). Note that it would be difficult to solve Equation (1.4) numerically, since for $x = 0$ we have the singularity $y'(0) = \infty$.

1.3   The Child and the Toy

Let us now solve a more general problem and suppose that a child is walking on the plane along a curve given by the two functions of time $X(t)$ and $Y(t)$.

Suppose now that the child is pulling or pushing some toy, by means of a rigid bar of length a. We are interested in computing the orbit of the toy when the child is walking around. Let $(x(t), y(t))$ be the position of the toy. From

FIGURE 1.2. *Velocities* $\mathbf{v}_C$ *and* $\mathbf{v}_T$.

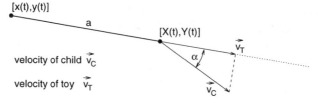

Figure 1.2 the following equations are obtained:

1. The distance between the points $(X(t), Y(t))$ and $(x(t), y(t))$ is always the length of the bar. Therefore

$$(X - x)^2 + (Y - y)^2 = a^2. \tag{1.5}$$

2. The toy is always moving in the direction of the bar. Therefore the difference vector of the two positions is a multiple of the velocity vector of the toy, $\mathbf{v}_T = (\dot{x}, \dot{y})^T$:

$$\begin{pmatrix} X - x \\ Y - y \end{pmatrix} = \lambda \begin{pmatrix} \dot{x} \\ \dot{y} \end{pmatrix} \quad \text{with} \quad \lambda > 0. \tag{1.6}$$

3. The speed of the toy depends on the direction of the velocity vector $\mathbf{v_C}$ of the child. Assume, e.g., that the child is walking on a circle of radius a (length of the bar). In this special case the toy will stay at the center of the circle and will not move at all (this is the final state of the first numerical example, see Figure 1.3).

   From Figure 1.2 we see that *the modulus of the velocity* $\mathbf{v_T}$ *of the toy is given by the modulus of the projection of the velocity* $\mathbf{v_C}$ *of the child onto the bar.*

Inserting Equation (1.6) into Equation (1.5), we obtain

$$a^2 = \lambda^2(\dot{x}^2 + \dot{y}^2) \quad \longrightarrow \quad \lambda = \frac{a}{\sqrt{\dot{x}^2 + \dot{y}^2}}.$$

Therefore

$$\frac{a}{\sqrt{\dot{x}^2 + \dot{y}^2}} \begin{pmatrix} \dot{x} \\ \dot{y} \end{pmatrix} = \begin{pmatrix} X - x \\ Y - y \end{pmatrix}. \tag{1.7}$$

We would like to solve Equation (1.7) for $\dot{x}$ and $\dot{y}$. Since we know the modulus of the velocity vector of the toy $|\mathbf{v}_T| = |\mathbf{v}_C| \cos \alpha$, see Figure 1.2, this can be done by the following steps:

- Normalize the difference vector $(X - x, Y - y)^T$ and obtain a vector $\mathbf{w}$ of unit length.

- Determine the projection of $\mathbf{v}_C = (\dot{X}, \dot{Y})^T$ onto the subspace generated by $\mathbf{w}$. This is simply the scalar product $\mathbf{v}_C^T \mathbf{w}$, since $\mathbf{v}_C^T \mathbf{w} = |\mathbf{v}_C||\mathbf{w}| \cos \alpha$ and $|\mathbf{w}| = 1$.

- $\mathbf{v}_T = (\dot{x}, \dot{y})^T = (\mathbf{v}_C^T \mathbf{w}) \mathbf{w}$.

Now we can write the function to evaluate the system of differential equations in MATLAB.

ALGORITHM 1.1. Function f.

```
function zs = f(t,z)
%
[X Xs Y Ys] = child(t);
v =[Xs; Ys];
w =[X-z(1); Y-z(2)];
w = w/norm(w);
zs = (v'*w)*w;
```

The function `f` calls the function `child` which returns the position $(X(t), Y(t))$ and velocity of the child $(Xs(t), Ys(t))$ for a given time `t`. As an example consider a child walking on the circle $X(t) = 5 \cos t; Y(t) = 5 \sin t$. The corresponding function `child` for this case is:

ALGORITHM 1.2. Function Child.

```
function [X, Xs, Y, Ys] = child(t);
%
X  = 5*cos(t);  Y  =  5*sin(t);
Xs = -5*sin(t);  Ys =  5*cos(t);
```

MATLAB offers two M-files `ode23` and `ode45` to integrate differential equations. In the following main program we will call one of these functions and also define the initial conditions (Note that for $t = 0$ the child is at the point $(5, 0)$ and the toy at $(10, 0)$):

```
>> % main1.m
>> y0 = [10 0]';
>> [t y] = ode45('f',[0 100],y0);
>> clf; hold on;
>> axis([-6 10 -6 10]);
>> axis('square');
>> plot(y(:,1),y(:,2));
```

FIGURE 1.3. *Child Walks on the Circle.*

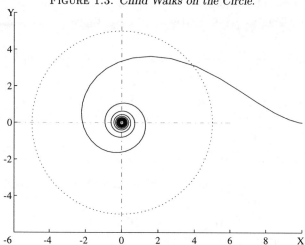

If we plot the two columns of y we obtain the orbit of the toy (cf. Figure 1.3). Furthermore we add the curve of the child in the same plot with the statements:

```
>> t = 0:0.05:6.3
>> [X, Xs, Y, Ys] = child(t);
>> plot(X,Y,':')
>> hold off;
```

Note that the length of the bar a does not appear explicitly in the programs; *it is defined implicitly by the position of the toy, (initial condition), and the position of the child (function* child*) for* $t = 0$.

We conclude this section with some more examples. Let the child be walking along the graph of a sine function: $X(t) = t$ and $Y(t) = 5\sin t$. The child's curve is again plotted with a dotted line. With the initial conditions $x(0) = 0$ and $y(0) = 10$ we obtain Figure 1.4.

In the next example, the child is again walking on the circle $X(t) = 5\cos t$, $Y(t) = 5\sin t$. With the initial condition $x(0) = 0$ and $y(0) = 10$, we obtain a nice flower-like orbit of the toy (cf. Figure 1.5).

1.4   The Jogger and the Dog

We consider the following problem: a jogger is running along his favorite trail on the plane in order to get his daily exercise. Suddenly, he is being attacked by a dog. The dog is running with constant speed w towards the jogger. Compute the orbit of the dog.

The orbit of the dog has the property that the velocity vector of the dog points at every time to its goal, the jogger. We assume that the jogger is running on some trail and that his motion is described by the two functions $X(t)$ and $Y(t)$.

FIGURE 1.4. *Example 2.*

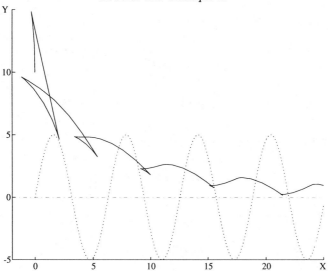

FIGURE 1.5. *Flower Orbit.*

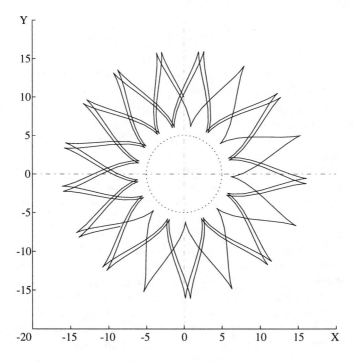

Let us assume that for $t = 0$ the dog is at the point (x_0, y_0), and that at time t his position will be $(x(t), y(t))$. The following equations hold:

1. $\dot{x}^2 + \dot{y}^2 = w^2$: The dog is running with constant speed.

2. The velocity vector of the dog is parallel to the difference vector between the position of the jogger and the dog:

$$\begin{pmatrix} \dot{x} \\ \dot{y} \end{pmatrix} = \lambda \begin{pmatrix} X - x \\ Y - y \end{pmatrix} \quad \text{with } \lambda > 0.$$

If we substitute this in the first equation we obtain

$$w^2 = \dot{x}^2 + \dot{y}^2 = \lambda^2 \left\| \begin{pmatrix} X - x \\ Y - y \end{pmatrix} \right\|^2.$$

This equation can be solved for λ:

$$\lambda = \frac{w}{\left\| \begin{pmatrix} X-x \\ Y-y \end{pmatrix} \right\|} > 0.$$

Finally, substitution of this expression for λ in the second equation yields the differential equation of the orbit of the dog:

$$\begin{pmatrix} \dot{x} \\ \dot{y} \end{pmatrix} = \frac{w}{\left\| \begin{pmatrix} X-x \\ Y-y \end{pmatrix} \right\|} \begin{pmatrix} X - x \\ Y - y \end{pmatrix}. \tag{1.8}$$

Again we will make use of one of the M-files `ode23.m` or `ode45.m` to integrate the system of differential equations. We notice that the system (1.8) has a singularity when the dog reaches the jogger. In this case the norm of the difference vector becomes zero and we have to stop the integration. The above mentioned MATLAB functions for integrating differential equations require as input an interval of the independent variable. MATLAB 5.0 provides now also the possibility to define another termination criterion for the integration, different from a given upper bound for the independent variable. It is possible to terminate the integration by checking zero crossings of a function. In our example one would like to terminate integration when the dog reaches the jogger, i.e. when $\|(X - x, Y - y)\|$ becomes small. In order to do so we have to add a third input and two more output parameters to the M-function `dog.m`. The integrator `ode23` or `ode45` calls the function in two ways: The first one consists of dropping the third parameter. The function then returns only the parameter `zs`: the speed of the dog. In the second way the keyword `'events'` is assigned to the parameter `flag`. This keyword tells the function to return the zero-crossing function in the first output `zs`. The second output `isterminal` is a logical vector that tells the integrator, which components of the first output force the procedure to stop when they become zero. Every component with this property is marked with a nonzero entry in `isterminal`. The third output parameter `direction` is also a vector that indicates for each component of `zs` if zero crossings shall only be regarded for increasing values (`direction = 1`), decreasing values (`direction = -1`) or in both cases (`direction = 0`). The condition for zero crossings is checked in the integrator. The speed w of the dog must be declared global in `dog` and in the main program. The orbit of the jogger is given by the M-function `jogger.m`.

ALGORITHM 1.3. *Function Dog.*

```
function [zs,isterminal,direction] = dog(t,z,flag);
%
global w   % w = speed of the dog
X= jogger(t);
h= X-z;
nh= norm(h);
if nargin < 3 | isempty(flag) % normal output
   zs= (w/nh)*h;
else
   switch(flag)
   case 'events'   % at norm(h)=0 there is a singularity
      zs= nh-1e-3;   % zero crossing at pos_dog=pos_jogger
      isterminal= 1; % this is a stopping event
      direction= 0;  % don't care if decrease or increase
   otherwise
      error(['Unknown flag: ' flag]);
   end
end
```

The main program `main2.m` defines the initial conditions and calls `ode23` for the integration. We have to provide an upper bound of the time t for the integration.

```
>> % main2.m
>> global w
>> y0 = [60;70];  % initial conditions, starting point of the dog
>> w = 10;        % w  speed of the dog
>> options= odeset('RelTol',1e-5,'Events','on');
>> [t,Y] = ode23('dog',[0,20],y0,options);
>> clf; hold on;
>> axis([-10,100,-10,70]);
>> plot(Y(:,1),Y(:,2));
>> J=[];

>> for h= 1: length(t),
>>    w  = jogger(t(h));
>>    J = [J; w'];
>> end;
>> plot(J(:,1), J(:,2),':');
```

The integration will stop either if the upper bound for the time t is reached or if the dog catches up with the jogger. For the latter case we set the flag Events of the ODE options to 'on'. This tells the integrator to check for zero crossings of the function dog called with flag = 'events'. After the call to ode23 the variable Y contains a table with the values of the two functions $x(t)$ and $y(t)$. We plot the orbit of the dog simply by the statement plot(Y(:,1),Y(:,2)). In order to show also the orbit of the jogger we have to compute it again using the vector t and the function jogger.

Let us now compute a few examples. First we let the jogger run along the x-axis:

ALGORITHM 1.4. *First Jogger Example.*

```
function s = jogger(t);
s   = [8*t; 0];
```

In the above main program we chose the speed of the dog as $w = 10$, and since here we have $X(t) = 8t$ the jogger is slower. As we can see in Figure 1.6 the dog is catching the poor jogger.

If we wish to indicate the position of the jogger's troubles, *(perhaps to build a small memorial)*, we can make use of the following file cross.m

ALGORITHM 1.5. *Drawing a Cross.*

```
function cross(Cx,Cy,v)
% draws at position  Cx,Cy  a cross of height 2.5v
% and width 2*v
Kx = [Cx Cx Cx Cx-v Cx+v];
Ky = [Cy Cy+2.5*v Cy+1.5*v Cy+1.5*v Cy+1.5*v];
plot(Kx,Ky);
plot(Cx,Cy,'o');
```

The cross in the plot was generated by appending the statements

```
>> p = max(size(Y));
>> cross(Y(p,1),Y(p,2),2)
>> hold off;
```

to the main program. The next example shows the situation where the jogger turns around and tries to run back home:

ALGORITHM 1.6. *Second Jogger Example.*

```
function s = jogger1(t);
%
if t<6, s = [8*t; 0];
else    s = [8*(12-t) ;0];
end
```

FIGURE 1.6. *Jogger Running on the Line* $y = 0$.

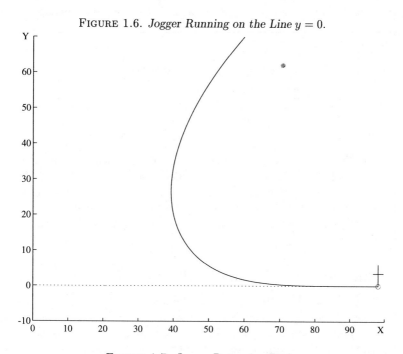

FIGURE 1.7. *Jogger Returning Back.*

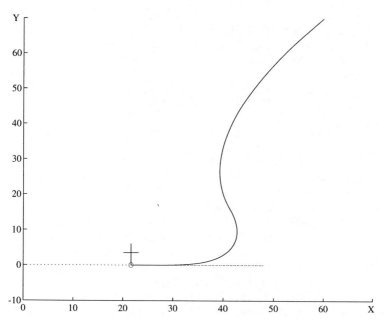

However, using the same main program as before the dog catches up with the jogger at time $t = 9.3$ (cf. Figure 1.7). Let us now consider a faster jogger running on an ellipse

ALGORITHM 1.7. *Third Jogger Example.*

```
function s = jogger2(t);
s   = [  10+20*cos(t)
         20 + 15*sin(t)];
```

If the dog also runs fast ($w = 19$), he manages to reach the jogger at time $t = 8.97$ (cf. Figure 1.8). We finally consider an old, slow dog ($w = 10$). He tries to catch a jogger running on a elliptic track. However, instead of waiting for the jogger somewhere on the ellipse, he runs (too slow) after his target, and we can see a steady state developing where the dog is running on a closed orbit inside the ellipse (cf. Figure 1.9).

1.5   Showing the Motions with MATLAB

It would be nice to show simultaneously the motions of the child and the toy or the dog and the jogger instead of just plotting statically their orbits. This is possible using the *handle graphics* commands in MATLAB. The main program for the child and its toy now looks as follows:

```
>> % main3.m
>> y0 = [0 20]';
>> options= odeset('RelTol',1e-10);
>> [t y] = ode45 ('f', [0 40], y0, options);
>> [X, Xs, Y, Ys] = child (t);

>> xmin = min (min (X), min (y (:, 1)));
>> xmax = max (max (X), max (y (:, 1)));
>> ymin = min (min (Y), min (y (:, 2)));
>> ymax = max (max (Y), max (y (:, 2)));

>> clf; hold on;
>> axis ([xmin xmax ymin ymax]);
>> % axis('equal');
>> title ('The Child and the Toy.');
>> stickhandle = line ('Color', 'yellow', 'EraseMode', 'xor', ...
>>                  'LineStyle', '-', 'XData', [], 'YData', []);

>> for k = 1:length(t)-1,
>>    plot ([X(k), X(k+1)], [Y(k), Y(k+1)], '-', ...
>>          'Color', 'yellow', 'EraseMode', 'none');
>>    plot ([y(k,1), y(k+1,1)], [y(k,2), y(k+1,2)], '-', ...
>>          'Color', 'green', 'EraseMode', 'none');
>>    set (stickhandle, 'XData', [X(k+1), y(k+1,1)], ...
```

FIGURE 1.8. *Jogger on an Ellipse.*

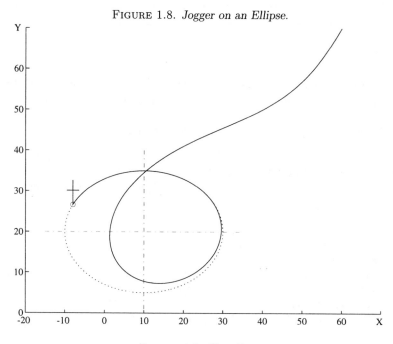

FIGURE 1.9. *Slow Dog.*

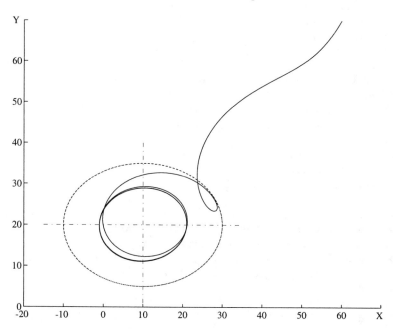

```
>>          'YData', [Y(k+1), y(k+1,2)]);
>> drawnow;
>> end;
>> hold off;
```

We define the variable stickhandle as a handle to a graphical object of type *line* associated with the stick. In the loop, we draw new segments of the child and toy orbits and move the position of the stick. The drawnow command forces these objects to be plotted instantaneously. Therefore, we can watch the two orbits and the stick being plotted simultaneously.

In the case of the jogger and the dog we do not even have to define a handle. All we have to do is to draw the segments of the two orbits in the proper sequence:

```
>> % main4.m
>> global w;
>> y0 = [60; 70];  % initial conditions, starting point of the dog
>> w = 10;           % w  speed of the dog
>> options= odeset('RelTol',1e-5,'Events','on');
>> [t,Y] = ode23 ('dog', [0 20], y0, options);

>> J=[];
>> for h= 1:length(t),
>>    w = jogger(t(h));
>>    J = [J; w'];
>> end

>> xmin = min (min (Y (:, 1)), min (J (:, 1)));
>> xmax = max (max (Y (:, 1)), max (J (:, 1)));
>> ymin = min (min (Y (:, 2)), min (J (:, 2)));
>> ymax = max (max (Y (:, 2)), max (J (:, 2)));
>> clf; hold on;
>> axis ([xmin xmax ymin ymax]);
>> % axis ('equal');
>> title ('The Jogger and the Dog.');

>> for h=1:length(t)-1,
>>    plot ([Y(h,1), Y(h+1,1)] , [Y(h,2), Y(h+1,2)], '-', ...
>>          'Color', 'yellow', 'EraseMode','none');
>>    plot ([J(h,1), J(h+1,1)] , [J(h,2), J(h+1,2)], ':', ...
>>          'Color', 'green', 'EraseMode','none');
>>    drawnow;
>>    pause(1);
>> end
>> hold off;
```

1.6   Jogger with Constant Velocity

We continue in this section with the elliptical orbit of the jogger. If we describe the ellipse as in Algorithm 1.7 and consider t as the time variable the velocity of the jogger will not be constant. Let $s(t)$ be the parameter for the description of the ellipse with center in (m_1, m_2) and main semi-axes a and b:

$$X(s) = m_1 + a\cos(s), \quad Y(s) = m_2 + b\sin(s)$$

We want to determine s as a monotonic increasing function of the time t such that for equidistant times t_i the points on the ellipse $X(s(t_i)), Y(s(t_i))$ are also equidistant on the border of the ellipse. An equivalent condition is that the velocity of the jogger $V(t)$ is constant:

$$V = \sqrt{\left(\frac{d\,X(s(t))}{d\,t}\right)^2 + \left(\frac{d\,Y(s(t))}{d\,t}\right)^2} = \text{const.} \tag{1.9}$$

Now $s(t)$ can be computed by first solving Equation (1.9) for $\mathtt{Ds} = ds/dt$:

```
> Xe := m1 + a*cos(s(t)): Ye := m2 + b*sin(s(t)):
> Ve2 := diff(Xe, t)^2 +diff(Ye, t)^2 =V^2;
```

$$Ve2 := a^2 \sin(s(t))^2 \left(\frac{\partial}{\partial t} s(t)\right)^2 + b^2 \cos(s(t))^2 \left(\frac{\partial}{\partial t} s(t)\right)^2 = V^2$$

```
> Ds := solve(Ve2, diff(s(t), t));
```

$$Ds :=$$
$$\frac{V}{\sqrt{a^2 \sin(s(t))^2 + b^2 - b^2 \sin(s(t))^2}}, \quad -\frac{V}{\sqrt{a^2 \sin(s(t))^2 + b^2 - b^2 \sin(s(t))^2}}$$

We have to select the first equation, because the the jogger moves counterclockwise. The differential equation has no analytical solution so we will solve it numerically with MATLAB. Together with the two differential equations (1.8) we obtain a system of three coupled differential equations:

$$
\begin{aligned}
X(t) &= m_1 + a\cos(s(t)) \\
Y(t) &= m_2 + b\sin(s(t)) \\
\dot{x}(t) &= c(X(t) - x(t)) \\
\dot{y}(t) &= c(Y(t) - y(t)) \\
\dot{s}(t) &= \frac{V}{\sqrt{a^2 \sin(s(t))^2 + b^2 \cos(s(t))^2}} \\
c &= \frac{w}{\left\| \begin{pmatrix} X-x \\ Y-y \end{pmatrix} \right\|}
\end{aligned}
$$

This system is implemented as function fkt (Algorithm 1.8). The corresponding main program is given as Algorithm 1.9.

ALGORITHM 1.8.

Function fkt for the Jogger with Constant Velocity.

```
function [ydot,isterminal,direction] = fkt(t,y,flag)
% system of differential equations
% for the jogger-dog problem
% where the jogger runs with constant
% velocity on an ellipse
global a b m c w
A =  cos(y(3)); B =  sin(y(3));
X = m(1) + a*A; Y = m(2) + b*B;
h = [X;Y] -y(1:2); nh = norm(h);
zs =  (w/nh)*h;

if nargin < 3 | isempty(flag) % normal output
    ydot = [zs;c/sqrt((a*B)^2+(b*A)^2)];
else
    switch(flag)
    case 'events' % at norm(h)=0 there is a singularity
        ydot= nh-(1e-3); % zero crossing at pos_dog=pos_jogger
        isterminal= 1;   % this is a stopping event
        direction= 0;    % don't care if decrease or increase
    otherwise
        error(['Unknown flag ''' flag '''.']);
    end
end
```

To compare the results with the previous computations we choose the constant in Equation (1.9) as the average jogger velocity of the example in Section 1.4:

$$V \equiv \overline{V(t)} = \frac{2}{\pi} \int_0^{\frac{\pi}{2}} V(t)\, dt \equiv \frac{L}{T}\ .$$

```
> T := 2*Pi:
> L := evalf(4*int(sqrt(20^2*sin(f)^2 + 15^2*cos(f)^2),
>         f = 0..Pi/2));
> V := evalf(L/T);
```

$$L := 110.5174608$$

$$V := 17.58940018$$

If we execute Algorithm 1.9 we notice that the dog catches the jogger at time $t = 8.22834$. This is a little earlier than in Section 1.4.

ALGORITHM 1.9.

Main Program for the Jogger with Constant Velocity.

```
>> % main5.m
>> global a b m c w
>> a = 20; b = 15; % semi-axes
>> m = [10;20];
>> c = 17.58940018; % constant velocity of jogger
>> w = 19; % velocity of dog
>> y0 = [60, 70, 0]'; % ini. cond., starting point of the dog

>> options= odeset('AbsTol',1e-5,'Events','on');
>> [t,Y]= ode23 ('fkt', [0 20], y0, options);
>> clf; hold on;
>> axis ([-10 70 -10 70]);
>> % axis ('equal');
>> title ('The Jogger Runs with Constant Velocity.');
>> p = length(t)-1;
>> for h=1:p
>>    plot ([Y(h,1), Y(h+1,1)] , [Y(h,2), Y(h+1,2)], '-', ...
>>          'Color', 'yellow', 'EraseMode','none');
>>    s1 = Y(h,3); s2 = Y(h+1,3);
>>    X1 = m(1) + a*cos(s1); Y1 = m(2) + b*sin(s1);
>>    X2 = m(1) + a*cos(s2); Y2 = m(2) + b*sin(s2);
>>    plot ([X1, X2] , [Y1, Y2], ':', ...
>>          'Color', 'green', 'EraseMode','none');
>>    drawnow;
>> end;
>> cross(Y(p,1),Y(p,2),2);
>> hold off;
```

1.7 Using a Moving Coordinate System

In this section we will use a Cartesian coordinate system to describe the position of the child respectively the jogger. The position of the toy respectively the dog will be described in a moving polar coordinate system, see Figure 1.10. The origin of the moving system is the current position of the child respectively jogger. The current positions of the toy or the dog are thus given by the distance $\rho(t)$ and the polar angle $\phi(t)$.

If $[X(t), Y(t)]$ describe the current child/toy position then the position of the jogger/dog in Cartesian coordinates $[x(t), y(t)]$ is

$$x(t) = X(t) + \rho(t)\cos(\phi(t)), \quad y(t) = Y(t) + \rho(t)\sin(\phi(t)). \qquad (1.10)$$

We want to express the system of differential equations (1.7) respectively (1.8) in the new variables $\rho(t)$ and $\phi(t)$. By doing so we will obtain a transformed system of differential equations for the functions $\rho(t)$ and $\phi(t)$.

FIGURE 1.10. *The moving Coordinate System,* $[\rho(t), \phi(t)]$

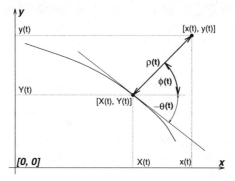

It is interesting to note that both systems are special cases of

$$\begin{pmatrix} \dot{x} \\ \dot{y} \end{pmatrix} = \frac{\sqrt{\dot{x}^2 + \dot{y}^2}}{\left\| \begin{pmatrix} X - x \\ Y - y \end{pmatrix} \right\|} \begin{pmatrix} X - x \\ Y - y \end{pmatrix}. \tag{1.11}$$

If we keep the velocity constant $\sqrt{\dot{x}^2 + \dot{y}^2} = w = const.$ then the system (1.11) describes the jogger/dog problem. We obtain on the other hand the equations for the child/toy problem if the distance is constant:

$$\rho(t) = \left\| \begin{pmatrix} X - x \\ Y - y \end{pmatrix} \right\| = a = const.$$

Since for the child/toy problem $\rho(t) = const.$, we expect that in polar coordinates the system will simplify to only one differential equation for $\phi(t)$. We will make this transformation with MAPLE.

1.7.1 Transformation for Jogger/Dog

We begin by defining the system (1.8):

```
> restart;
> S(t) := sqrt((X(t) - x(t))^2 + (Y(t) - y(t))^2):
> rx := diff(x(t), t) = W*(X(t) - x(t))/S(t);
> ry := diff(y(t), t) = W*(Y(t) - y(t))/S(t);
```

$$rx := \frac{\partial}{\partial t} x(t) = \frac{W\,(X(t) - x(t))}{\sqrt{(X(t) - x(t))^2 + (Y(t) - y(t))^2}}$$

$$ry := \frac{\partial}{\partial t} y(t) = \frac{W\,(Y(t) - y(t))}{\sqrt{(X(t) - x(t))^2 + (Y(t) - y(t))^2}}$$

Now we introduce the transformation of the functions and their derivatives:

```
> Rx := x(t) = X(t) + rho(t)*cos(phi(t)):
> Ry := y(t) = Y(t) + rho(t)*sin(phi(t)):
> Vx := diff(Rx, t);  Vy := diff(Ry, t);
```

$$Vx := \frac{\partial}{\partial t} x(t) = (\frac{\partial}{\partial t} X(t)) + (\frac{\partial}{\partial t} \rho(t)) \cos(\phi(t)) - \rho(t) \sin(\phi(t)) (\frac{\partial}{\partial t} \phi(t))$$

$$Vy := \frac{\partial}{\partial t} y(t) = (\frac{\partial}{\partial t} Y(t)) + (\frac{\partial}{\partial t} \rho(t)) \sin(\phi(t)) + \rho(t) \cos(\phi(t)) (\frac{\partial}{\partial t} \phi(t))$$

We substitute and simplify the result:

```
> qx := subs(Vx, rx):   qy := subs(Vy, ry):
> qx := simplify(subs(Rx, Ry, qx), symbolic);
> qy := simplify(subs(Rx, Ry, qy), symbolic);
```

$qx :=$

$$(\frac{\partial}{\partial t} X(t)) + (\frac{\partial}{\partial t} \rho(t)) \cos(\phi(t)) - \rho(t) \sin(\phi(t)) (\frac{\partial}{\partial t} \phi(t)) = -W \cos(\phi(t))$$

$qy :=$

$$(\frac{\partial}{\partial t} Y(t)) + (\frac{\partial}{\partial t} \rho(t)) \sin(\phi(t)) + \rho(t) \cos(\phi(t)) (\frac{\partial}{\partial t} \phi(t)) = -W \sin(\phi(t))$$

Finally we solve for the derivatives of $\rho(t)$ and $\phi(t)$:

```
> Dsys := solve({qx, qy}, {diff(rho(t), t), diff(phi(t), t)});
```

$$Dsys := \left\{ \frac{\partial}{\partial t} \phi(t) = \frac{-\cos(\phi(t)) (\frac{\partial}{\partial t} Y(t)) + (\frac{\partial}{\partial t} X(t)) \sin(\phi(t))}{\rho(t)}, \right.$$

$$\left. \frac{\partial}{\partial t} \rho(t) = -(\frac{\partial}{\partial t} Y(t)) \sin(\phi(t)) - \cos(\phi(t)) (\frac{\partial}{\partial t} X(t)) - W \right\}$$

```
> Dradial := Dsys[2]; Daxial := Dsys[1];
```

$$Dradial := \frac{\partial}{\partial t} \rho(t) = -(\frac{\partial}{\partial t} Y(t)) \sin(\phi(t)) - \cos(\phi(t)) (\frac{\partial}{\partial t} X(t)) - W \quad (1.12)$$

$$Daxial := \frac{\partial}{\partial t} \phi(t) = \frac{-\cos(\phi(t)) (\frac{\partial}{\partial t} Y(t)) + (\frac{\partial}{\partial t} X(t)) \sin(\phi(t))}{\rho(t)} \quad (1.13)$$

The resulting system of differential equations is somewhat simpler but cannot be solved analytically. We will therefore not continue the discussion.

1.7.2   Transformation for Child/Toy

As before we define the system of differential equations:
```
> Rx := x(t) = X(t) + a*cos(phi(t)):
> Ry := y(t) = Y(t) + a*sin(phi(t)):
> Vx := diff(Rx, t): Vy := diff(Ry, t):
```
We will introduce the following substitution RW for the velocity of the toy:
```
> RW := W(t) = subs(Vx, Vy, sqrt(diff(x(t), t)^2
>              + diff(y(t), t)^2));
```

$$RW := W(t) =$$
$$\sqrt{((\frac{\partial}{\partial t}X(t)) - a\sin(\phi(t))(\frac{\partial}{\partial t}\phi(t)))^2 + ((\frac{\partial}{\partial t}Y(t)) + a\cos(\phi(t))(\frac{\partial}{\partial t}\phi(t)))^2}$$

The following statements generate the system (1.7)
```
> qx := diff(x(t), t) = W(t)/a*(X(t) - x(t));
> qy := diff(y(t), t) = W(t)/a*(Y(t) - y(t));
```

$$qx := \frac{\partial}{\partial t}x(t) = \frac{W(t)(X(t) - x(t))}{a}$$
$$qy := \frac{\partial}{\partial t}y(t) = \frac{W(t)(Y(t) - y(t))}{a}$$

We introduce the new variables and the derivatives by eliminating $x(t)$ and $y(t)$. Furthermore we squared both equations to get rid of the square root.
```
> qx := subs(RW, Vx, Rx, map(u -> u^2, qx));
> qy := subs(RW, Vy, Ry, map(u -> u^2, qy));
```

$$qx := ((\frac{\partial}{\partial t}X(t)) - a\sin(\phi(t))(\frac{\partial}{\partial t}\phi(t)))^2 =$$
$$(((\frac{\partial}{\partial t}X(t)) - a\sin(\phi(t))(\frac{\partial}{\partial t}\phi(t)))^2 + ((\frac{\partial}{\partial t}Y(t)) + a\cos(\phi(t))(\frac{\partial}{\partial t}\phi(t)))^2)$$
$$\cos(\phi(t))^2$$

$$qy := ((\frac{\partial}{\partial t}Y(t)) + a\cos(\phi(t))(\frac{\partial}{\partial t}\phi(t)))^2 =$$
$$(((\frac{\partial}{\partial t}X(t)) - a\sin(\phi(t))(\frac{\partial}{\partial t}\phi(t)))^2 + ((\frac{\partial}{\partial t}Y(t)) + a\cos(\phi(t))(\frac{\partial}{\partial t}\phi(t)))^2)$$
$$\sin(\phi(t))^2$$

We want to show that both equations are the same, and that the system reduces to only one differential equation for $\phi(t)$. To do this we make the following substitution:
```
> Subst := [rhs(Vx) = A, rhs(Vy) = B, phi(t) = F];
```

$$Subst := [(\frac{\partial}{\partial t}X(t)) - a\sin(\phi(t))(\frac{\partial}{\partial t}\phi(t)) = A,$$
$$(\frac{\partial}{\partial t}Y(t)) + a\cos(\phi(t))(\frac{\partial}{\partial t}\phi(t)) = B, \phi(t) = F]$$

```
> q1x := expand(subs(Subst, qx));
> q1y := expand(subs(Subst, qy));
```

$$q1x := A^2 = \cos(F)^2\,A^2 + \cos(F)^2\,B^2$$

$$q1y := B^2 = \sin(F)^2\,A^2 + \sin(F)^2\,B^2$$

In order to see that indeed the equations are the same we simplify them by collecting A and B terms.

```
> q1x := A^2*sin(F)^2 = B^2*cos(F)^2;
> q1y := B^2*cos(F)^2 = A^2*sin(F)^2;
```

$$q1x := \sin(F)^2\,A^2 = \cos(F)^2\,B^2$$

$$q1y := \cos(F)^2\,B^2 = \sin(F)^2\,A^2$$

Now we see that they are identical. We continue the computation with the first one and remove the squares and the substitution. By solving for the derivative of $\phi(t)$ we obtain the desired differential equation:

```
> Q1 := map(u -> simplify(sqrt(u), symbolic), q1x);
```

$$Q1 := \sin(F)\,A = \cos(F)\,B$$

```
> BackSubst := [seq(rhs(op(i, Subst)) =
>                    lhs(op(i, Subst)), i = 1..nops(Subst))];
```

$$BackSubst := [A = (\frac{\partial}{\partial t}\,X(t)) - a\sin(\phi(t))\,(\frac{\partial}{\partial t}\,\phi(t)),$$
$$B = (\frac{\partial}{\partial t}\,Y(t)) + a\cos(\phi(t))\,(\frac{\partial}{\partial t}\,\phi(t)),\; F = \phi(t)]$$

```
> Q2 := subs(BackSubst, Q1);
```

$$Q2 := \sin(\phi(t))\,((\frac{\partial}{\partial t}\,X(t)) - a\sin(\phi(t))\,(\frac{\partial}{\partial t}\,\phi(t))) =$$
$$\cos(\phi(t))\,((\frac{\partial}{\partial t}\,Y(t)) + a\cos(\phi(t))\,(\frac{\partial}{\partial t}\,\phi(t)))$$

```
> Daxial := diff(phi(t), t) =
>                 simplify(solve(Q2, diff(phi(t), t)));
```

$$Daxial := \frac{\partial}{\partial t}\,\phi(t) = \frac{\sin(\phi(t))\,(\frac{\partial}{\partial t}X(t)) - \cos(\phi(t))\,(\frac{\partial}{\partial t}Y(t))}{a}$$

It is interesting to note that we would obtain the same equation if we would replace the function $\rho(t)$ by the constant a in the differential equation for $\phi(t)$ for the jogger/dog problem.

1.8   Examples

Since the system of differential equations has simplified into one equation we may hope to obtain analytical solution for certain cases. We begin with the first example where the child is walking on a circle.

```
> Child_Subst := X(t) = R*cos(omega*t), Y(t) = R*sin(omega*t);
```
$$Child\_Subst := X(t) = R\cos(\omega\,t),\ Y(t) = R\sin(\omega\,t)$$

```
> Das := subs(Child_Subst, Daxial);
```
$$Das := \frac{\partial}{\partial t}\,\phi(t) = \frac{\sin(\phi(t))\,(\frac{\partial}{\partial t}R\cos(\omega\,t)) - \cos(\phi(t))\,(\frac{\partial}{\partial t}R\sin(\omega\,t))}{a}$$

```
> Das := combine(Das, trig);
```
$$Das := \frac{\partial}{\partial t}\,\phi(t) = -\frac{R\,\omega\cos(-\phi(t)+\omega\,t)}{a}$$

```
> Sol := dsolve(Das, phi(t));
```
$$Sol := 2\,\frac{\arctan\left(\dfrac{(a-R)\tan(-\frac{1}{2}\phi(t)+\frac{1}{2}\omega\,t)}{\sqrt{(a+R)\,(a-R)}}\right)a}{\sqrt{a^2-R^2\,\omega^2}} - \frac{t}{\omega} = \_C1$$

We received an analytical expression! The function $\phi(t)$ is given by an implicit equation though. Therefore MAPLE would not solve the differential equation with given initial condition. Though the `solve` command cannot solve in the equation above for $\phi(t)$ it is possible to do it by hand. We will not do it right here because it is simpler to determine the constant of integration for special cases in this form.

To reproduce the orbit given in Figure 1.3 we introduce the initial condition:

```
> IniSol := eval(subs(t = 0, phi(0) = alpha, Sol));
```
$$IniSol := -2\,\frac{\arctan\left(\dfrac{(a-R)\tan(\frac{1}{2}\alpha)}{\sqrt{(a+R)\,(a-R)}}\right)a}{\sqrt{a^2-R^2\,\omega^2}} = \_C1$$

Now we solve for the special solution (by helping MAPLE):

```
> Sol := map(u -> (u + t/omega)*sqrt(a^2 - R^2)*omega^2/a/2, Sol):
> Sol := map(u -> tan(u)*sqrt((a + R)*(a - R))/(a - R), Sol):
> Sol := simplify(map(u -> arctan(u), Sol), symbolic):
> _C1 := lhs(IniSol):
> Sol := map(u -> (u - omega*t/2)*(-2), Sol):
```

For Figure 1.3 we now specify the initial angle α:

```
> Sol0 := eval(subs(alpha = 0, Sol));
```
$$Sol0 := \phi(t) = -2\arctan\left(\frac{\tan(\frac{1}{2}\frac{t\,\omega\,\sqrt{a^2-R^2}}{a})\sqrt{(a+R)\,(a-R)}}{a-R}\right) + \omega\,t$$

and we want to set the length of the bar a equal to the radius of the circle R. This is not possible by simple substitution—we rather need to compute the limit $a \to R$:

```
> Sol00 := lhs(Sol0) = limit(rhs(Sol0), a = R, 'left'):
> Sol00 := symplify (S00, symbolic);
```

$$Sol00 := \phi(t) = -2\arctan(\omega\,t) + \omega\,t$$

```
> Toy := [x(t), y(t)]:
> Toy_Plot := subs(Rx, Ry, Child_Subst, Sol00, omega = 1,
>                   a = R, R = 5, [Toy[], t = 0..40]);
```

$$Toy\_Plot := [5\cos(t) + 5\cos(-2\arctan(t) + t),$$
$$5\sin(t) + 5\sin(-2\arctan(t) + t),\ t = 0..40]$$

```
> plot(Toy_Plot, scaling = constrained, color = black);
```

We obtain the same plot as in Figure 1.3. Figure 1.5 is obtained with the following statements:

```
> Sol15 := eval(subs(alpha = arctan(-2)+Pi, Sol)):
> Toy_Plot := subs(Rx, Ry, Child_Subst, Sol15, omega = 1,
>                   a = sqrt(125), R = 5, [Toy[], t = 0..200]):
> plot(Toy_Plot, scaling = constrained, color = black);
```

Finally we compute the orbit of a variant of the first example. We will consider the case where the bar is shorter than the radius of the circle:

```
> Sol1 := evalc(Sol0):
> Sol1 := evalc(subs(signum(a^2 - R^2) = -1, Sol1)):
> Sol1 := subs(abs(a^2 - R^2) = R^2 - a^2, Sol1);
```

$$Sol1 := \phi(t) =$$
$$2\arctan\left(\frac{\sinh(\frac{1}{2}\frac{\omega\,t\,\sqrt{R^2 - a^2}}{a})\cosh(\frac{1}{2}\frac{\omega\,t\,\sqrt{R^2 - a^2}}{a})\sqrt{R^2 - a^2}}{(1 + \sinh(\frac{1}{2}\frac{\omega\,t\,\sqrt{R^2 - a^2}}{a})^2)\,(a - R)}\right) + \omega\,t$$

```
> Toy_Plot := subs(Rx, Ry, Child_Subst, Sol1, omega = 1,
>                   a = 4.95, R = 5, [Toy[], t = 0..40]):
> plot(Toy_Plot, scaling = constrained, color = black);
```

The orbit is given in Figure 1.11.

FIGURE 1.11. *Toy Orbit for $a < R$*

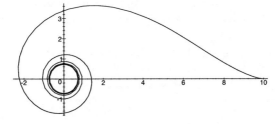

Note that for the child walking on a straight line there is an analytical solution, the tractrix, as shown in the first section of this chapter. Let (X_0, Y_0) be the initial position and (V_x, V_y) the constant velocity vector of the child:

```
> a := 'a': Vx := 'Vx': Vy := 'Vy':
> Child_Subst := X(t) = Xo + Vx*t, Y(t) = Yo + Vy*t:
> Das := combine(subs(Child_Subst, Daxial));
```

$$Das := \frac{\partial}{\partial t} \phi(t) = \frac{\sin(\phi(t))\ Vx - \cos(\phi(t))\ Vy}{a}$$

```
> Sol := dsolve(Das, phi(t));
```

$$Sol := 2\ \frac{\text{arctanh}\left(\dfrac{1}{2}\ \dfrac{2\ Vy \tan(\frac{1}{2}\phi(t)) + 2\ Vx}{\sqrt{Vy^2 + Vx^2}}\right)\ a}{\sqrt{Vy^2 + Vx^2}} + t = \_C2$$

Repeating the same process as before in the case of the child walking on a circle, we obtain again an implicit function for $\phi(t)$.

In general it will not be possible to obtain an analytic solution for the orbit of the toy. If we consider the child walking on the sine curve, we have to compute the orbit of the toy numerically.

```
> X := t -> t:   Y := t -> 5*sin(t):   a:= 10:
> Daxial;
```

$$Daxial := \frac{\partial}{\partial t} \phi(t) = \frac{1}{10}\sin(\phi(t)) - \frac{1}{2}\cos(\phi(t))\cos(t)$$

```
> F := dsolve({Daxial, phi(0) = Pi/2}, phi(t), numeric);
```

$$F := \mathbf{proc}(rkf45\_x)\ \dots\ \mathbf{end}$$

Because the plot is the same as the Fig. 1.4 we will not print it here. We will, however, add a few commands to show the movements dynamically on the screen. Note that it is not possible to use the `animate` command for the same purpose.

We would like to see the movement of the bar, and the trajectory of both ends of the bar. One end has to move along the sine curve, the second will describe the computed orbit. We would like to see a movie describing the process during T seconds, as a sequence of N partial plots.

```
> T := 6*Pi: N := 200: L := a:
> SF := [evalf(seq(rhs(F(T*i/N)[2]), i = 0..N))]:
> ST := [evalf(seq(T*i/N, i= 0..N))]:
> plot([seq([X(ST[i]) + L*cos(SF[i]), Y(ST[i]) + L*sin(SF[i])],
>       i =1..N)]);
```

The last MAPLE command will display the orbit. For the movie we have to prepare and store the sequences of plots with increasing number of points.

```
> TS := [seq(op(1, plot([seq([X(ST[i]) + L*cos(SF[i]),
>         Y(ST[i]) + L*sin(SF[i])], i = 1..j)])), j = 1..N)]:
> BS := [seq(op(1, plot([[X(ST[i]), Y(ST[i])],
>         [X(ST[i]) + L*cos(SF[i]), Y(ST[i]) + L*sin(SF[i])]])),
>         i=1..N)]:
> PS := [seq(op(1, plot([seq([X(ST[i]), Y(ST[i])],
>         i = 1..j)])), j = 1..N)]:
```

The sequence TS contains information about the orbit. It stores N partial plots containing 1 to N points. The sequence BS describes the bar position for each frame. Finally PS describes the *sine* trajectory. Its structure is similar to TS.

The whole process can now be viewed with the command

```
> PLOT(ANIMATE(seq([TS[i], BS[i], PS[i]], i=1..N)),
>         AXESLABELS(x, y), VIEW(-2..X(ST[N]), DEFAULT));
```

References

[1] E. HAIRER, S.P. NØRSETT and G. WANNER, *Solving Ordinary Differential Equations I*, Springer-Verlag Berlin Heidelberg, 1987.

[2] H. HEUSER, *Gewöhnliche Differentialgleichungen*, B. G. Teubner, Stuttgart, 1989.

Chapter 2. Trajectory of a Spinning Tennis Ball

F. Klvaňa

2.1 Introduction

Consider a tennis ball with mass m and diameter d, moving in air near the earth surface. The ball is spinning with angular velocity $\vec{\omega}$ (the vector $\vec{\omega}$ has the direction of the axis of rotation and magnitude $\omega = d\varphi(t)/dt = \dot{\varphi}(t)$, where $\varphi(t)$ is an angle of rotation). We will impose a Cartesian coordinates system (xyz) on the surface of the earth with the z axis directed vertically.

FIGURE 2.1. *Spinning Ball Moving in Air.*

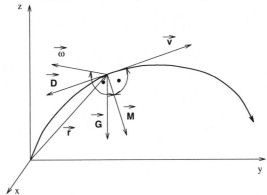

As a model of the ball we can then take a mass point, which is moving under influence of the following forces (cf. Figure 2.1)

- The weight force $\vec{G} = m\vec{g}$, where $\vec{g} = (0, 0, -g)$ is a vector of the gravitational acceleration.

- The drag force $\vec{D} = -D_L(v)\vec{v}/v$, which has opposite direction to the velocity $\vec{v}$.

- The Magnus force $\vec{M} = M_L\vec{\omega}/\omega \times \vec{v}/v$; this force is orthogonal to $\vec{v}$ and $\vec{\omega}$.

The magnitudes of the drag force $D_L(v)$ and the Magnus force $M_L(v)$ are usually supposed to have a form given by the theory of ideal fluids [1]:

$$D_L(v) = C_D \frac{1}{2} \frac{\pi d^2}{4} \rho v^2$$

$$M_L(v) = C_M \frac{1}{2} \frac{\pi d^2}{4} \rho v^2$$

(2.1)

where ρ is air density. The coefficients C_D and C_M depend for real fluids (air) on the velocity v, the ball revolution and the material of the surface of the ball. Usually we have to find these coefficients experimentally.

In [2] results of experiments with a spinning tennis ball are reported. It is shown that for a tennis ball in the regions of velocity $v \in [13.6, 28]$ $[m \, s^{-1}]$ and a ball revolution $n \in [800, 3250]$ rpm, the coefficients C_D and C_M depend on v/w only, where $w = d/2 \cdot |\, \vec{\omega} \times \vec{v}/v \,|$ is in some sense the projection of the equatorial velocity $wd/2$ of the spinning ball onto the velocity vector $\vec{v}$. The following expressions for the coefficients were obtained:

$$C_D = 0.508 + \left(\frac{1}{22.053 + 4.196 \left(\frac{v}{w} \right)^{5/2}} \right)^{2/5}$$

$$C_M = \frac{1}{2.022 + 0.981 \left(\frac{v}{w} \right)}$$

(2.2)

For a tennis ball we can neglect the deceleration of the ball revolution, so w is constant.

The trajectory of the ball is then defined by Newton's equations for the position vector $\vec{r}(t)$

$$m \frac{d^2 \vec{r}(t)}{dt^2} = -m\vec{g} - D_L \frac{\vec{v}}{v} + M_L \frac{\vec{\omega}}{w} \times \frac{\vec{v}}{v}$$

(2.3)

with initial conditions

$$\vec{r}(0) = \vec{r}_0 \qquad \text{and} \qquad \frac{d\vec{r}}{dt}(0) = \vec{v}_0.$$

The Equation (2.3) is a nonlinear system of three differential equations and no analytical solution exists for it, so we have to solve it numerically.

In practice, the most usual case is the topspin lob, for which the vector of angular velocity lies in a horizontal plane and is orthogonal to the vector $\vec{v}_0$, and, as follows from Equation (2.3), is also orthogonal to $\vec{v}(t)$ for $t \geq 0$; so the trajectory lies in the vertical plane. Let us choose the x axis in this plane. Then the final form of Equations (2.3) is

$$\ddot{x} = -C_D \alpha v \dot{x} + \eta C_M \alpha v \dot{z}$$

$$\ddot{z} = -g - C_D \alpha v \dot{z} - \eta C_M \alpha v \dot{x}$$

(2.4)

where $v = \sqrt{\dot{x}^2 + \dot{z}^2}$ and $\alpha = (\rho \pi d^2)/(8m)$. The parameter $\eta = \pm 1$ describes direction of rotation (for topspin $\eta = 1$).

As initial conditions for $t = 0$ we take

$$x(0) = 0, \quad z(0) = h, \quad \dot{x}(0) = v_0 \cos(\vartheta), \quad \dot{z}(0) = v_0 \sin(\vartheta) \qquad (2.5)$$

where v_0 is the magnitude of the initial velocity vector $\vec{v_0}$ and ϑ is an angle between $\vec{v_0}$ and $x-$axis.

2.2 MAPLE Solution

In order to demonstrate the influence of drag and Magnus forces on a trajectory of a ball, we will consider three models: a ball in vacuum, a ball in air without spin and a spinning ball.

Ball in Vacuum

In vacuum, the only force acting on the ball is gravity and the Equations (2.4) have a very simple form

$$\ddot{x} = 0, \quad \ddot{z} = -g \qquad (2.6)$$

MAPLE can find a general solution of (2.6) using function `dsolve`:

```
>                       # motion of ball in vacuum
> eqnid := diff(x(t), t$2) = 0, diff(z(t),t$2) = -g:
> varid := {x(t), z(t)}:
> initcid := x(0) = 0, z(0) = h,
>                 D(x)(0) = v0*cos(theta), D(z)(0) = v0*sin(theta):
>                 #solution for ball in vacuum
> resid := dsolve({eqnid, initcid}, varid);
```

$$resid := \{\mathrm{x}(t) = t\,v0\,\cos(\theta),\ \mathrm{z}(t) = h + t\,v0\,\sin(\theta) - \frac{1}{2}t^2\,g\}$$

Notice that function `dsolve` returns the solution $< result >$ in the form of a set of equations

$$< varname >=< expression >,$$

so there is no defined order of these equations. In order to access individual solutions corresponding to $< varname >$ we will use the function

$$subs(< result >, < varname >).$$

Ball in the Air

For the models of a ball in the air we cannot solve Equations (2.4) analytically and we have to use a numeric solution. Since we want to use Newton's method later to calculate a flight time, we need the derivative of $z(t)$ and hence we transform the differential equation system of second order (2.4) into a differential equation system of first order, (as the velocities $\dot{x}$ and $\dot{z}$ we use variables v_x and v_z),

$$
\begin{aligned}
\dot{x} &= v_x \\
\dot{v}_x &= -C_L\,\alpha\,v\cdot v_x + \eta\,C_M\,\alpha\,v\cdot v_z \\
\dot{z} &= v_z \\
\dot{v}_z &= -g - C_L\,\alpha\,v\cdot v_z - \eta\,C_M\,\alpha\,v\cdot v_x
\end{aligned}
\tag{2.7}
$$

where $v = \sqrt{v_x^2 + v_z^2}$. The initial conditions (2.5) have the form

$$
x(0) = 0, \quad z(0) = h, \quad v_x(0) = v_0\cos(\vartheta), \quad v_z(0) = v_0\sin(\vartheta). \tag{2.8}
$$

We will use the SI system of units. The numeric values of the parameters are $g = 9.81\,[m\,s^{-1}]$, diameter of the tennis ball $d = 0.063\,[m]$, its mass $m = 0.05\,[kg]$, and air density $\rho = 1.29\,[kg\,m^{-3}]$. As initial conditions we choose

$$
h = 1\,[m], \quad v_0 = 25\,[m\,s^{-1}], \quad \vartheta = 15°
$$

and as the parameters of the spin of a ball we take $w = 20\,[m\,s^{-1}]$ and $\eta = 1$ (top spin). The MAPLE statements to solve the problem follow[1].

```
>                  # numeric solution of ball in air
>                  # Basic constants in SI units
> g := 9.81:    # gravitational acceleration:
> d := 0.063: m := 0.05: # diameter and mass of ball
> rov := 1.29:  # density of air
> alpha := evalf(Pi*d^2/(8*m)*rov):
> v := (dx^2 + dz^2)^(1/2): # definition of velocity
> Cd := .508+1/(22.503 + 4.196/(w/v(t))^(5/2))^(2/5):
> Cm := eta/(2.202 + .981/(w/v(t))):
>                  # eta =+-1  defines direction of rotation,
>                  # for top spinning eta = 1
> var := {x(t), z(t), dx(t), dz(t)}:
>                  # initial conditions
> initc := x(0) = 0, z(0) = h,
>          dx(0) = v0*cos(theta), dz(0) = v0*sin(theta):
>                  # equations of motion of ball in air
>                  # rotation (drag force only)
> eqnt0:= diff(x(t),t) = dx(t),
>         diff(dx(t),t)= -0.508*alpha*dx(t)*v(t),
>         diff(z(t),t) = dz(t),
>         diff(dz(t),t)= -g-0.508*alpha*dz(t)*v(t):
>                  # equations of motion of rotating ball in air
>                  # (influence of Magnus effect)
> eqnt1:= diff(x(t),t) = dx(t),
>         diff(dx(t),t)= (-Cd*dx(t) + Cm*dz(t))*alpha*v(t),
>         diff(z(t),t) = dz(t),
>         diff(dz(t),t)= -g-(Cd*dz(t) + Cm*dx(t))*alpha*v(t):
```

---

[1]This example has been tested using MAPLE V Release 3.

```
>                     #numeric values of initial parameters
> h := 1: v0 := 25:   theta := Pi/180*15: # theta= 15 degrees
> w := 20:  eta := 1:
>                     # solution for non rotating ball
> res0 := dsolve({eqnt0,initc},var,numeric);
```

$$res0 := \mathbf{proc}(rkf45\_x) \dots \mathbf{end}$$

```
>                     # result is a set of equations
> res0(0.5);
```

$$[t = .5,\, \mathrm{x}(t) = 10.76902243898168,\, \mathrm{z}(t) = 2.745476721738618,$$
$$\mathrm{dx}(t) = 19.31511754217100,\, \mathrm{dz}(t) = .7584804887012169]$$

```
>                     # solution for rotating ball
> res1 := dsolve({eqnt1, initc}, var, numeric);
```

$$res1 := \mathbf{proc}(rkf45\_x) \dots \mathbf{end}$$

Function `dsolve(...,numeric)` returns as result a function with one parameter (independent variable). The result of a call of this function is a set of equations

$$< variable >=< value >$$

(in our case the variables are $t, x(t), z(t), v_x(t), v_z(t)$). For an access of numeric values we can again use the function `subs`.

Plotting the Graphs

As the final part of our solution we will plot the graphs of the trajectories for all three models in the same picture. For this we have to solve the subproblem of finding the flight time (solution of equation $z(t) = 0$).

To compute the flight time for the ball in vacuum t_{maxid}, we can use the symbolic solution of (2.6) in res_{id} and the function `fsolve`.

```
>                 # calculation of tmax - time of falling to earth
>                 # for ball in vacuum
> tmaxid := fsolve(subs(resid, z(t)) = 0, t, t = 0..5);
```

$$tmaxid := 1.458903640$$

In the other cases we can easily use Newton's method to solve equation $z(t) = 0$, because by integrating the differential equations we obtain also the derivative of z in variable v_z. Then the recurrence relation for the sequence of approximations of the solution of equation $z(t) = 0$ will be

$$t_{n+1} = t_n - \frac{z(t_n)}{v_z(t_n)}. \tag{2.9}$$

ALGORITHM 2.1. *Function* zzero.

```
zzero := proc (u, t0, z, dz) local tn, ts, up;
        # find root of z(t) = subs(u(t), z) = 0
        # using Newton method
        # using diff(z, t) = subs(u(t), dz)
    tn := t0; ts := 0.0;
    while abs((tn - ts)/tn) > 10^(-4) do;
            ts := tn;
            up := u(ts);
            tn := ts - subs(up, z)/subs(up, dz);
        od;
        tn;
    end;
```

As the initial approximation t_0 we can use t_{maxid}. See the MAPLE function zzero, which implements this Newton iteration (cf. Algorithm 2.1).

```
>               # calculation of the flight time for the other models
> tmax0:= zzero(res0, tmaxid, z(t), dz(t));
> tmax1:= zzero(res1, tmaxid, z(t), dz(t));
```

$$tmax0 := 1.362013689$$

$$tmax1 := .9472277855$$

The simplest, (and probably the fastest) method for creating the graphs of our ball trajectories in air (from a numeric solution) is to use an array $[x(t_i), z(t_i)]$ as an input parameter for the function plot. To create this array from the result of dsolve(..., numeric) for a time interval with constant time step we define the simple function *tablepar*. The rest of the MAPLE statements for the creation of the graphs in red, blue and black color will then be

```
>               # making graphs:
> Gid := plot([subs(resid, x(t)), subs(resid, z(t)),
>               t = 0..tmaxid], linestyle = 2):
>               # for models with numeric solution
>               # calculation of tables [x(t),z(t)]  for plotting
> tablepar := proc(u, x, y, xmin, xmax, npoints) local i,Step;
>       Step := (xmax - xmin)/npoints;
>       [seq([subs(u(xmin + i*Step), x), subs(u(xmin + i*Step) ,y)],
>           i = 0 .. npoints)]
> end:
> G0 := plot(tablepar(res0, x(t), z(t), 0, tmax0, 15),
>           linestyle = 3):
> G1 := plot(tablepar(res1, x(t), z(t), 0, tmax1, 15),
>           linestyle = 1):
>               # plotting of all graphs
> plots[display]({Gid, G0, G1});
```

The different graphs are the same as the trajectories computed with MAT-LAB (cf. Figure 2.2).

ALGORITHM 2.2. *A Set of Needed M-functions.*

```
function xdot= tennisip(t,x,flag)
  global g
  xdot = [ x(3)
           x(4)
           0
           -g   ];

function xdot= tennis0p(t,x)
  global g alpha
  v= sqrt(x(3)^2+x(4)^2);
  xdot = [ x(3)
           x(4)
           -alpha*0.508*x(3)*v
           -g-alpha*0.508*x(4)*v ];

function xdot= tennis1p(t,x)
  global g alpha w etha
  v = sqrt(x(3)^2 + x(4)^2);
  Cd = (0.508 + 1/(22.503 + 4.196*(v/w)^0.4))*alpha*v;
  Cm = etha*w/(2.022*w + 0.981*v)*alpha*v;
  xdot = [ x(3)
           x(4)
           -Cd*x(3) + Cm*x(4)
           -g-Cd*x(4) - Cm*x(3) ];
```

2.3 MATLAB Solution

Because of the nonlinear nature of our problem we have to use numerical methods to solve it. A numerical system like MATLAB appears to be more adequate than a symbolic one. So we will now try to solve our problem in MATLAB.

To solve an initial value problem of a system of n differential equations

$$\frac{d\vec{x}}{dt} = \vec{f}(t, \vec{x}), \tag{2.10}$$

where $\vec{x} = (x_1, x_2, ..., x_n)$, we can use the MATLAB function ode23 (or ode45), which implements an embedded Runge-Kutta method of order 2 and 3 (respectively 4 and 5).

We have to define a M-function for each model to implement the system of differential Equations (2.10). To transfer the parameters g, α, w and η into the M-files we define them as global in our program. Let us use the following mapping of variables

$$x \longrightarrow x(1), \quad z \longrightarrow x(2), \quad v_x \longrightarrow x(3), \quad v_z \longrightarrow x(4).$$

Needed M-files are in Algorithm (2.2).

In the following main program we compute and plot the trajectories for all three models (cf. Figure 2.2). To compute a sufficient dense table $[x_i, z_i]$ to plot the ball trajectories for the models in the air, we interpolate by a spline function 100 points of the solution $z = z(x)$, using the functions spline.

```
>>           % Trajectory of spinning tennis ball
>>           % initialization
>> global g alpha w etha
>>           % basic constants im MKS units
>> g = 9.81; d = 0.063; m = 0.05; rho = 1.29;
>> alpha=pi*d^2/(8*m)*rho;
>> etha = 1;
>> w = 20;
>>           % initial conditions
>> h = 1; v0 = 25; theta = pi/180*15;
>> xin = [0, h, v0*cos(theta), v0*sin(theta)];
>>           % flight time for vacuum
>> tmaxid = (xin(4) + sqrt(xin(4)^2 + 2*g*xin(2)))/g;
>>           % solution in vacuum
>> [tid, xid] = ode23('tennisip', [0 tmaxid], xin);
>>           % solution without spin
>> [t0, x0] = ode23('tennis0p', [0 tmaxid], xin);
>>           % solution with spin
>> [t1, x1] = ode23('tennis1p', [0 tmaxid], xin);
>> N = max(xid(:, 1)); x = 0:N/100:N;
>> axis([0,max(xid(:,1)), 0, max(xid(:,2))])
>> hold on;
>> plot(x, spline(xid(:,1), xid(:, 2), x), ':r');
>> plot(x, spline(x0(:,1), x0(:, 2), x), '--b');
>> plot(x, spline(x1(:,1), x1(:, 2), x), '-w');
>> hold off;
```

Note that we did not have to compute the flight time for the two models in the air. By using the axis statement for the largest trajectory (the ball in vacuum) followed by a hold statement the trajectories are truncated and not plotted below the x-axis.

If we wish to actually compute the flight times we could proceed similarly as shown in the MAPLE solution. This solution is rather costly in terms of computing time, since for each evaluation of the function and the derivative we have to integrate numerically the system of differential equations from the beginning.

It is simpler, however, to compute an approximation of the flight time using inverse interpolation. In order to do so we need to find a time interval in which the function is invertible. For this we take those values in the table where the z-values of the trajectory are monotonically decreasing. This can be done with the min and max function:

```
>>           % Determine flight time  by inverse interpolation
>> [z, j] = min(x0(:, 2)); [y, i] = max(x0(:, 2));
>> tmax0 = spline(x0(i + 1:j, 2), t0(i + 1:j), 0)
>> [z, j] = min(x1(:, 2)); [y, i] = max(x1(:, 2));
>> tmax1 = spline(x1(i+1:j,2), t1(i+1:j), 0)
```

We have to take as the first point of the monotonic region of z the point with the index $(i+1)$ in order to ensure the region $z(i+1:j)$ is monotonic. We then obtain for the flight times the approximate values $t_{max0} = 1.3620$ and $t_{max1} = 0.9193$ which compare well with the MAPLE results.

FIGURE 2.2.

Trajectories of a Ball: $\cdots\cdots$ *in Vacuum,*

$- - - -$ *in Air without Spin,*

———   *with Topspin.*

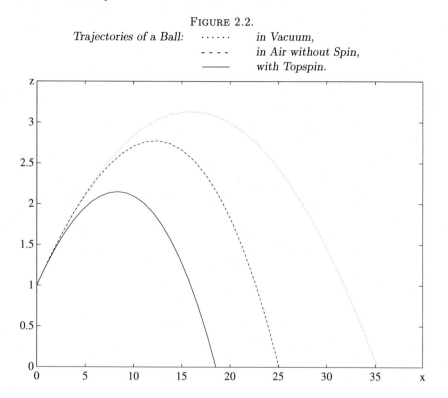

2.4   Simpler Solution for MATLAB 5

Version 5 of MATLAB allows us to integrate the differential equation until the resulting trajectory crosses the x-axis[2]. Therefore it is no longer necessary to interpolate with splines. According to the explanations made in Chapter 1 on page 9, the M-functions defined in Algorithm (2.2) have to be adapted to check for zero crossings; the updated M-functions are shown in Algorithm 2.3. The main program for calculating the three trajectories becomes shorter: The option Events is set to on, which causes the integrator to check for zero crossings of the z-coordinate and to stop the integration if there is one. Therefore the time for the trajectories can be taken from the last element of the time vectors t0 and t1 returned by the integrator.

---

[2]This solution was added by Leonhard Jaschke

ALGORITHM 2.3.

The M-Files of the last section rewritten for MATLAB 5.

```
function [xdot,isterminal,direction]= tennisip2(t,x,flag)
global g
if nargin < 3 | isempty(flag) % normal output
   xdot = [x(3), x(4), 0, -g]';
else
   switch(flag)
   case 'events'  % at ||h||=0 there is a singularity
      xdot= x(2);
      isterminal= 1;
      direction= -1;
   otherwise
      error(['Unknown flag ''' flag '''.']);
   end
end

function [xdot,isterminal,direction]= tennis0p2(t,x,flag)
global g alpha
if nargin < 3 | isempty(flag) % normal output
   v= sqrt(x(3)^2+x(4)^2);
   xdot = [x(3), x(4), ...
           -alpha*0.508*x(3)*v, -g-alpha*0.508*x(4)*v]';
else
   switch(flag)
   case 'events'  % at ||h||=0 there is a singularity
      xdot= x(2);
      isterminal= 1;
      direction= -1;
   otherwise
      error(['Unknown flag ''' flag '''.']);
   end
end

function [xdot,isterminal,direction]= tennis1p2(t,x,flag)
global g alpha w etha
if nargin < 3 | isempty(flag) % normal output
   v= sqrt(x(3)^2+x(4)^2);
   Cd=(0.508+1/(22.503+4.196*(v/w)^0.4))*alpha*v;
   Cm=etha*w/(2.022*w+0.981*v)*alpha*v;
   xdot = [x(3), x(4), ...
           -Cd*x(3)+Cm*x(4), -g-Cd*x(4)-Cm*x(3)]';
else
   switch(flag)
   case 'events'  % at ||h||=0 there is a singularity
      xdot= x(2);
      isterminal= 1;
      direction= -1;
   otherwise
      error(['Unknown flag ''' flag '''.']);
   end
end
```

The integrators take rather large integration steps, which makes the plots of the trajectories not as smooth as those obtained by interpolating with splines. To smoothen these trajectories we must take smaller steps. This can be done by setting an upper bound for the step width using the option Maxstep. Usually, this value is set to one tenth of the length of the time interval. In the following algorithm we set this value to tmaxid/100 and get the same smooth trajectories as shown in Figure 2.2:

```
>>        % Trajectory of rotating tennis ball
>>        % initialization
>> global g alpha w etha
>> g= 9.81; d= 0.063; m= 0.05;  rho= 1.29;
>> alpha= pi*d^2/(8*m)*rho;
>> etha= 1;
>> w=20;
>>        %initial condition
>> h=1; v0=25; theta=pi/180*15;
>> xin=[0,h,v0*cos(theta),v0*sin(theta)]
>>        % flight time for vacuum
>> tmaxid=(xin(4)+sqrt(xin(4)^2+2*g*xin(2)))/g
>>        % setting options for the integrator
>> options= odeset('Events','on','Maxstep',tmaxid/100);
>>        % solution in vacuum
>> [tid,xid]= ode23('tennisip2',[0 tmaxid],xin,options);
>>        % solution without spin
>> [t0,x0]= ode23('tennis0p2',[0 tmaxid],xin,options);
>>        % solution with spin
>> [t1,x1]= ode23('tennis1p2',[0 tmaxid],xin,options);
>> clf;
>> axis([0,max(xid(:,1))*1.13,0,max(xid(:,2))*1.13]);
>> hold on;
>> plot(xid(:,1),xid(:,2),':r');
>> plot(x0(:,1),x0(:,2),'--b');
>> plot(x1(:,1),x1(:,2),'-g');
>> hold off;
>>        % flight time for the trajectory without spin
>> tmax0= t0(length(t0))
>>        % flight time for the trajectory with spin
>> tmax1= t1(length(t1))
```

References

[1] E. G. RICHARDSON, *Dynamics of Real Fluids*, Edward Arnold, 1961.

[2] A. ŠTĚPÁNEK, *The Aerodynamics of Tennis Balls - The Topspin Lob*, American Journal of Physics, 56, 1988, pp. 138 – 142.

Chapter 3. The Illumination Problem

S. Bartoň and D. Gruntz

3.1 Introduction

In this article we consider a horizontal road illuminated by two lights, where P_i is the illumination power and h_i the height of a lamp. The coordinates of the lamps are $(0, h_1)$ and (s, h_2) where s is the horizontal distance between the two light sources. Let $X = (x, 0)$ be a point on the road somewhere between the two lights. In this chapter we will look for a point X which is minimally illuminated. In Figure 3.1 we have made a sketch of the situation we will refer to later in this chapter.

FIGURE 3.1. *Illumination Problem Description.*

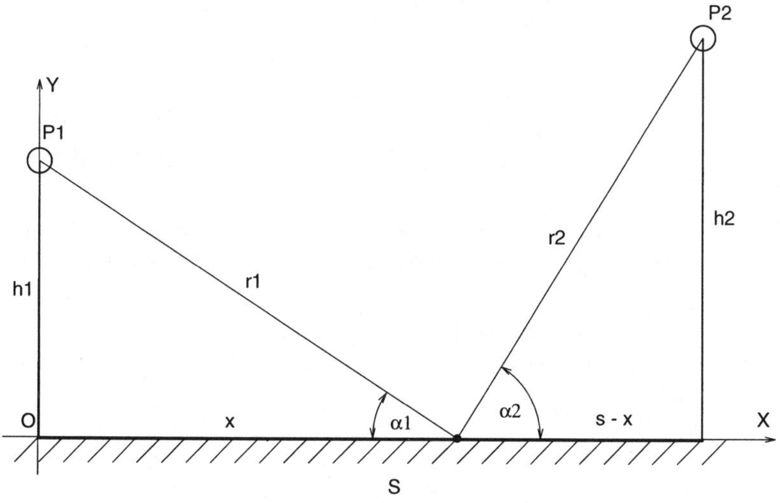

In the first section we will find X given the height and the intensity of both lamps. In the second section we will maximize the illumination at X by varying the height of the second lamp. In the last section we will go even further and optimize the illumination at X with respect to the heights of both lamps. In effect, we will optimize the illumination of the road by varying the heights of the lamps.

3.2   Finding the Minimal Illumination Point on a Road

In this section we look for the minimally illuminated point X between the two light sources. It is known from physics, that the light intensity depends on the inverse value of the square of the distance to the light source and on the impact angle of the light rays, see [1, 2].

The distance from X to the first light source is x and to the second one $s-x$. Using the Pythagorean theorem we can determine r_1 and r_2, the distances from X to the two light sources,

$$r_1{}^2 = h_1{}^2 + x^2, \quad r_2{}^2 = h_2{}^2 + (s-x)^2.$$

The light intensities from the two lamps at X are given by

$$I_1(x) = \frac{P_1}{r_1{}^2} = \frac{P_1}{h_1{}^2 + x^2}, \quad I_2(x) = \frac{P_2}{r_2{}^2} = \frac{P_2}{h_2{}^2 + (s-x)^2}.$$

If the impact angles of the light rays are α_1 and α_2, the road illumination depends on $\sin\alpha_1$ and $\sin\alpha_2$ which are given by

$$\sin\alpha_1 = \frac{h_1}{\sqrt{h_1{}^2 + x^2}}, \quad \sin\alpha_2 = \frac{h_2}{\sqrt{h_2{}^2 + (s-x)^2}}.$$

Thus the total illumination $C(x)$ at the point X is

$$C(x) = I_1(x)\sin\alpha_1 + I_2(x)\sin\alpha_2 = \frac{P_1 h_1}{\sqrt{\left(h_1{}^2 + x^2\right)^3}} + \frac{P_2 h_2}{\sqrt{\left(h_2{}^2 + (s-x)^2\right)^3}}. \tag{3.1}$$

By minimizing $C(x)$ for $0 \le x \le s$ we get the coordinates of X. To find the minimum we can simply take the derivative of $C(x)$ and find its roots. We will attempt this using MAPLE.

```
> S[1] := P[1]*h[1]/(h[1]^2 + x^2)^(3/2):
> S[2] := P[2]*h[2]/(h[2]^2 + (s - x)^2)^(3/2):
> C :=  S[1] + S[2]:
> dC := diff(C, x);
```

$$dC := -3\,\frac{P_1\,h_1\,x}{(h_1{}^2 + x^2)^{5/2}} - \frac{3}{2}\,\frac{P_2\,h_2\,(-2\,s + 2\,x)}{(h_2{}^2 + (s-x)^2)^{5/2}}$$

```
> solve(dC=0,x):
```

If you try this command, you will see that MAPLE will never return an answer. Using algebraic manipulations, the equation $dC = 0$ can be transformed into a polynomial equation in x. In particular we will move one of the terms of the equation to the right hand side, square both sides, move the right hand side back to the left and write the expression over a common denominator. We observe that the numerator must be zero.

```
> eq := diff(S[1], x)^2 - diff(S[2], x)^2;
```

$$eq := 9\,\frac{P_1{}^2\,h_1{}^2\,x^2}{(h_1{}^2 + x^2)^5} - \frac{9}{4}\,\frac{P_2{}^2\,h_2{}^2\,(-2\,s + 2\,x)^2}{(h_2{}^2 + (s - x)^2)^5}$$

```
> eq := collect(primpart(numer(eq)), x);
```

The result of the last command is a degree-12 polynomial in x,

$$(P_1^2 h_1^2 - P_2^2 h_2^2)x^{12} + (2P_2^2 h_2^2 s - 10P_1^2 h_1^2 s)x^{11} + \cdots - P_2^2 h_2^2 h_1^{10} s^2 = 0. \quad (3.2)$$

This polynomial is difficult to solve in closed form without specifying the constants.

We consider the following numerical values: $P_1 = 2000\ [W]$, $P_2 = 3000\ [W]$, $h_1 = 5\ [m]$, $h_2 = 6\ [m]$ and $s = 20\ [m]$. The functions $C(x)$, $C'(x) \equiv dC$, $S_1(x)$ and $S_2(x)$ (the illumination intensities on the road implied by each lamp separately) can be plotted using MAPLE, see Figure 3.2. An interval containing the zero of the function $C'(x)$ can be picked off from this graph. We will use this interval to determine X using `fsolve`.

FIGURE 3.2. *Illumination as a Function of the Position.*

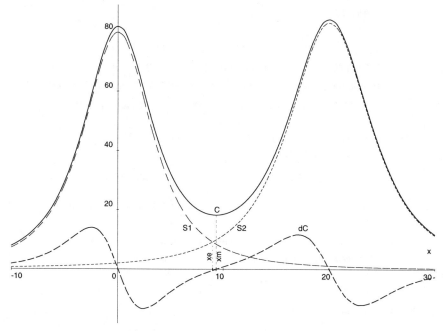

```
> P[1] := 2000:  P[2] := 3000: s := 20: h[1] := 5: h[2] := 6:
> plot({C, S[1], S[2], dC}, x = -s/2..s*3/2);
```

```
> Xm := fsolve(dC = 0, x, 5..10);
```

$$Xm := 9.338299136$$

```
> Cmin := subs(x = Xm, C);
```

$$Cmin := 18.24392572$$

```
> Xe := fsolve(S[1] = S[2], x, 0..s);
```

$$Xe := 9.003061731$$

```
> dX := Xm - Xe;
```

$$dx := .335237405$$

It is interesting to note, that the position X differs from the point x_e which is equally illuminated from both lamps.

For this numerical example we can also determine the point X directly. We can apply the `solve` command to Equation (3.2) and obtain an algebraic number defined by a degree-12 polynomial with integer coefficients. Real (physical) solutions correspond to points between the two light sources.

```
> solve(eq, x);
```

$$
\begin{aligned}
\text{RootOf}(&56\,\_Z^{12} + 1760\,\_Z^{11} - 411975\,\_Z^{10} + 24315000\,\_Z^{9} \\
&- 886167750\,\_Z^{8} + 22194630000\,\_Z^{7} - 388140507750\,\_Z^{6} \\
&+ 4698666870000\,\_Z^{5} - 37664582848875\,\_Z^{4} + 180676117955000\,\_Z^{3} \\
&- 393823660188775\,\_Z^{2} - 31640625000\,\_Z + 316406250000)
\end{aligned}
$$

```
> select(t -> type(t, numeric) and t > 0 and t < s,
>        [allvalues(")]);
```

$$[.02848997038, 9.338299136, 19.97669581]$$

MAPLE returned three extremes, and it is necessary to determine which solution is a minimum.

```
> map(t -> if subs(x = t, diff(dC, x)) < 0
>       then max else min fi, ");
```

$$[max, min, max]$$

```
> map(t -> subs(x = t, C), "");
```

$$[81.98104008, 18.24392572, 84.47655488]$$

As can be seen the same result was obtained for X, namely $x = 9.338299136$. Notice that the maximal illuminated points are located near the two lamps and not immediately beneath them.

3.3   Varying h_2 to Maximize the Illumination

In this section we use the same numerical values as in the previous one but consider the height of the second light source as a variable and maximize the illumination at X. Hence $C(x, h_2)$ is a function of two variables.

As a first step, we find the function $x(h_2)$ such that

$$C(x(h_2), h_2) = \min_{0 \le x \le s} C(x, h_2).$$

To do so, we vary h_2 from 3 $[m]$ to 9 $[m]$ and resolve the problem for each value of h_2 as in the last section.

```
> h[2] := 'h[2]':
> H2 := array(0..30):  # array for the values of h[2]
> X  := array(0..30):  # array for the values of x(h[2])
> for i from 0 to 30 do
>     H2[i] := 3 + 6*i/30:
>     X[i] := fsolve(subs(h[2]=H2[i], dC), x, 0..s):
> od:
> H2 := convert(H2, list):
> X  := convert(X, list):
```

Figure 3.3 is a plot of $x(h_2)$ generated with the following command:

```
> plot(zip((h2, x) -> [h2, x], H2, X), 3..9);
```

FIGURE 3.3.
x-Coordinate of the Minimal Illuminated Point for
$3 \le h_2 \le 9.$

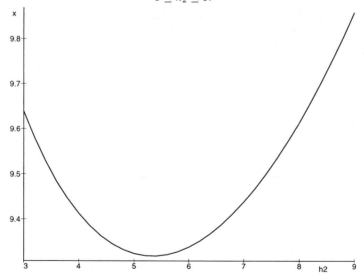

We will plot $C(x, h_2)$ as a 3-dimensional plot and overlay it with the space curve $C(x(h_2), h_2)$ of the points with minimal illumination. As expected, the space curve lies in the valley of $C(x, h_2)$.

```
> f := unapply(C, x, h[2]):
> Cu := [seq([X[i], H2[i], f(X[i], H2[i])], i = 1..31)]:
> with(plots):
> PL1 := spacecurve(Cu, thickness = 2):
> PL2 := plot3d(subs(h[2] = h2 ,C), x = -s/2..3*s/2, h2 = 3..9,
>                    style = wireframe):

> display({PL1, PL2});
```

FIGURE 3.4. *Illumination Function* $C(x, h_2)$.

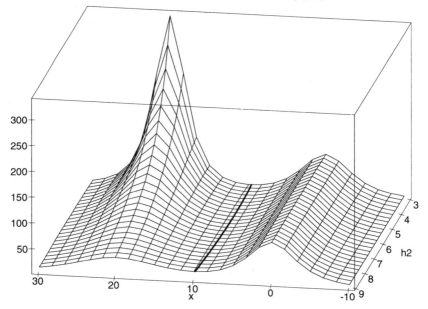

As a second step we find the point of maximal illumination on the curve $x(h_2)$. This point is among the stationary points of the function $C(x, h_2)$, i.e. among the points where the gradient of $C(x, h_2)$ becomes zero.

```
> with(linalg):
> g := grad(C, [x, h[2]]);
```

$$g := \left[-30000\,\frac{x}{(25 + x^2)^{5/2}} - 4500\,\frac{h_2\,(-40 + 2\,x)}{(h_2{}^2 + (20 - x)^2)^{5/2}},\right.$$
$$\left. \frac{3000}{(h_2{}^2 + (20 - x)^2)^{3/2}} - 9000\,\frac{h_2{}^2}{(h_2{}^2 + (20 - x)^2)^{5/2}} \right]$$

```
> Sol := fsolve({g[1] = 0, g[2] = 0}, {x, h[2]},
>               {x = 0..s, h[2] = 3..9});
```

$$Sol := \{x = 9.503151310, h_2 = 7.422392890\}$$

In order to verify that this solution is a maximum we look at the eigenvalues of the Hessian of $C(x, h_2)$ at the above point.

```
> H := subs(Sol, hessian(C, [x, h[2]]));
```

$$H := \begin{bmatrix} 1.056539134 & -.1793442077 \\ -.1793442077 & -.2536310107 \end{bmatrix}$$

```
> eigenvals(H);
```

$$-.2777372198, \ 1.080645343$$

The different signs of the eigenvalues tell us that we have found a saddle point. Since $H[1,1] = \partial^2 C/\partial x^2 \doteq 1.06 > 0$ we have a minimum in x-direction, and since $H[2,2] = \partial^2 C/\partial h_2{}^2 \doteq -0.25 < 0$ we have a maximum in h_2-direction, hence we found what we were looking for, namely the maximally illuminated point among all minimally illuminated points $(x(h_2), 0)$.

Note that we can analytically solve the second equation $g_2 = 0$ for h_2 in our numerical example.

```
> {solve(g[2] = 0, h[2])};
```

$$\{-\frac{1}{2}\sqrt{2}\,(-20 + x), \ \frac{1}{2}\sqrt{2}\,(-20 + x)\}$$

This means that for the optimal height of the second lamp, the impact angle α_2 is given by

```
> tan(alpha[2]) = normal("[1]/(s-x));
```

$$\tan(\alpha_2) = \frac{1}{2}\sqrt{2}$$

```
> evalf(arctan(rhs(")));
```

$$.6154797085$$

```
> evalf(convert(", degrees));
```

$$35.26438965 \ degrees$$

or $\alpha_2 = 35° \ 15' \ 51.8028''$.

3.4 Optimal Illumination

It is very important to have a homogeneous illumination on a road. This problem cannot be solved using point-light sources. We will always obtain a maximal illumination beneath the light sources and a minimum somewhere between them. But for lamps of a given illumination and a given separation we can maximize the illumination of the minimally illuminated point by adjusting the heights of the light sources. We will consider this problem in this section.

The total illumination $C(x, h_1, h_2)$ is now a function of three variables. The minimal illuminated point is again, as in lower dimensions, determined by the roots of the gradient of C. We try to find a general solution; in MAPLE this means unassigning the variables we used for the numerical example.

```
> P[1] := 'P[1]': P[2] := 'P[2]': h[1] := 'h[1]':
> h[2] := 'h[2]': s := 's':
> g := grad(C, [x, h[1], h[2]]);
```

$$g := \left[-3\,\frac{P_1\,h_1\,x}{(h_1{}^2+x^2)^{5/2}} - \frac{3}{2}\,\frac{P_2\,h_2\,(-2\,s+2\,x)}{(h_2{}^2+(s-x)^2)^{5/2}}, \right.$$

$$\left. \frac{P_1}{(h_1{}^2+x^2)^{3/2}} - 3\,\frac{P_1\,h_1{}^2}{(h_1{}^2+x^2)^{5/2}},\; \frac{P_2}{(h_2{}^2+(s-x)^2)^{3/2}} - 3\,\frac{P_2\,h_2{}^2}{(h_2{}^2+(s-x)^2)^{5/2}} \right]$$

MAPLE cannot analytically solve for the roots of equation $g = 0$. However, we can solve the second equation for h_1 and the third for h_2.

```
> sh1 := {solve(g[2] = 0, h[1])};
```

$$sh1 := \{-\frac{1}{2}\,\sqrt{2}\,x,\; \frac{1}{2}\,\sqrt{2}\,x\}$$

```
> sh2 := {solve(g[3] = 0, h[2])};
```

$$sh2 := \{-\frac{1}{2}\,\sqrt{2}\,(s-x),\; \frac{1}{2}\,\sqrt{2}\,(s-x)\}$$

We are interested in positive values only since it is rather unusual to illuminate a road from below.

```
> ho[1] := sh1[2];
```

$$ho_1 := \frac{1}{2}\,\sqrt{2}\,x$$

```
> ho[2]:= sh2[2];
```

$$ho_2 := \frac{1}{2}\,\sqrt{2}\,(s-x)$$

Note that the optimal height for each lamp is independent of the illumination powers. This result defines the geometry! Hence the impact angles are the same as the one we computed in Section 3.3, namely

$$\tan\alpha_1 = \tan\alpha_2 = \frac{\sqrt{2}}{2} \quad \Rightarrow \alpha_1 = \alpha_2 = 35°\ 15'\ 51.8028''.$$

By substituting the optimal heights into g_1 we can find the overall solution. We assign the real solution to the variable Xo since it is the only physically realistic solution.

```
> Q := subs(h[1] = ho[1], h[2] = ho[2], g[1]);
```

$$Q := -\frac{4}{9}\,\frac{P_1\,x^2\,\sqrt{3}}{(x^2)^{5/2}} - \frac{2}{9}\,\frac{P_2\,(s-x)\,\sqrt{3}\,(-2\,s+2\,x)}{((s-x)^2)^{5/2}}$$

```
> G := simplify(Q, symbolic);
```

$$G := -\frac{4}{9}\,\frac{\sqrt{3}\,(P_1\,s^3 - 3\,P_1\,x\,s^2 + 3\,P_1\,x^2\,s - P_2\,x^3 - P_1\,x^3)}{x^3\,(s-x)^3}$$

The equation $G = 0$ can be solved directly using the commands

```
> Xsols := [solve(G = 0, x)];
```

$$Xsols := [(\%2 - \%1 + \frac{P_1}{P_2 + P_1}) s,$$

$$(-\frac{1}{2}\%2 + \frac{1}{2}\%1 + \frac{P_1}{P_2 + P_1} + \frac{1}{2} I \sqrt{3} (\%2 + \%1)) s,$$

$$(-\frac{1}{2}\%2 + \frac{1}{2}\%1 + \frac{P_1}{P_2 + P_1} - \frac{1}{2} I \sqrt{3} (\%2 + \%1)) s]$$

$$\%1 := \frac{P_1 P_2}{(P_2 + P_1)(P_1 P_2{}^2)^{1/3}}$$

$$\%2 := \frac{(P_1 P_2{}^2)^{1/3}}{P_2 + P_1}$$

```
> Xsol := remove(has, Xsols, I);
```

$$Xsol := \left[\left(\frac{(P_1 P_2{}^2)^{1/3}}{P_2 + P_1} - \frac{P_1 P_2}{(P_2 + P_1)(P_1 P_2{}^2)^{1/3}} + \frac{P_1}{P_2 + P_1}\right) s\right]$$

```
> Xsol := simplify(Xsol, symbolic);
```

$$Xsol := [\frac{(P_1{}^{1/3} P_2{}^{2/3} - P_1{}^{2/3} P_2{}^{1/3} + P_1) s}{P_2 + P_1}]$$

```
> assume(p1>0, p2>0);
> Xo := subs(P[1] = p1^3, P[2] = p2^3, Xsol[1]):
> Xo := simplify(Xo);
```

$$Xo := \frac{s\, p1\tilde{\ }}{p1\tilde{\ } + p2\tilde{\ }}$$

```
> Xo := subs(p1=P[1]^(1/3), p2=P[2]^(1/3), Xo);
```

$$Xo := \frac{P_1{}^{1/3} s}{P_1{}^{1/3} + P_2{}^{1/3}} .$$

A geometrical determination of the optimal illumination and its physical interpretation for given values of P_1, P_2 and s can be found with some simple manipulations.

```
> eq1 := op(1, Q)^2 = normal(op(2, Q))^2;
```

$$eq1 := \frac{16}{27} \frac{P_1{}^2}{x^6} = \frac{16}{27} \frac{P_2{}^2}{(s - x)^6}$$

```
> eq2 := simplify(eq1*27*x^6*(s - x)^6/16, symbolic);
```

$$eq2 := (s - x)^6 P_1{}^2 = x^6 P_2{}^2$$

```
> eq3 := simplify(map(t -> t^(1/6), eq2), symbolic);
```

$$eq3 := (s - x) P_1{}^{1/3} = x P_2{}^{1/3}$$

The last simplification is valid since we know that all the values P_1, P_2, x are positive and $x < s$. The obtained equation can be interpreted as follows: The maximum illumination of the minimally illuminated point is obtained if the quotient of the distances x and $s - x$ is equal to the quotient of the cube roots of the light powers. In other words: The maximum possible illumination of the minimally illuminated point is reached if the volume densities of the illuminating powers are equal. The solution of eq3 is equal to Xo.

Let us plot the optimally illuminated point Xo as a function of the power of the light sources for the case $s = 20$. We will then determine the optimal heights for the case $P_1 = 2000$ and $P_2 = 3000$

```
> s := 20:
> plot3d(subs(P[1] = p1, P[2] = p2, Xo), p1 = 0..2000,
>        p2 = 0..3000, orientation = [110, 60], axes=BOXED);
> P[1] := 2000: P[2] := 3000: x := Xo:
> h[1] := ho[1];
```

$$h_1 := 10\,\frac{\sqrt{2}\,2000^{1/3}}{2000^{1/3} + 3000^{1/3}}$$

```
> evalf(");
```

$$6.593948668$$

```
> h[2] := ho[2];
```

$$h_2 := -\frac{1}{2}\,\sqrt{2}\,(-20 + 20\,\frac{2000^{1/3}}{2000^{1/3} + 3000^{1/3}})$$

```
> evalf(");
```

$$7.548186950$$

As we can see from Figure 3.5, for a wide range of power values the optimal illuminated point Xo is around $s/2 = 10$. This can also be seen by comparing the values of Xo and Xm, the position of minimal illumination computed in the first section. The relative difference is

```
> evalf( (Xm - Xo)/Xm );
```

$$.0013972025$$

but the relative difference of the illuminations is

```
> evalf( (C - Cmin)/Cmin );
```

$$.040658247$$

which is about 30 times larger.

Finally in Figure 3.6 we will compare the optimal solution Cop with that of Section 3.2 where we used the fixed heights $h_1 = 5$ and $h_2 = 6$.

```
> Cop := evalf(C);
```

$$Cop := 18.98569176$$

```
> x := 'x': Cop := C: h[1] := 5: h[2] := 6:
> plot({C, Cop}, x = -s/2..3*s/2);
```

FIGURE 3.5. *Xo as a Function of P_1 and P_2.*

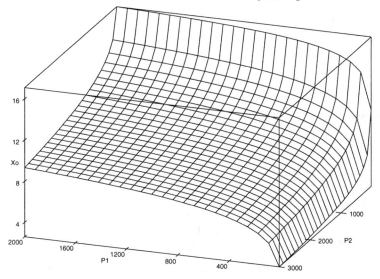

FIGURE 3.6. *The Optimal Illumination.*

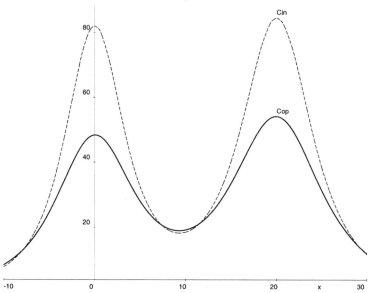

3.5   Conclusion

Based on our computations we can make some recommendations to persons involved with illumination.

- If we insert Xo into the optimal heights then they depend on the light powers P_1, P_2 and the distance s only.

$$ho_1 = \frac{1}{2} \frac{\sqrt{2}\, s\, P_1^{1/3}}{P_1^{1/3} + P_2^{1/3}} \; , \qquad ho_2 = \frac{1}{2} \frac{\sqrt{2}\, s\, P_2^{1/3}}{P_1^{1/3} + P_2^{1/3}} \; .$$

- For the special case $P_1 = P_2$ the optimal height for both lamps is

$$ho_2 = ho_2 = \frac{s}{\sqrt{8}} \; .$$

References

[1] M. E. Gettys and F. J. Keller, *Classical and modern Physics*, Mc. Graw Hill, 1989.

[2] R. G. Lerner and G. L. Trigg, *Encyclopedia of physics*, VCH Publishers, New York, 1991.

Chapter 4. Orbits in the Planar Three-Body Problem

D. Gruntz and J. Waldvogel

4.1 Introduction

The planar three-body problem is the problem of describing the motion of three point masses in the plane under their mutual Newtonian gravitation. It is a popular application of numerical integration of systems of ordinary differential equations since most solutions are too complex to be described in terms of known functions.

In addition, the three-body problem is a classical problem with a long history and many applications (see e.g. the exhaustive accounts by Szebehely [9] or by Marchal [4]). Nevertheless, the sometimes complicated interplay of the three bodies can often be described in terms of two-body interactions and is therefore qualitatively simple to understand. About 100 years ago the French Academy of Sciences set out a prize for the solution of the problem which was awarded to Sundman [7] for a series solution convergent at all times. However, owing to the excessively slow convergence of Sundman's series it is of no practical value for discussing orbits.

In this article we will demonstrate how MAPLE and MATLAB can be used efficiently to construct and display numerical solutions of the planar three-body problem. In Section 4.2 we will straight forwardly use the differential equations of motion and the numerical integrator of MATLAB. Although for most initial conditions this approach will quickly produce an initial segment of the solution, it will usually fail at a sufficiently close encounter of two bodies, owing to the singularity at the corresponding collision.

In classical celestial mechanics the regularizing transformation by T. Levi-Civita [3] is an efficient technique to overcome the problems of numerically integrating over a collision or near-collision between two bodies. Since three different pairs can be formed with three bodies it was suggested by Szebehely and Peters [8] to apply Levi-Civita's transformation to the closest pair if the mutual distance becomes smaller than a certain limit.

In Section 4.3 we will use a set of variables suggested by Waldvogel [10] that amounts to automatically regularizing each of the three types of close encounters whenever they occur. Owing to the complexity of the transformed equations of motion, the Hamiltonian formalism will be used for deriving these equations. Then MAPLE's capability of differentiating algorithms (automatic differentiation) will be used to generate the regularized equations of motion.

4.2   Equations of Motion in Physical Coordinates

Let $m_j > 0$ $(j = 0, 1, 2)$ be the masses of the three bodies, and let $x_j \in \mathbf{R}^2$ and $\dot{x}_j \in \mathbf{R}^2$ be their position and velocity (column) vectors in an inertial coordinate system (dots denoting derivatives with respect to time t). For the mutual distances of the bodies the notation of Figure 4.1 will be used:

$$r_0 = |x_2 - x_1|, \quad r_1 = |x_0 - x_2|, \quad r_2 = |x_1 - x_0|. \tag{4.1}$$

FIGURE 4.1. *The Three-Body Problem in Physical Coordinates.*

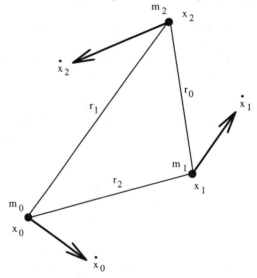

Next we notice that the Newtonian gravitational force F exerted onto m_0 by m_1 is given by

$$F = -m_0\, m_1\, \frac{x_0 - x_1}{|x_0 - x_1|^3}$$

if the units of length, time and mass are chosen such that the gravitational constant has the value 1. Therefore the Newtonian equations of motion (in their most primitive form) become

$$\ddot{x}_0 = m_1\, \frac{x_1 - x_0}{r_2{}^3} + m_2\, \frac{x_2 - x_0}{r_1{}^3}$$

$$\ddot{x}_1 = m_2\, \frac{x_2 - x_1}{r_0{}^3} + m_0\, \frac{x_0 - x_1}{r_2{}^3} \tag{4.2}$$

$$\ddot{x}_2 = m_0\, \frac{x_0 - x_2}{r_1{}^3} + m_1\, \frac{x_1 - x_2}{r_0{}^3}.$$

This system of 12 degrees of freedom may be integrated by MATLAB in a straight forward manner by defining the (column) vector $Y \in \mathbf{R}^{12}$ of dependent variables as, e.g.,

$$Y = [x_0; \ \dot{x}_0; \ x_1; \ \dot{x}_1; \ x_2; \ \dot{x}_2].$$

The system (4.2) may then be coded by the MATLAB function Ydot = f(t,Y) presented as Algorithm 4.1.

ALGORITHM 4.1. *Function* Ydot.

```
function Ydot = f(t, Y)

global m0 m1 m2 % masses of the three bodies

x0 = Y(1:2); x1 = Y(5:6); x2 = Y(9:10);
d0 = (x2-x1)/norm(x2-x1)^3;
d1 = (x0-x2)/norm(x0-x2)^3;
d2 = (x1-x0)/norm(x1-x0)^3;

Ydot( 1: 2) = Y( 3: 4);
Ydot( 5: 6) = Y( 7: 8);
Ydot( 9:10) = Y(11:12);
Ydot( 3: 4) = m1*d2 - m2*d1;
Ydot( 7: 8) = m2*d0 - m0*d2;
Ydot(11:12) = m0*d1 - m1*d0;
Ydot = Ydot(:);
```

A call to MATLAB's integrator ode113 together with a few additional lines of MATLAB code produces the orbits of the three bodies.

```
>> global m0 m1 m2;
>> m0 = 5; m1 = 3; m2 = 4;
>> x00 = [1;-1]; x10 = [1;3]; x20 = [-2;-1]; xp0 = [0;0];
>> options= odeset('RelTol',1e-10,'AbsTol',1e-10);
>> [T1,Y1] = ode113('f', [0 63], [x00;xp0;x10;xp0;x20;xp0],
                    options);
```

In the above example the so-called Pythagorean initial data

$$m_0 = 5, \quad m_1 = 3, \quad m_2 = 4,$$

$$x_0 = \begin{pmatrix} 1 \\ -1 \end{pmatrix}, \quad x_1 = \begin{pmatrix} 1 \\ 3 \end{pmatrix}, \quad x_2 = \begin{pmatrix} -2 \\ -1 \end{pmatrix}, \tag{4.3}$$

$$\dot{x}_0 = 0, \quad \dot{x}_1 = 0, \quad \dot{x}_2 = 0$$

were used for historical reasons (see also Figure 4.2). These data were first considered in 1913 by Burrau [1], but the final evolution of the system was only settled in 1967 by Szebehely and Peters [8] by careful numerical integration. An account of the history of this problem, which has no direct astronomical or physical significance, is also given in [4].

FIGURE 4.2. *Initial Configuration of the Pythagorean Problem.*

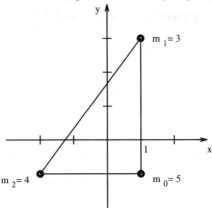

For the time interval $0 \le t \le 63$, 11016 integration steps were carried out in about three minutes of CPU time on a PC with a 200 MHz Intel Pentium Pro Processor. In Figures 4.3 and 4.4 we show the orbits of the three bodies in the intervals $0 \le t \le 10$ and $10 \le t \le 20$, and finally in Figure 4.5 the orbits in the interval $50 \le t \le 63$ are shown.

```
>> R1 = 1:1900;
>> plot(Y1(R1,1),Y1(R1,2),'-', ...
         Y1(R1,5),Y1(R1,6),':', Y1(R1,9),Y1(R1,10),'-.')
>> R2 = 1901:3633;
>> plot(Y1(R2,1),Y1(R2,2),'-', ...
         Y1(R2,5),Y1(R2,6),':', Y1(R2,9),Y1(R2,10),'-.')
>> R3 = 8515:11016;
>> plot(Y1(R3,1),Y1(R3,2),'-', ...
         Y1(R3,5),Y1(R3,6),':', Y1(R3,9),Y1(R3,10),'-.')
```

In the above example the smallest step size used was

```
>> [m,k] = min(diff(T1));
>> m

m =

    8.8727e-008

>> T1(k)

ans =

    15.8299
```

This small step was needed at $t = 15.8299$ where a near-collision between m_0 and m_2 occurred. The integrator barely managed to overcome this near-collision.

However, the accuracy of the orbit for $t > 15.83$ is rather poor, as is seen by comparing Figure 4.5 with Figure 4.9 and with the results in [8], in spite of the small error tolerance of $1e-10$.

FIGURE 4.3.
Orbits in $0 \leq t \leq 10$. Solid Line: m_0, Dotted Line: m_1, Dashdotted Line: m_2.

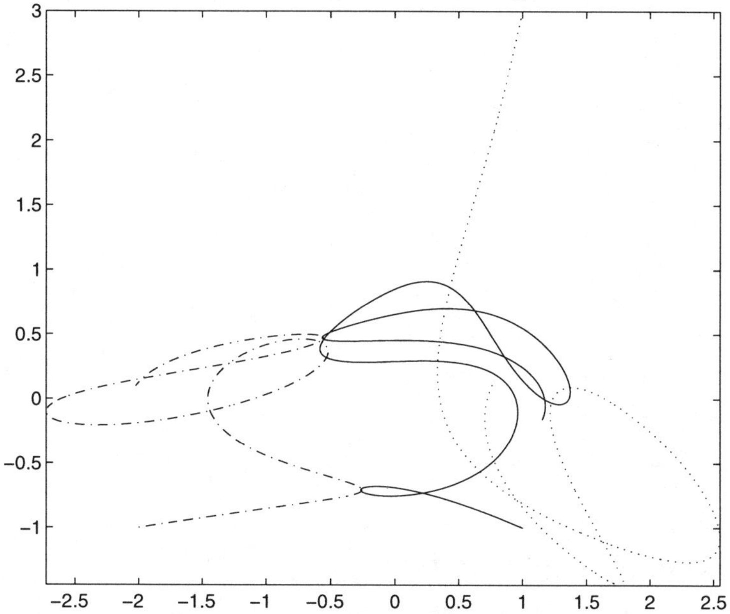

FIGURE 4.4. *Orbits in $10 \leq t \leq 20$.*

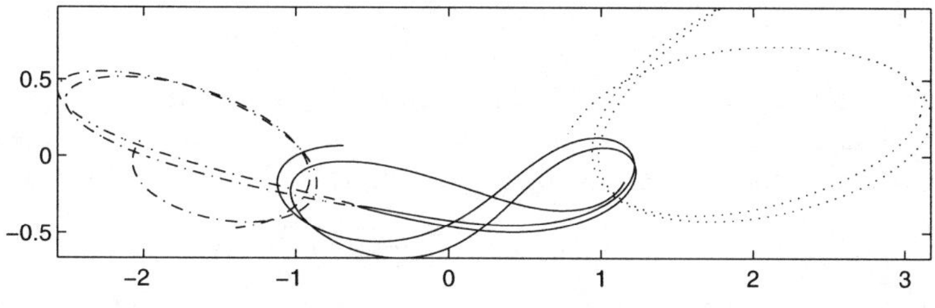

In fact, the final evolution predicted by this numerical integration is incorrect. To compensate for the high velocities at the near-collisions the integrator has to dramatically reduce the step size, sometimes rather near the level of the accuracy tolerance leading to inaccurate results. These problems will be overcome in the next section by introducing new variables such that the singularities due to all possible binary collisions are regularized.

FIGURE 4.5. *Orbits in* $50 \leq t \leq 63$.

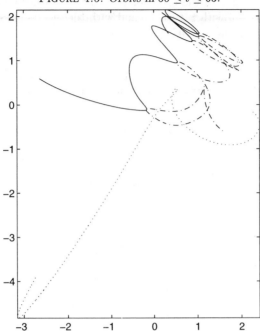

4.3   Global Regularization

For simplicity we will assume that the center of mass is initially at rest at the origin, i.e.

$$\sum_{j=0}^{2} m_j\, x_j = 0, \quad \sum_{j=0}^{2} m_j\, \dot{x}_j = 0; \tag{4.4}$$

then the equations of motion (4.2) imply that (4.4) is satisfied throughout the motion. This fact will be used to eliminate one of the variables x_j. This is best done in Hamiltonian formalism by introducing the relative coordinates

$$X = x_1 - x_0, \quad Y = x_2 - x_0 \tag{4.5}$$

with respect to x_0 as well as the canonically conjugated momenta [6]

$$P = m_1\, \dot{x}_1, \quad Q = m_2\, \dot{x}_2. \tag{4.6}$$

The mutual distances between the point masses expressed in the relative coordinates become

$$r_0 = |Y - X|, \quad r_1 = |Y|, \quad r_2 = |X|. \tag{4.7}$$

The equations of motion may then be derived from the Hamiltonian

$$H = \frac{|P + Q|^2}{2m_0} + \frac{|P|^2}{2m_1} + \frac{|Q|^2}{2m_2} - \frac{m_1\, m_2}{|Y - X|} - \frac{m_0\, m_1}{|X|} - \frac{m_0\, m_2}{|Y|} \tag{4.8}$$

as

$$\dot{X} = \frac{\partial H}{\partial P}, \quad \dot{Y} = \frac{\partial H}{\partial Q}, \quad \dot{P} = -\frac{\partial H}{\partial X}, \quad \dot{Q} = -\frac{\partial H}{\partial Y}. \tag{4.9}$$

The coordinates x_0, x_1, x_2 of the three bodies may be recovered by

$$x_0 = -\frac{m_1 X + m_2 Y}{m_0 + m_1 + m_2}, \quad x_1 = x_0 + X, \quad x_2 = x_0 + Y, \tag{4.10}$$

as follows from Equations (4.4) and (4.5).

In the following the complex number $v_1 + i v_2$ associated with a vector $v = (v_1, v_2)^T \in \mathbf{R}^2$ will be denoted by the same symbol $v \in \mathbf{C}$ for convenience. Incidentally, the conventional symbol for the modulus, $|v| = \sqrt{v_1{}^2 + v_2{}^2}$, agrees in both notations.

To regularize the equations of motion at the collision between m_0 and m_1 Levi-Civita's method calls for introducing the new complex coordinate $x \in \mathbf{C}$ instead of $X \in \mathbf{C}$ according to the conformal map $X = x^2$. Furthermore, a new (fictitious) time s must be introduced according to the relation $dt = |X| \, ds$ between the differentials [3].

A generalization of this transformation due to Waldvogel [10] will be used in order to simultaneously regularize all three types of collisions in the three-body problem. Instead of the complex coordinates X, Y and time t we will use new coordinates $x \in \mathbf{C}, y \in \mathbf{C}$ and the fictitious time s that relate to X, Y, t by

$$X = \left(\frac{x^2 - y^2}{2}\right)^2, \quad Y = \left(\frac{x^2 + y^2}{2}\right)^2, \quad dt = r_0 \, r_1 \, r_2 \, ds. \tag{4.11}$$

The key for the regularizing effect of the transformation (4.11) is the relation

$$Y - X = (x \, y)^2; \tag{4.12}$$

hence all three complex relative coordinates, $X, Y, Y - X$, are written as complete squares in the new complex coordinates.

The mechanism for transforming the equations of motion (4.9) to the new variables calls for introducing new momenta $p \in \mathbf{C}, q \in \mathbf{C}$ such that the transformation from the set (X, Y, P, Q) of variables to (x, y, p, q) is canonical. The relations defining the new momenta turn out to be [10]

$$\begin{pmatrix} p \\ q \end{pmatrix} = \overline{A}^T \begin{pmatrix} P \\ Q \end{pmatrix} \tag{4.13}$$

where

$$A = \begin{pmatrix} \frac{\partial X}{\partial x} & \frac{\partial X}{\partial y} \\ \frac{\partial Y}{\partial x} & \frac{\partial Y}{\partial y} \end{pmatrix} = \begin{pmatrix} x\,(x^2 - y^2) & -y\,(x^2 - y^2) \\ x\,(x^2 + y^2) & y\,(x^2 + y^2) \end{pmatrix} \tag{4.14}$$

and $\overline{A}^T$ denotes the complex conjugate transpose of A.

Then the regularized equations of motion are

$$\frac{dx}{ds} = \frac{\partial K}{\partial p}, \quad \frac{dy}{ds} = \frac{\partial K}{\partial q}, \quad \frac{dp}{ds} = -\frac{\partial K}{\partial x}, \quad \frac{dq}{ds} = -\frac{\partial K}{\partial y}, \quad \frac{dt}{ds} = r_0\, r_1\, r_2, \quad (4.15)$$

where

$$K = r_0\, r_1\, r_2\, (H - E) \tag{4.16}$$

is the regularized Hamiltonian, expressed in terms of the new variables, and E is the initial value of H, i.e. the total energy. Since $H(t) = E = \text{constant}$ on an orbit there follows $K(s) = 0$ on this orbit.

For the further manipulations the capabilities of MAPLE will be used as much as possible. However, the use of symbolic differentiation to form the gradient of K will produce too many terms resulting in an inefficient evaluation of the right-hand sides of (4.15).

The following method proved successful. In a first step the expression for K is written as elegantly as possible by using naturally occurring auxiliary quantities such as r_0, r_1, r_2, etc.. Then the gradient of K is evaluated by means of automatic differentiation thus generating efficient code for it.

Let $x = x_1 + i\, x_2$, $y = y_1 + i\, y_2$ with $x_1 \in \mathbf{R}, x_2 \in \mathbf{R}$ in a new meaning. Then we obtain according to (4.11) and (4.7)

```
> x  := x1 + I*x2:
> y  := y1 + I*y2:
> X  := ((x^2-y^2)/2)^2:
> Y  := ((x^2+y^2)/2)^2:
> r0 := factor(evalc(abs(Y-X))):
> r0 := simplify(r0, power, symbolic);
```

$$r0 := (y2^2 + y1^2)\,(x1^2 + x2^2)$$

```
> r1 := factor(evalc(abs(Y)));
```

$$r1 := \frac{1}{4}(x2^2 + x1^2 + 2\,x2\,y1 + y1^2 - 2\,x1\,y2 + y2^2)$$
$$(x2^2 + x1^2 - 2\,x2\,y1 + y1^2 + 2\,x1\,y2 + y2^2)$$

```
> r2 := factor(evalc(abs(X)));
```

$$r2 := \frac{1}{4}(x2^2 + x1^2 + 2\,x1\,y1 + y1^2 + 2\,x2\,y2 + y2^2)$$
$$(x2^2 + x1^2 - 2\,x1\,y1 + y1^2 - 2\,x2\,y2 + y2^2)$$

It is easy to see that the factors of r_1 and r_2 are sums of squares. We have written a small MAPLE procedure **reduce** (see Algorithm 4.4 on page 71) to simplify such expressions, since there exists no direct MAPLE command to perform this operation.

```
> r1 := reduce(r1);
```

$$r1 := \frac{1}{4}\left((x2 + y1)^2 + (x1 - y2)^2\right)\left((x2 - y1)^2 + (x1 + y2)^2\right)$$

```
> r2 := reduce(r2);
```

$$r2 := \frac{1}{4}\left((x2 + y2)^2 + (x1 + y1)^2\right)\left((x2 - y2)^2 + (x1 - y1)^2\right)$$

According to (4.16), (4.8) and (4.7) we obtain for K

$$K = \frac{L_0}{2m_0} + \frac{L_1}{2m_1} + \frac{L_2}{2m_2} - (m_0 m_1 r_0 r_1 + m_1 m_2 r_1 r_2 + m_2 m_0 r_2 r_0) - E r_0 r_1 r_2.$$
(4.17)

For the auxiliary quantities L_j a simple calculation yields

$$
\begin{aligned}
L_0 &= r_0 r_1 r_2 |P + Q|^2 = \frac{r_0}{16} |\overline{x} p - \overline{y} q|^2 \\
L_1 &= r_0 r_1 r_2 |P|^2 = \frac{r_1}{16} |\overline{y} p - \overline{x} q|^2 \\
L_2 &= r_0 r_1 r_2 |Q|^2 = \frac{r_2}{16} |\overline{y} p + \overline{x} q|^2.
\end{aligned}
$$
(4.18)

One way of writing these expressions in mostly factored form while using real notation with $p = p_1 + i p_2$, $q = q_1 + i q_2$ is

$$
\begin{aligned}
L_0 &= \frac{r_0}{16}\left(|p|^2 |x|^2 + |q|^2 |y|^2 - A - B\right) \\
L_1 &= \frac{r_1}{16}\left(|p|^2 |y|^2 + |q|^2 |x|^2 - A + B\right) \\
L_2 &= \frac{r_2}{16}\left(|p|^2 |y|^2 + |q|^2 |x|^2 + A - B\right)
\end{aligned}
$$

where

$$
\begin{aligned}
A &= 2(x_1 y_1 + x_2 y_2)(p_1 q_1 + p_2 q_2) \\
B &= 2(x_2 y_1 - x_1 y_2)(p_2 q_1 - p_1 q_2).
\end{aligned}
$$

In Algorithm 4.2 a MAPLE procedure computing the regularized Hamiltonian K is given, where the meaning of L_j has been slightly modified and EE is used for the total energy E.

To compute the partial derivatives of K with respect to all its arguments we use the automatic differentiation capability of MAPLE. For an introduction to automatic (or algorithmic) differentiation we refer to [2, 5]. It is well known that the so-called *reverse mode* of automatic differentiation is best suited for computing gradients. Applying the *forward mode* to K leads to a procedure with about 300 multiplications (after optimization), whereas the reverse mode leads to a procedure with only about 200 multiplications. This number can still

ALGORITHM 4.2. *Procedure* K.

```
K := proc(x1,x2,y1,y2,p1,p2,q1,q2)
     local xx,yy,pp,qq,r0,r1,r2,
           A,B,L0,L1,L2,m01,m12,m20,apb,amb;

     xx:=x1^2+x2^2; yy:=y1^2+y2^2;
     pp:=p1^2+p2^2; qq:=q1^2+q2^2;

     r0:=xx*yy;
     r1:=((x1-y2)^2+(x2+y1)^2)*((x1+y2)^2+(x2-y1)^2)/4;
     r2:=((x1+y1)^2+(x2+y2)^2)*((x1-y1)^2+(x2-y2)^2)/4;

     A:=2*(p1*q1+p2*q2)*(x1*y1+x2*y2);
     B:=2*(p2*q1-p1*q2)*(x2*y1-x1*y2);

     apb:=A+B:
     amb:=A-B:
     L0:=r0*(pp*xx+qq*yy-apb);
     L1:=r1*(pp*yy+qq*xx-amb);
     L2:=r2*(pp*yy+qq*xx+amb);

     m01 := m0*m1;
     m12 := m1*m2;
     m20 := m2*m0;

     L0/32/m0+L1/32/m1+L2/32/m2-m01*r0*r1-m12*r1*r2-m20*r2*r0
        -EE*r0*r1*r2;
   end:
```

be reduced if we first split up the products in the procedure K in order to avoid the generation of common subexpressions when computing the derivatives.

```
> SK   := SPLIT(K):
> RSK  := REVERSEMODE(SK):
> ORSK := readlib(optimize)(RSK):
> COST(ORSK);
```

138 *multiplications* + 129 *assignments* + 81 *subscripts* + 86 *additions*

+ 3 *divisions* + *functions*

The procedures we used here can be obtained using anonymous FTP from `ftp.inf.ethz.ch`[1]. The final result is the MAPLE procedure ORSK to evaluate the right-hand sides of Equations (4.15) with only 144 multiplicative operations (multiplications and divisions). One of the authors [10] has written a hand-optimized procedure which requires roughly 100 multiplicative operations and 50 additions, so the result obtained by MAPLE is almost optimal.

---

[1]URL: `ftp://ftp.inf.ethz.ch/pub/SolvingProblems/ed3/chap04/`

Notice that MAPLE can also be used to *prove* that the procedure ORSK is in fact correct. For that we compute the gradient using the `gradient` function of the `linalg` package and compare it element by element with the result generated by the procedure ORSK on symbolic input arguments.

```
> G1 := linalg[grad](K(x1,x2,y1,y2,p1,p2,q1,q2),
>                          [x1,x2,y1,y2,p1,p2,q1,q2]):
> G2 := [ORSK(x1,x2,y1,y2,p1,p2,q1,q2)]:
> zip((g1,g2)->expand(g1-g2), convert(G1, list), G2);
```

$$[0,\ 0,\ 0,\ 0,\ 0,\ 0,\ 0,\ 0]$$

For convenience numerical integration and graphical output of the orbits will again be done in MATLAB. To convert the MAPLE procedure ORSK into a MAT-LAB procedure we have to make some syntactical changes which can be done with the help of a simple editor. First, the MAPLE assignments ":=" must be converted to the MATLAB notation "=", and array references must be converted from square brackets (`a[1]`) to round parentheses (`a(1)`). Additionally a procedure head must be added and the result must be stored in an array. This leads to a MATLAB procedure of about 150 lines which is shown in Algorithm 4.3.

ALGORITHM 4.3. *Function* threebp.

```
function yprime = threebp(s, y)

    global m0 m1 m2 EE

    x1 = y(1); x2 = y(2); y1 = y(3); y2 = y(4);
    p1 = y(5); p2 = y(6); q1 = y(7); q2 = y(8);

    % here comes the Maple generated code
    t1 = x1^2;
    ...
    ...
    t137 = t43*xx+t44*xx+t45*yy;

    grd(1) = -y2*t96+y1*t98+t100+t101+t102+t103+2*x1*t104;
    grd(2) = y1*t96+y2*t98+t109+t110+t111+t112+2*x2*t104;
    grd(3) = x2*t96+x1*t98-t100+t101-t111+t112+2*y1*t117;
    grd(4) = -x1*t96+x2*t98-t109+t110+t102-t103+2*y2*t117;
    grd(5) = -q2*t124+q1*t126+2*p1*t128;
    grd(6) = q1*t124+q2*t126+2*p2*t128;
    grd(7) = p2*t124+p1*t126+2*q1*t137;
    grd(8) = -p1*t124+p2*t126+2*q2*t137;

    yprime(1:4)= grd(5:8);
    yprime(5:8)=-grd(1:4);
    yprime(9)=r0*r1*r2;
end
```

Since in regularized variables time t is a dependent variable the orbit itself is obtained from the first 8 differential equations of the system (4.15), whereas the temporal evolution of the three-body system may be obtained from the last equation.

When integrating this system of differential equations the time-critical part is the computation of the derivative, i.e. the evaluation of the above procedure, which is executed for every step at least once. Since procedures in MATLAB are interpreted it is a good idea to write a C program (again some syntactical changes) to perform this computation and to link it dynamically to MATLAB. This way a speed-up factor of about 10 is achieved. For a more detailed description of the mechanism of linking C code to MAPLE we refer to Section 9.3.4 on page 146.

4.4 The Pythagorean Three-Body Problem

For integrating an orbit in regularized variables the initial values of x, y, p, q at $s = t = 0$ must be calculated by inverting the transformation (4.11),

$$x = \sqrt{\sqrt{Y} + \sqrt{X}}, \quad y = \sqrt{\sqrt{Y} - \sqrt{X}},$$

and by applying (4.13). For the Pythagorean initial data (see Equation (4.3) and Figure 4.2) we may again use MAPLE to obtain these initial values. The function evalc splits a complex expression into its real and imaginary parts, and the function radnormal simplifies radicals.

```
> Digits := 20: readlib(radnormal):
> X := 4*I:
> Y := -3:
> x := sqrt(sqrt(Y)+sqrt(X));
```

$$x := \sqrt{I\sqrt{3} + \sqrt{2} + I\sqrt{2}}$$

```
> x := map(t->map(radnormal,t),evalc(x));
```

$$x := \frac{1}{2}\sqrt{2 + 2\sqrt{3}\sqrt{2} + 2\sqrt{2}} + \frac{1}{2}I\sqrt{2 + 2\sqrt{3}\sqrt{2} - 2\sqrt{2}}$$

```
> evalf(");
```

$$1.5594395315555318362 + 1.0087804965427521214\,I$$

```
> y := sqrt(sqrt(Y)-sqrt(X));
```

$$y := \sqrt{I\sqrt{3} - \sqrt{2} - I\sqrt{2}}$$

```
> y := map(t->map(radnormal,t),evalc(y));
```

$$y := \frac{1}{2}\sqrt{-2 + 2\sqrt{3}\sqrt{2} - 2\sqrt{2}} + \frac{1}{2}I\sqrt{-2 + 2\sqrt{3}\sqrt{2} + 2\sqrt{2}}$$

```
> evalf(");
```

$$.13280847188730666477 + 1.1966000386838271257\,I$$

Since the bodies are at rest initially, $p = q = 0$. The total energy E can be obtained from the condition $K(0) = 0$.

```
> m0 := 5: m1 := 3: m2 := 4:
> radnormal(solve(K(Re(x), Im(x), Re(y), Im(y), 0, 0, 0, 0), EE));
```

$$\frac{-769}{60}$$

We now switch to MATLAB to integrate the system of differential equation using ode45.

```
>> global m0 m1 m2 EE;
>> m0 = 5; m1 = 3; m2 = 4;
>> p10 = 0; q10 = 0; p20 = 0; q20 = 0;
>> x10 = 1.5594395315555318362;  x20 = 1.0087804965427521214;
>> y10 = 0.13280847188730666477; y20 = 1.1966000386838271256;
>> EE = -769/60;
>> options= odeset('RelTol',1e-10,'AbsTol',1e-10);
>> [S,Z] = ode113('threebp', [0 7.98], ...
                  [x10,x20,y10,y20,p10,p20,q10,q20,0], options);
```

For the above integration with the accuracy tolerance 10^{-10}, 2768 integration steps were needed (about 1 minute on a Pentium Pro 200). The interval $0 \leq s \leq 7.98$ in fictitious time corresponds to $0 \leq t \leq t_f = 63.6127$.

```
>> size(Z)

ans =

        2768        9

>> Z(size(Z))

ans =

    63.6127
```

Next we recover the coordinates x_0, x_1, x_2 of the three bodies by means of Equations (4.11) and (4.10).

```
>> x = Z(:,1)+i*Z(:,2);
>> y = Z(:,3)+i*Z(:,4);
```

```
>> X = (x.^2-y.^2).^2/4;
>> Y = (x.^2+y.^2).^2/4;
>> x0 = - (m1*X+m2*Y) / (m0+m1+m2);
>> x1 = x0 + X;
>> x2 = x0 + Y;
```

We now may look at the orbits of the three bodies in animated graphics with
the following commands.

```
>> clf
>> axis([-3.3361 3.4907 -4.9904 8.5989])
>> hold on
>> [n,e]=size(x1);
>> for k=1:n-1,
>>     plot(x0(k:k+1),'r-','EraseMode','none');
>>     plot(x1(k:k+1),'g:','EraseMode','none');
>>     plot(x2(k:k+1),'b-.','EraseMode','none');
>>     drawnow
>> end
>> hold off
```

FIGURE 4.6. *Orbits in* $20 \leq t \leq 30$.

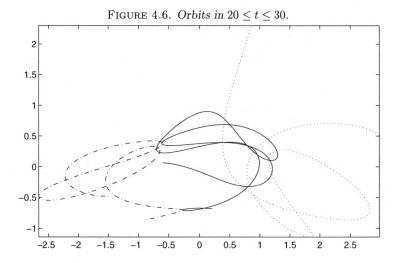

A few snapshots of this animation will be given in the following. The orbits for
$0 \leq t \leq 20$ are the same as those shown in Figures 4.3 and 4.4. The remaining
time interval is covered by Figure 4.6, Figure 4.8 and Figure 4.9. In these figures
the motion of the mass m_1 is shown by a dotted line, the dashes represent the
orbit of m_2 and the solid line illustrates the motion of the body with the largest
mass m_0.

Since we have all data of the orbits available, it is possible to discuss the
orbits further. We may for example find the times of near-collisions of two
bodies. This can be visualized by plotting the distances r_2 and r_1 which are the

FIGURE 4.7. Distances r_2 and r_1.

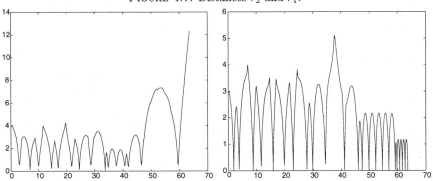

absolute values of X and Y according to (4.7). We plot these distances versus
the physical time t:

```
>> T2=Z(:,9);
>> plot(T2, abs(X))
>> plot(T2, abs(Y))
```

The smallest distance between any two bodies at any time $t \geq 0$ during the
evolution of this three-body system occurs at $t = 15.8299$ between m_0 and m_2:
$r_1 = 9.1094e{-}04$. This near-collision can be seen in Figure 4.4 and in more
detail in Figure 4.10.

```
>> [m,k]=min(abs(Y))


m =


    4.4744e-004


k =


   655

>> T2(k)

ans =

   15.8299
```

A simple way to discuss the velocity vectors of the three masses is to use the
forward difference quotients between two consecutive integration steps as an
approximation, e.g.

$$\dot{x}_0 = \texttt{diff(x0) ./ diff(T)}.$$

FIGURE 4.8. *Orbits in the Intervals* $30 \leq t \leq 40$ *and* $40 \leq t \leq 50$.

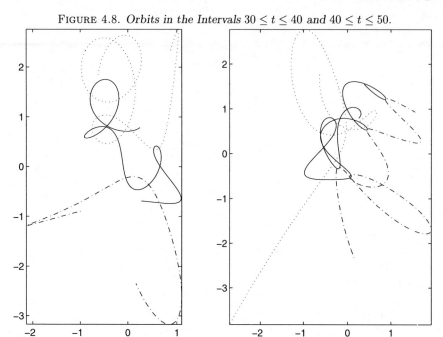

However, it is also possible to obtain the true velocity vectors from the orbital data by using Equation (4.13) and (4.6). First we prepare the symbolic expressions for P and Q by means of MAPLE (suppressing the complex conjugations for the moment):

```
> x := 'x': y := 'y':
> B := linalg[matrix](2,2,[[x*(x^2-y^2), -y*(x^2-y^2)],
>                          [x*(x^2+y^2), y*(x^2+y^2)]]);
```

$$B := \begin{bmatrix} x\,(x^2 - y^2) & -y\,(x^2 - y^2) \\ x\,(x^2 + y^2) & y\,(x^2 + y^2) \end{bmatrix}$$

```
> invB := linalg[inverse](transpose(B)):
> pq := linalg[vector]([p,q]):
> PQ := map(normal,linalg[multiply](invB,pq));
```

$$PQ := \left[-\frac{1}{2}\,\frac{p\,y - q\,x}{x\,(-x^2 + y^2)\,y},\ \frac{1}{2}\,\frac{p\,y + q\,x}{x\,(x^2 + y^2)\,y} \right]$$

FIGURE 4.9. *Orbits in* $50 \leq t \leq 63$.

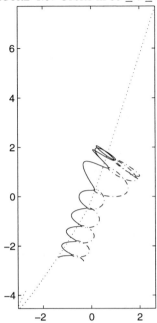

This agrees, apart from the bars, with the results given in [10], namely

$$P = \frac{p\,\bar{y} - q\,\bar{x}}{2\bar{x}\,\bar{y}\,(\bar{x}^2 - \bar{y}^2)}, \quad Q = \frac{p\,\bar{y} + q\,\bar{x}}{2\bar{x}\,\bar{y}\,(\bar{x}^2 + \bar{y}^2)}. \tag{4.19}$$

Again we may look for the maximum speeds. It turns out that m_0 and m_2 have their maximum speeds at $t = 15.8299$, the near-collision we found in Section 4.2. At this time, the velocity of m_1 is very small. Note that these maximum speeds are the maxima on the discrete set of points generated by the numerical integration. Therefore the maxima may vary considerably if the integrator is changed

```
>> xbar = Z(:,1)-i*Z(:,2);
>> ybar = Z(:,3)-i*Z(:,4);
>> p = Z(:,5)+i*Z(:,6);
>> q = Z(:,7)+i*Z(:,8);
>> P = (p ./ xbar - q ./ ybar) ./ (2*(xbar.^2 - ybar.^2));
>> Q = (p ./ xbar + q ./ ybar) ./ (2*(xbar.^2 + ybar.^2));
>> vx1 = P/m1;
>> vx2 = Q/m2;
>> vx0 = - (m1*vx1+m2*vx2)/m0;
```

```
>> [v,k]=max(abs(vx0))

v =

   89.1372

k =

   655

>> [v,k]=max(abs(vx2))

v =

   111.4273

k =

   655

>> abs(vx1(k))

ans =

   0.0404
```

The near-collision at $t = 15.8299$ has a special significance. Notice that the
three bodies describe approximately the same orbits before and after this close
encounter (see Figure 4.4). This trend continues through Figure 4.6, and at $t \approx$
31.66 the three bodies approximately occupy their initial positions with small
velocities (see left plot in Figure 4.8). This means that near the Pythagorean
initial conditions a periodic solution exists. In Figure 4.10 we show a detail of
Figure 4.4 which was generated from the data obtained in Section 4.2 by the
following commands:

```
>> [m,k] = min(diff(T1));
>> R = k-200:k+200;
>> plot(Y1(R,1),Y1(R,2),'-', ...
        Y1(R,5),Y1(R,6),':', Y1(R,9),Y1(R,10),'-.')
>> hold on
>> plot(Y1(k,1), Y1(k,2), '+')
>> plot(Y1(k,9), Y1(k,10), '+')
>> axis([-0.60 -0.59 -0.33 -0.32])
>> hold off
```

Finally, Figure 4.9 shows the asymptotic behavior of the three orbits after a
close encounter of all three bodies. Notice that the final evolution obtained in
Section 4.2 (Figure 4.5) with the direct approach is incorrect.

FIGURE 4.10. *Orbits of m_2 and m_0 near $t = 15.8299$.*

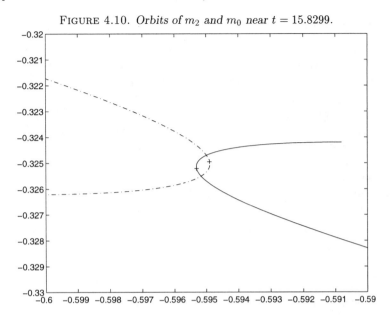

Another interesting plot is the size of the steps used during the integration. The maximum step used with the global regularization was 0.0264 at $t = 59.4421$ and the minimum step, except for the initial and the final phase of the motion, was $3.7553e{-}004$ at $t = 52.8367$. For the direct approach described in Section 4.2 the minimum step used was $8.8727e{-}008$ (see page 54). In Figure 4.11 we show the step size used for the integrations with the direct approach (left) and with the global regularization (right). Notice that the two pictures are scaled differently.

```
>> plot(diff(T1)); axis([0,11015,0,0.07])
>> plot(diff(S)); axis([0,2767,0,0.03])
```

FIGURE 4.11. *Step Sizes During the Integrations.*

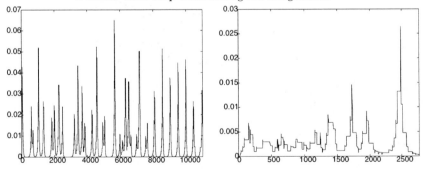

4.5   Conclusions

Experiments show that the long-term evolution of the three-body system is
extremely sensitive to the initial data and to the accuracy of the numerical
integration. Typically, neighboring orbits deviate exponentially in time. This
means that systems like the one at hand are not predictable over a long time
span (like the weather).

Nevertheless, the mathematical problem has a well defined, unique solution
which can be approximated over a finite time interval $0 \leq t \leq t_f$ if the numerical
integration is done with sufficient accuracy.

A rough check of the achieved accuracy in the interval $[0, t_f]$ is provided by
the Hamiltonian $K(s)$ which theoretically vanishes on the entire orbit. In our
example the maximum value of $K(s)$ for $s \in [0, 8]$ is

```
>> for k=1:n, KK(k) = K(Z(k,1:8)); end;
>> norm(KK,inf)

ans =

    1.9638e-006
```

A more reliable check is obtained by integrating the same orbit with different
accuracy tolerances. The orbit is valid at least as long as the different approximations agree to sufficient accuracy. Alternatively, the validity of an orbit may
be checked by integrating it backwards.

ALGORITHM 4.4. *Function* reduce.

```
reduce := proc(a) local p,P,S,c,i,j,f;
    if type(a,{name,constant}) then a
    elif type(a,'+') then p := a; P := convert(p,list);
        S := map(t -> if (type(t,'^') and type(op(2,t),even))
                        then op(1,t)^(op(2,t)/2) fi, P);
        for i to nops(S) do
            for j from i+1 to nops(S) do
                if    has(p, 2*S[i]*S[j]) then
                    p := p - (S[i]^2+2*S[i]*S[j]+S[j]^2)
                        + (S[i]+S[j])^2
                elif has(p,-2*S[i]*S[j]) then
                    p := p - (S[i]^2-2*S[i]*S[j]+S[j]^2)
                        + (S[i]-S[j])^2
                fi
            od;
        od;
        p
    else map(reduce, a)
    fi
end:
```

In the Pythagorean problem of three bodies it is found that the motion begins with a long interplay involving many close encounters of all three types. At time $t = 46.6$ the smallest mass, m_1, shoots through a newly formed binary of m_0 and m_2 and is accelerated to a near-escape in the third quadrant. However, m_1 comes back, just to receive another kick at $t = 59.4$ after which it is definitely expelled in the first quadrant, whereas the binary escapes in the opposite direction. Possibly, this is one of the mechanisms at work in the formation of binaries and of halos in galaxies.

Maple Code of the procedure reduce

The procedure reduce in Algorithm 4.4 tries to write a given expression a as a sum of squares.

References

[1] C. BURRAU, *Numerische Berechnung eines Spezialfalles des Dreikörperproblems,* Astron. Nachr. 195, 1913, p. 113.

[2] A. GRIEWANK and G.F. CORLISS editors, *Automatic Differentiation of Algorithms: Theory, Implementation, and Application,* Proceedings of the SIAM Workshop on Automatic Differentiation, held in Breckenridge, Colorado,1991, SIAM Philadelphia, 1991.

[3] T. LEVI-CIVITA, *Sur la régularisation du problème des trois corps,* Acta Math. 42, 1920, pp. 99-144.

[4] CH. MARCHAL, *The Three-Body Problem,* Elsevier, 1990.

[5] M.B. MONAGAN and W.M. NEUENSCHWANDER, *GRADIENT: Algorithmic Differentiation in Maple,* Proceedings of the 1993 International Symposium on Symbolic and Algebraic Computation, ISSAC'93, 1993, p. 68.

[6] C.L. SIEGEL and J.K. MOSER, *Lectures on Celestial Mechanics,* Springer, 1971.

[7] K.F. SUNDMAN, *Mémoire sur le problème des trois corps,* Acta Math. 36, 1912, pp. 105-179.

[8] V. SZEBEHELY and C.F. PETERS, *Complete Solution of a General Problem of Three Bodies,* Astron. J. 72, 1967, pp. 876-883.

[9] V. SZEBEHELY, *Theory of Orbits,* Academic Press, 1967.

[10] J. WALDVOGEL, *A New Regularization of the Planar Problem of Three Bodies,* Celest. Mech. 6, 1972, pp. 221-231.

Chapter 5. The Internal Field in Semiconductors

F. Klvaňa

5.1 Introduction

Let us consider a semiconductor of length l in x-direction, which is doped with a concentration of electrically active impurities $C(x) = C_D^+(x) - C_A^-(x)$. The C_A^-, C_D^+ are the acceptor and donor impurity concentrations respectively and are independent of y and z. Let this semiconductor be connected to an external potential $U(x)$ with $U(0) = U_0$ and $U(l) = 0$. Then, if the semiconductor has sufficiently large dimensions in y- and z-directions, all physical properties will depend only on x, and we can study it as a one-dimensional object.

We consider the problem of finding the internal potential in this semiconductor for a given carrier (electrons and holes) concentration $C(x)$. To simplify the problem we will assume that we can neglect recombination and generation of carriers.

Under the standard conditions we can assume the system of electrons and holes to be a classical system. Then the equilibrium concentrations of electrons $n(x)$ and of holes $p(x)$ in an internal potential field $\psi(x)$ are given by the Boltzmann statistics ([2])

$$n(x) = n_i \, e^{\frac{q}{\theta}(\psi(x) - \varphi_F)}, \qquad p(x) = n_i \, e^{\frac{-q}{\theta}(\psi(x) - \varphi_F)},$$

where n_i is an intrinsic concentration of electrons. φ_F is the so-called Fermi potential, which is constant over the whole semiconductor, and is taken as a reference level for the potential ($\varphi_F = 0$). q is the charge of an electron, $\theta = kT$ is the statistical temperature.

Assuming that the above conditions hold, the internal potential $\psi(x)$ is given as a solution of the Poisson equation, which for our one-dimensional problem has the form

$$\frac{d^2\psi(x)}{dx^2} = \frac{q}{\varepsilon} \, (n(x) - p(x) - C(x)), \tag{5.1}$$

where ε is the dielectric constant.

It is useful to represent the concentration $C(x)$ by the so-called builtin potential $\psi_D(x)$ using the definition

$$C(x) = n_i \, (e^{\frac{q}{\theta}\psi_D(x)} - e^{-\frac{q}{\theta}\psi_D(x)}). \tag{5.2}$$

The boundary conditions for Equation (5.1) are of Dirichlet type and have the form

$$\psi(0) = \psi_D(0) + U_0 \quad , \qquad \psi(l) = \psi_D(l). \tag{5.3}$$

Finally we introduce the dimensionless quantities

$$\varphi(X) = \frac{q}{\theta}\psi(x) \,, \quad \varphi_D(X) = \frac{q}{\theta}\psi_D(x) \,, \quad u_0 = \frac{q}{\theta}U_0 \,, \quad c(X) = \frac{C(x)}{n_i} \tag{5.4}$$

with the scaling transformation $x = \lambda_D \cdot X$, where $\lambda_D = \sqrt{\varepsilon\,\theta/q^2\,n_i}$ is the Debye length. From Equations (5.2) and (5.4) it follows that

$$\varphi_D(X) = \operatorname{arcsinh}\left(\frac{c(X)}{2}\right) = \ln\left(\frac{1}{2}\left(\sqrt{c(X)^2 + 4} + c(X)\right)\right) \,.$$

Equation (5.1) is transformed to

$$\frac{d^2\varphi(X)}{dX^2} = e^{\varphi(X)} - e^{-\varphi(X)} - c(X), \tag{5.5}$$

with the boundary conditions

$$\varphi(0) = \varphi_D(0) + u_0 \,, \qquad \varphi(L) = \varphi_D(L) \,, \quad \text{where } L = \frac{l}{\lambda_D}. \tag{5.6}$$

5.2   Solving a Nonlinear Poisson Equation Using MAPLE

Equation (5.5) together with boundary conditions (5.6) represents a nonlinear boundary value problem which has to be solved numerically. MAPLE cannot solve a boundary value problem directly, an appropriate numerical method must be implemented for its solution. We choose the method of finite differences (see e.g. [1]).

Let us define a mesh of $N + 1$ points $X_0 = 0, X_1, ..., X_N = L$ on the interval $[0, L]$ and denote $h_i = X_{i+1} - X_i$ ($i = 0, 1, ..., N - 1$) and $\varphi_i = \varphi(X_i)$, $\varphi_{Di} = \varphi_D(X_i)$, $c_i = c(X_i)$. Then the discretized form of Equation (5.5) in interior points is

$$\frac{1}{h_i^*}\left(\frac{\varphi_{i+1} - \varphi_i}{h_i} - \frac{\varphi_i - \varphi_{i-1}}{h_{i-1}}\right) = e^{\varphi_i} - e^{-\varphi_i} - c_i \,, \qquad i = 1, ..., N - 1, \tag{5.7}$$

where $h_i^* = \frac{1}{2}(h_{i-1} + h_i)$. Together with the boundary conditions

$$\varphi_0 = \varphi_{D0} + u_0, \quad \varphi_N = \varphi_{DN},$$

this is a system of nonlinear equations for the unknowns $\varphi_1, ..., \varphi_{N-1}$:

$$\begin{aligned} a_1\,\varphi_1 + b_1\,\varphi_2 &= F_1(\varphi_1) \\ b_{i-1}\,\varphi_{i-1} + a_i\,\varphi_i + b_i\,\varphi_{i+1} &= F_i(\varphi_i) \qquad \text{for} \quad i = 2, ..., N - 2 \\ b_{N-2}\,\varphi_{N-2} + a_{N-1}\,\varphi_{N-1} &= F_{N-1}(\varphi_{N-1}) \end{aligned}$$

where

$$
\begin{aligned}
b_i &= \tfrac{1}{h_i}, \quad a_i = -(b_{i-1} + b_i), \\
F_1 &= h_1^* \left(e^{\varphi_1} - e^{-\varphi_1} - c_1 \right) - b_0\, \varphi_0, \\
F_{N-1} &= h_{N-1}^* \left(e^{\varphi_{N-1}} - e^{-\varphi_{N-1}} - c_{N-1} \right) - b_{N-1}\, \varphi_N, \\
F_i &= h_i^* \left(e^{\varphi_i} - e^{-\varphi_i} - c_i \right) \qquad \text{for} \quad i = 2, \ldots, N-2.
\end{aligned}
$$

We rewrite the above system of Equations (5.8) in matrix form as

$$
\hat{A} \cdot \hat{\varphi} = \hat{F}(\hat{\varphi})
$$

where $\hat{A}$ is a symmetric tridiagonal matrix with the nonzero elements

$$
A_{ii} = a_i, \quad A_{i,i-1} = b_{i-1}, \quad A_{i,i+1} = b_i,
$$

and $\hat{\varphi}$ is the column vector with the elements $\varphi_1, \ldots, \varphi_{N-1}$. To solve this system of nonlinear equations we use Newton's method (see for example [1]).

Let us define the vector function

$$
\hat{G}(\hat{\varphi}) = \hat{A} \cdot \hat{\varphi} - \hat{F}(\hat{\varphi}).
$$

Newton's method defines a sequence

$$
\hat{\varphi}^{(0)}, \, \hat{\varphi}^{(1)}, \, \hat{\varphi}^{(2)}, \cdots
$$

of approximate solutions of the system $\hat{G}(\hat{\varphi}) = 0$ by the recurrence relation

$$
\hat{\varphi}^{(k+1)} = \hat{\varphi}^{(k)} + \hat{H}^{(k)}.
$$

The correction $\hat{H}^{(k)}$ is the solution of the system of equations

$$
\hat{G}(\hat{\varphi}^{(k)}) + \hat{J}(\hat{\varphi}^{(k)})\hat{H}^{(k)} = 0,
$$

where the Jacobian $\hat{J}(\hat{\varphi}^{(k)})$ contains the partial derivatives of the vector function $\hat{G}(\hat{\varphi}^{(k)})$. The i-th equation of the system is given by

$$
G_i(\hat{\varphi}^{(k)}) + \sum_{j=1}^{N-1} \frac{\partial G_i(\hat{\varphi}^{(k)})}{\partial \varphi_j^{(k)}} \cdot H_j^{(k)} = 0 \qquad \text{for} \quad i = 1, \ldots, N-1. \tag{5.8}
$$

Since

$$
\frac{\partial G_i}{\partial \varphi_j} = A_{ij} - \frac{\partial F_i}{\partial \varphi_j} = A_{ij} - h_i^* \left(e^{\varphi_i} + e^{-\varphi_i} \right) \cdot \delta_{ij},
$$

the Jacobian splits into a constant matrix $\hat{A}$ and a diagonal matrix $\hat{D}(\varphi_j^{(k)})$ depending on the current iterate. The system of linear equations for the correction becomes

$$
(\hat{A} - \hat{D}(\varphi_j^{(k)}))\hat{H}^{(k)} = \hat{F}(\hat{\varphi}^{(k)}) - \hat{A} \cdot \hat{\varphi}^{(k)}, \tag{5.9}
$$

where $D_{ii} = h_i^* \left(e^{\varphi_i} + e^{-\varphi_i} \right)$. As an initial guess $\hat{\varphi}^{(0)}$, it is recommended for physical reasons to use the builtin potential $\hat{\varphi}_D$. The iteration process is stopped if

$$\|\hat{\varphi}^{(k+1)} - \hat{\varphi}^{(k)}\|_{max} = \|\hat{H}^{(k)}\|_{max} = \max_i |H_i^{(k)}| < \epsilon$$

where ϵ is an error tolerance for the result.

It is a complicated task to prove convergence of our iteration process. But for physically realistic values of the parameters ([2]), Newton's method gives a unique solution of the problem and has good convergence properties.

For this problem in MAPLE we will take a model semiconductor with a P-N jump in the middle of the region $[0, L]$, where $L = 10$, with the impurity distribution

$$c(X) = \tanh\left(20 \left(\frac{X}{L} - \frac{1}{2}\right)\right).$$

For simplicity, we choose a uniform grid with step $h_i = h = 1/N$, where N is the number of points of the mesh. For the quantities $\hat{\varphi}$, $\hat{\varphi}_D$, $\hat{h}^*$ and $\hat{F}$ we will use arrays phi, phiD, hp and Fp. Tridiagonal matrices will be represented by the vectors of the diagonal and off-diagonal elements A, Ap and B, where

$$A_i = a_i, \quad Ap_i = a_i - h_i^* \left(e^{\varphi_i} + e^{-\varphi_i} \right), \quad B_i = b_i.$$

The first part of the MAPLE program follows:

```
>         #dimensionless formulation, x,L in units of Debye's length
> N:=20: L:=10.0: h:=L/N:
> U0:=0: #normalized potential on the boundary
>         #concentration of impurities - region of N-P jump
> c:=tanh(20.0*(x/L-0.5)):
>         #mesh of N+1 point for region 0..L
> xp:=array(0..N):
> for i from 0 to N do xp[i]:=i*h od:
>         #array of concentrations Ca and builtin potential phiD
> phiD:=array(0..N):
> Ca:=array(0..N):
> for i from 0 to N do
>     Ca[i]:=evalf(subs(x=xp[i],c));
>     phiD[i]:=arcsinh(Ca[i]/2)
> od:
```

To solve the tridiagonal System (5.9) we could use the standard library function linsolve, but it appears to be slow, since it does not take into account the sparsity of the matrix. So we implement the standard LU decomposition for tridiagonal systems without pivoting, which is very fast and economical in storage. The function solvetridiag(n,A,B,F) solves the symmetric tridiagonal system of n linear equations; A is the vector of the main diagonal, B contains the off-diagonal elements, and F is the vector of the right-hand sides. This function returns the solution as an array [0..n+1]. The indices 0 and n+1 refer to the corresponding elements of the vector F (see Algorithm (5.1)).

ALGORITHM 5.1.

MAPLE *Functions* solvetridiag *and* norma.

```
solvetridiag:=proc(N, A, B, F) local alfa, beta, pom, x, i;
    # solution of a linear system of  N equations
    # with tridiagonal symmetric matrix, where
    # A is a vector diagonal of elements,
    # B is a vector of off-diagonal elements and
    # F is a vector of right hand site,
    #result is a vector (0..N+1)
    x:=array(0..N+1,[(0)=F[0],(N+1)=F[N+1]]);
    alfa[1]:= -B[1]/A[1]; beta[1]:= F[1]/A[1];
    for i from 2 to N-1 do
        pom:= A[i]+B[i-1]*alfa[i-1];
        alfa[i]:= -B[i]/pom;
        beta[i]:= (F[i]-B[i-1]*beta[i-1])/pom
    od;
    x[N]:= (F[N]-B[N-1]*beta[N-1])/(A[N]+B[N-1]*alfa[N-1]);
    for i from N-1 by -1 to 1 do
        x[i]:= alfa[i]*x[i+1]+beta[i]
    od;
    eval(x)
end;  #solvetridiag

norma:= proc(N, X) local mx,i,pom;
    #calculation of a infinity norm for 1-D array X, i=1..N
    mx:=0;
    for i to N do
        pom:=abs(X[i]);
        if pom > mx then mx := pom fi
    od;
    mx
end;
```

The implementation of the Newton method is now straightforward. We define the function iterate(), which performs the iterations calculating the new potential phi from the old one, until the norm of the correction H is less than a tolerance epsilon. iterate() also returns the norm of the difference of the last two iterations. We will choose epsilon = 0.001.

```
>     #generation of vectors A,B and average step hp
> B := array(0..N-1): A := array(1..N): hp := array(1..N-1):
> for i from 0 to N-1 do B[i] := 1/(xp[i+1]-xp[i]) od:
> for i from 1 to N-1 do A[i] := -(B[i-1]+B[i]) od:
> for i from 1 to N-1 do hp[i] := (xp[i+1]-xp[i-1])/2 od:
> phi := array(0..N):

>     #initial approximation of phi
> phi := eval(phiD):
```

ALGORITHM 5.2. *Definition of the Iteration Function.*

```
iterate:=proc(phi) local i, nor, H, ex, Ap, Fp;
    #iterations for phi begin
    nor:=1;
    while nor > epsilon do
        for i to N-1 do
            ex := exp(phi[i]);
            Ap[i] := A[i] - hp[i]*(ex+1/ex);
            Fp[i] := Fpoc[i] + hp[i]*(ex-1/ex) - A[i]*phi[i];
            if i = 1 then Fp[1] := Fp[1] - B[1]*phi[2]
                elif i = N-1 then
                    Fp[N-1] := Fp[N-1] - B[N-2]*phi[N-2]
                else
                    Fp[i] := Fp[i] - B[i-1]*phi[i-1] - B[i]*phi[i+1]
            fi;
        od;
        H:=solvetridiag(N-1,Ap,B,Fp);
        for i to N-1 do
            phi[i] := H[i] + phi[i];
            nor := norma(N-1,H);
        od;
    od;
end; #iterate
```

```
>       #introducing boundary condition
> phi[0] := phi[0]+U0:
>       #generation of vectors of right hand side Fpoc and Fp
> Fpoc := array(0..N): Fp := array(0..N):
> for i to N-1 do Fpoc[i] := -hp[i]*Ca[i] od:
> Fpoc[1] := Fpoc[1] - B[0]*phi[0]:
> Fpoc[N-1] := Fpoc[N-1] - B[N-1]*phi[N]:
>       #introducing the boundary conditions into Fp
> Fp[0] := phi[0]: Fp[N] := phi[N]:
> epsilon := 0.001:    #error of the result
>       #solution for  phi
> iterate(phi);  #returns a value of norm for iteruj:=proc(phi)
```

$$.1899345025 \, 10^{-6}$$

```
>       #graph of builtin potential phiD
> GphiD:=plot(arcsinh(c/2),x=0..L, color=black, linestyle=2):
>       # plot of resulting phi and phiD
> Gphi:= plot([seq([xp[i],phi[i]],i=0..N)],color=black):
> plots[display]({Gphi,GphiD});    #end
```

The graph of the internal and builtin potentials is in Figure 5.1.

FIGURE 5.1. *Internal and Builtin Potentials for $u_0 = 0$.*

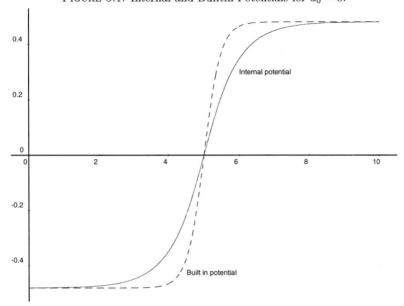

FIGURE 5.2. *Internal and Builtin Potentials for $u_0 = 0.2$.*

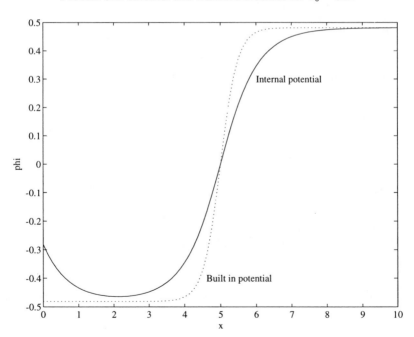

5.3   MATLAB Solution

In the previous section a relatively fast MAPLE algorithm was developed for
solving a pure numerical problem, by making use of special properties of the
problem (e.g. the tridiagonality of the matrix of the System (5.9)). A numerical
system such as MATLAB is of course more adequate for solving pure numerical
problems. By using the standard mechanism for dealing with sparse matrices
in MATLAB, we can formulate a very simple program for our problem.

MATLAB allows the definition of sparse matrices using the function spdiags.
The standard matrix operations then deal very efficiently with the sparsity of
such a matrix. We can therefore use the standard notation for the solution of
the linear system $A \cdot x = f$:

```
>> x = A\f
```

We again use Newton's method for solving the discretized form of the nonlinear
Poisson Equation (5.7). In the definitions of the tridiagonal matrix $\hat{A}$ and the
diagonal matrix $\hat{D}$ we use the function spdiagsx. The MATLAB program which
results is straightforward:

```
%program for computing the internal potential in a semiconductor

l = 10; n = 80;
u0 = 0.2                        %external potential
xl = (0:l/n:l)'                 %mesh
cl = tanh(20.0.*(xl./l-0.5))    %concentration of impurities
phid = asinh(cl*0.5)            %builtin potential
h = xl(2:n+1)-xl(1:n)           %steps
b = 1 ./h
                                %tridiagonal matrix A
A = spdiags([b(2:n) -(b(1:n-1)+b(2:n)) b(1:n-1)],-1:1,n-1,n-1)

hs = (h(1:n-1)+h(2:n))*0.5      %average steps
c = cl(2:n)

f0 = -hs.*c
f0(1) = f0(1)-(phid(1)+u0)*b(1);
f0(n-1) = f0(n-1)-phid(n+1)*b(n);
phi =  phid(2:n)                %initial approximation

nor = 1                         %iteration
while nor > 0.001
  ex = exp(phi); f =  f0+hs.*(ex-1./ex)-A*phi
  Dp = spdiags(hs.*(ex+1./ex),0,n-1,n-1)
  H = (A-Dp)\f
  nor = norm(H,inf)
  phi = phi+H
end
phiv = [phid(1)+u0; phi; phid(n+1)] %with boundary conditions

clf;
```

```
plot(xl,phiv,'-r', xl,phid,':g')
hold;
text(4.6,-0.4,'Builtin potential');
text(6,0.3,'Internal potential');
xlabel('x'); ylabel('phi');
```
                              %end

The plot of the results for the external potential $u_0 = 0.3$ are given in Figure 5.2.

References

[1] W.H. PRESS, B.P. FLANNERY, S.A. TEUKOLSKY
    and W.T. VETTERLING, *Numerical Recipes*, Cambridge University Press, 1988.

[2] S. SELBERHERR, *Analysis and Simulation of Semiconductor Devices*, Springer,
    1984.

Chapter 6. Some Least Squares Problems

W. Gander and U. von Matt

6.1 Introduction

In this chapter we will consider some least squares problems, that arise in quality control in manufacturing using coordinate measurement techniques [4], [5]. In mass production machines are producing parts, and it is important to know, if the output satisfies the quality requirements. Therefore some sample parts are usually taken out of the production line, measured carefully and compared with the nominal features. If they do not fit the specified tolerances, the machine may have to be serviced and adjusted to produce better parts.

6.2 Fitting Lines, Rectangles and Squares in the Plane

Fitting a line to a set of points in such a way that the sum of squares of the distances of the given points to the line is minimized, is known to be related to the computation of the main axes of an inertia tensor. In [1] this fact is used to fit a line and a plane to given points in the 3d space by solving an eigenvalue problem for a 3×3 matrix.

In this section we will develop a different algorithm, based on the singular value decomposition, that will allow us to fit lines, rectangles, and squares to measured points in the plane and that will also be useful for some problems in 3d space.

Let us first consider the problem of fitting a straight line to a set of given points $P_1, P_2, \ldots, P_m$ in the plane. We denote their coordinates with (x_{P_1}, y_{P_1}), (x_{P_2}, y_{P_2}), $\ldots$, (x_{P_m}, y_{P_m}). Sometimes it is useful to define the vectors of all x- and y-coordinates. We will use $\mathbf{x}_P$ for the vector $(x_{P_1}, x_{P_2}, \ldots, x_{P_m})$ and similarly $\mathbf{y}_P$ for the y coordinates.

The problem we want to solve is not *linear regression*. Linear regression means to fit the linear model $y = ax + b$ to the given points, i.e. to determine the two parameters a and b such that the sum of squares of the residual is minimized:

$$\sum_{i=1}^{m} r_i^2 = \min, \quad \text{where} \quad r_i = y_{P_i} - ax_{P_i} - b.$$

This simple linear least squares problem is solved in MATLAB by the following statements (assuming that $\mathbf{x} = \mathbf{x}_P$ and $\mathbf{y} = \mathbf{y}_P$):

```
>> p = [ x ones(size(x))]\y;
>> a = p(1); b = p(2);
```

In the case of the linear regression the sum of squares of the differences of the y coordinates of the points P_i to the fitted linear function is minimized. What we would like to minimize now, however, is the *sum of squares of the distances of the points from the fitted straight line.*

In the plane we can represent a straight line uniquely by the equations

$$c + n_1 x + n_2 y = 0, \quad n_1^2 + n_2^2 = 1. \tag{6.1}$$

The unit vector (n_1, n_2) is orthogonal to the line. A point is on the line if its coordinates (x, y) satisfy the first equation. On the other hand if $P = (x_P, y_P)$ is some point not on the line and we compute

$$r = c + n_1 x_P + n_2 y_P$$

then $|r|$ is its distance from the line. Therefore if we want to determine the line for which the sum of squares of the distances to given points is minimal, we have to solve the constrained least squares problem

$$||\mathbf{r}|| = \sum_{i=1}^{m} r_i^2 = \min$$

subject to

$$\begin{pmatrix} 1 & x_{P_1} & y_{P_1} \\ 1 & x_{P_2} & y_{P_2} \\ \vdots & \vdots & \vdots \\ 1 & x_{P_m} & y_{P_m} \end{pmatrix} \begin{pmatrix} c \\ n_1 \\ n_2 \end{pmatrix} = \begin{pmatrix} r_1 \\ r_2 \\ \vdots \\ r_m \end{pmatrix} \quad \text{and } n_1^2 + n_2^2 = 1. \tag{6.2}$$

Let $\mathbf{A}$ be the matrix of the linear system (6.2), $\mathbf{x}$ denote the vector of unknowns $(c, n_1, n_2)^T$ and $\mathbf{r}$ the right hand side. Since orthogonal transformations $\mathbf{y} = \mathbf{Q}^T \mathbf{r}$ leave the norm invariant ($||\mathbf{y}||_2 = ||\mathbf{r}||_2$ for an orthogonal matrix $\mathbf{Q}$), we can proceed as follows to solve problem (6.2).

First we compute the QR decomposition of $\mathbf{A}$ and reduce our problem to solving a small system:

$$\mathbf{A} = \mathbf{QR} \quad \Longrightarrow \quad \mathbf{Q}^T \mathbf{A} \mathbf{x} = \begin{pmatrix} r_{11} & r_{12} & r_{13} \\ 0 & r_{22} & r_{23} \\ 0 & 0 & r_{33} \\ 0 & 0 & 0 \\ \vdots & \vdots & \vdots \\ 0 & 0 & 0 \end{pmatrix} \begin{pmatrix} c \\ n_1 \\ n_2 \end{pmatrix} = \mathbf{Q}^T \mathbf{r} \tag{6.3}$$

Since the nonlinear constraint only involves two unknowns we now have to solve

$$\begin{pmatrix} r_{22} & r_{23} \\ 0 & r_{33} \end{pmatrix} \begin{pmatrix} n_1 \\ n_2 \end{pmatrix} \approx \begin{pmatrix} 0 \\ 0 \end{pmatrix}, \quad \text{subject to } n_1^2 + n_2^2 = 1. \tag{6.4}$$

ALGORITHM 6.1. *Function* `cslq`.

```
function [c,n] = clsq(A,dim);
% solves the constrained least squares Problem
% A (c n)' ~ 0 subject to norm(n,2)=1
% length(n) = dim
% [c,n] = clsq(A,dim)
[m,p] = size(A);
if p < dim+1, error ('not enough unknowns'); end;
if m < dim, error ('not enough equations'); end;
m = min (m, p);
R = triu (qr (A));
[U,S,V] = svd(R(p-dim+1:m,p-dim+1:p));
n = V(:,dim);
c = -R(1:p-dim,1:p-dim)\R(1:p-dim,p-dim+1:p)*n;
```

Problem (6.4) is of the form $||\mathbf{B}\mathbf{x}||_2 = \min$, subject to $||\mathbf{x}||_2 = 1$. The value of the minimum is the *smallest singular value of* $\mathbf{B}$, and the solution is given by the corresponding singular vector [2]. Thus we can determine n_1 and n_2 by a singular value decomposition of a 2-by-2 matrix. Inserting the values into the first component of (6.3) and setting it to zero, we then can compute c. As a slight generalization we denote the dimension of the normal vector $\mathbf{n}$ by `dim`. Then, the MATLAB function to solve problem (6.2) is given by Algorithm 6.1. Let us test the function `clsq` with the following main program:

```
>> % mainline.m
>> Px = [1:10]'
>> Py = [ 0.2 1.0 2.6 3.6 4.9 5.3 6.5 7.8 8.0 9.0]'
>> A = [ones(size(Px)) Px Py]
>> [c, n] = clsq(A,2)
```

The line computed by the program `mainline` has the equation $0.4162 - 0.7057x + 0.7086y = 0$. We would now like to plot the points and the fitted line. For this we need the function `plotline`, Algorithm 6.2. The picture is generated by adding the commands

```
>> clf; hold on;
>> axis([-1, 11 -1, 11])
>> plotline(Px,Py,'o',c,n,'-')
>> hold off;
```

to the program `mainline`. The result is shown as Figure 6.1.

Fitting two Parallel Lines

To fit two parallel lines, we must have two sets of points. We denote those two sets by $\{P_i\}, i = 1, \ldots, p$, and $\{Q_j\}, j = 1, \ldots, q$. Since the lines are parallel, their normal vector must be the same. Thus the equations for the lines are

$$
\begin{aligned}
c_1 + n_1 x + n_2 y &= 0, \\
c_2 + n_1 x + n_2 y &= 0, \\
n_1^2 + n_2^2 &= 1.
\end{aligned}
$$

ALGORITHM 6.2. *Function* plotline.

```
function plotline(x,y,s,c,n,t)
% plots the set of points (x,y) using the symbol s
% and plots the straight line c+n1*x+n2*y=0 using
% the line type defined by t
plot(x,y,s)
xrange = [min(x) max(x)];
yrange = [min(y) max(y)];
if n(1)==0, % c+n2*y=0  => y = -c/n(2)
   x1=xrange(1); y1 = -c/n(2);
   x2=xrange(2); y2 = y1
elseif n(2) == 0, % c+n1*x=0  => x = -c/n(1)
   y1=yrange(1); x1 = -c/n(1);
   y2=yrange(2); x2 = x1;
elseif xrange(2)-xrange(1)> yrange(2)-yrange(1),
   x1=xrange(1); y1 = -(c+n(1)*x1)/n(2);
   x2=xrange(2); y2 = -(c+n(1)*x2)/n(2);
else
   y1=yrange(1); x1 = -(c+n(2)*y1)/n(1);
   y2=yrange(2); x2 = -(c+n(2)*y2)/n(1);
end
plot([x1, x2], [y1,y2],t)
```

FIGURE 6.1. *Fitting a Straight Line.*

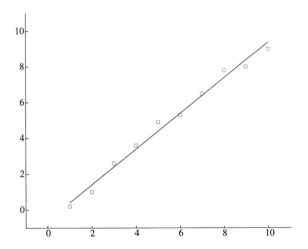

If we insert the coordinates of the two sets of points into these equations we get the following constrained least squares problem:

$$||\mathbf{r}|| = \sum_{i=1}^{m} r_i^2 = \min$$

subject to

$$\begin{pmatrix} 1 & 0 & x_{P_1} & y_{P_1} \\ 1 & 0 & x_{P_2} & y_{P_2} \\ \vdots & \vdots & \vdots & \vdots \\ 1 & 0 & x_{P_p} & y_{P_p} \\ 0 & 1 & x_{Q_1} & y_{Q_1} \\ 0 & 1 & x_{Q_2} & y_{Q_2} \\ \vdots & \vdots & \vdots & \vdots \\ 0 & 1 & x_{Q_q} & y_{Q_q} \end{pmatrix} \begin{pmatrix} c_1 \\ c_2 \\ n_1 \\ n_2 \end{pmatrix} = \begin{pmatrix} r_1 \\ r_2 \\ \vdots \\ r_{p+q} \end{pmatrix} \quad \text{and } n_1^2 + n_2^2 = 1. \qquad (6.5)$$

Again, we can use our function `clsq` to solve this problem:

```
>> % mainparallel.m
>> Px = [1:10]'
>> Py = [ 0.2 1.0 2.6 3.6 4.9 5.3 6.5 7.8 8.0 9.0]'
>> Qx = [ 1.5 2.6 3.0 4.3 5.0 6.4 7.6 8.5 9.9 ]'
>> Qy = [ 5.8 7.2 9.1 10.5 10.6 10.7 13.4 14.2 14.5]'
>> A = [ones(size(Px))   zeros(size(Px)) Px Py
>>      zeros(size(Qx))  ones(size(Qx)) Qx Qy ]
>> [c, n] = clsq(A,2)
>> clf; hold on;
>> axis([-1 11 -1 17])
>> plotline(Px,Py,'o',c(1),n,'-')
>> plotline(Qx,Qy,'+',c(2),n,'-')
>> hold off;
```

The results obtained by the program `mainparallel` are the two lines

$$\begin{aligned} 0.5091 - 0.7146x + 0.6996y &= 0, \\ -3.5877 - 0.7146x + 0.6996y &= 0, \end{aligned}$$

which are plotted as Figure 6.2.

Fitting Orthogonal Lines

To fit two orthogonal lines we can proceed very similar as in the the case of the parallel lines. If (n_1, n_2) is the normal vector of the first line, then the second line must have the normal vector $(-n_2, n_1)$ in order to be orthogonal. Therefore again we will have four unknowns: c_1, c_2, n_1 and n_2.

FIGURE 6.2. *Fitting Two Parallel Lines.*

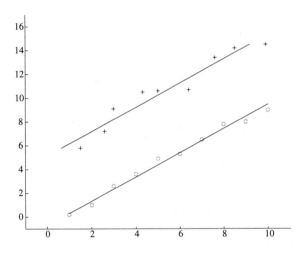

FIGURE 6.3. *Fitting Two Orthogonal Lines.*

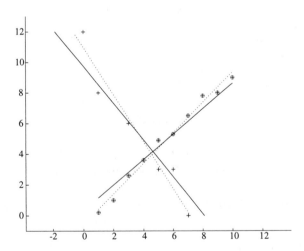

If the P_i's are the points associated with the first line and the Q_j's are the points associated with the second line, we obtain the following constrained least squares problem:

$$||\mathbf{r}|| = \sum_{i=1}^{m} r_i^2 = \min$$

subject to

$$\begin{pmatrix} 1 & 0 & x_{P_1} & y_{P_1} \\ 1 & 0 & x_{P_2} & y_{P_2} \\ \vdots & \vdots & \vdots & \vdots \\ 1 & 0 & x_{P_p} & y_{P_p} \\ 0 & 1 & y_{Q_1} & -x_{Q_1} \\ 0 & 1 & y_{Q_2} & -x_{Q_2} \\ \vdots & \vdots & \vdots & \vdots \\ 0 & 1 & y_{Q_q} & -x_{Q_q} \end{pmatrix} \begin{pmatrix} c_1 \\ c_2 \\ n_1 \\ n_2 \end{pmatrix} = \begin{pmatrix} r_1 \\ r_2 \\ \vdots \\ r_{p+q} \end{pmatrix} \quad \text{and } n_1^2 + n_2^2 = 1. \qquad (6.6)$$

We only have to change the definition of the matrix A in `mainparallel` in order to compute the equations of the two orthogonal lines. To obtain a nicer plot we also chose different values for the second set of points.

```
>> % mainorthogonal.m
>> Px = [1:10]'
>> Py = [ 0.2 1.0 2.6 3.6 4.9 5.3 6.5 7.8 8.0 9.0]'
>> Qx = [ 0 1 3 5 6 7]'
>> Qy = [12 8 6 3 3 0]'
>> A = [ones(size(Px))    zeros(size(Px)) Px  Py
>>        zeros(size(Qx)) ones(size(Qx))  Qy -Qx ]
>> [c, n] = clsq(A,2)
>> clf; hold on;
>> axis([-1 11 -1 13])
>> axis('equal')
>> plotline(Px,Py,'o',c(1),n,'-')
>> n2(1) =-n(2); n2(2) = n(1)
>> plotline(Qx,Qy,'+',c(2),n2,'-')
```

The Program `mainorthogonal` computes the two orthogonal lines

$$\begin{aligned} -0.2527 - 0.6384x + 0.7697y &= 0, \\ 6.2271 - 0.7697x - 0.6384y &= 0. \end{aligned}$$

We can also fit two lines individually to each set of points:

```
>> [c, n] = clsq([ones(size(Px)) Px Py],2)
>> plotline(Px,Py,'+',c,n,':')
>> [c, n] = clsq([ones(size(Qx)) Qx Qy],2)
>> plotline(Qx,Qy,'+',c,n,':')
>> hold off;
```

The individually computed lines are indicated by the dotted lines in Figure 6.3.

Fitting a Rectangle

Fitting a rectangle requires four sets of points:

$$P_i, i = 1, \ldots, p, \quad Q_j, j = 1, \ldots, q, \quad R_k, k = 1, \ldots, r, \quad S_l, l = 1, \ldots, s.$$

Since the sides of the rectangle are parallel and orthogonal we can proceed very similarly as before. The four sides will have the equations

$$
\begin{aligned}
a: \quad c_1 + n_1 x + n_2 y &= 0 \\
b: \quad c_2 - n_2 x + n_1 y &= 0 \\
c: \quad c_3 + n_1 x + n_2 y &= 0 \\
d: \quad c_4 - n_2 x + n_1 y &= 0 \\
n_1^2 + n_2^2 &= 1.
\end{aligned}
$$

Inserting the sets of points we get the following constrained least squares problem:

$$\|\mathbf{r}\| = \sum_{i=1}^m r_i^2 = \min$$

subject to

$$
\begin{pmatrix}
1 & 0 & 0 & 0 & x_{P_1} & y_{P_1} \\
\vdots & \vdots & \vdots & \vdots & \vdots & \vdots \\
1 & 0 & 0 & 0 & x_{P_p} & y_{P_p} \\
0 & 1 & 0 & 0 & y_{Q_1} & -x_{Q_1} \\
\vdots & \vdots & \vdots & \vdots & \vdots & \vdots \\
0 & 1 & 0 & 0 & y_{Q_q} & -x_{Q_q} \\
0 & 0 & 1 & 0 & x_{R_1} & y_{R_1} \\
\vdots & \vdots & \vdots & \vdots & \vdots & \vdots \\
0 & 0 & 1 & 0 & x_{R_r} & y_{R_r} \\
0 & 0 & 0 & 1 & y_{S_1} & -x_{S_1} \\
\vdots & \vdots & \vdots & \vdots & \vdots & \vdots \\
0 & 0 & 0 & 1 & y_{S_s} & -x_{S_s}
\end{pmatrix}
\begin{pmatrix} c_1 \\ c_2 \\ c_3 \\ c_4 \\ n_1 \\ n_2 \end{pmatrix}
=
\begin{pmatrix} r_1 \\ r_2 \\ \vdots \\ r_{p+q+r+s} \end{pmatrix}
\quad \text{and } n_1^2 + n_2^2 = 1.
$$

$$(6.7)$$

Instead of explicitly giving the coordinates of the four sets of points, we will now enter the points with the mouse using the ginput function in MATLAB. [X,Y] = ginput(N) gets N points from the current axes and returns the x- and y-coordinates in the vectors X and Y of length N. The points have to be entered clock- or counter clock wise in the same order as the sides of the rectangle: the next side is always orthogonal to the previous.

```
>> % rectangle.m
>> clf; hold on;
>> axis([0 10 0 10])
>> axis('equal')
>> p=100; q=100; r=100; s=100;

>> disp('enter points P_i belonging to side A')
```

```
>> disp('by clicking the mouse in the graphical window.')
>> disp('Finish the input by pressing the Return key')
>> [Px,Py] = ginput(p); plot(Px,Py,'o')
>> disp('enter points Q_i for side B ')
>> [Qx,Qy] = ginput(q); plot(Qx,Qy,'x')
>> disp('enter points R_i for side C ')
>> [Rx,Ry] = ginput(r); plot(Rx,Ry,'*')
>> disp('enter points S_i for side D ')
>> [Sx,Sy] = ginput(s); plot(Sx,Sy,'+')

>> zp = zeros(size(Px)); op =  ones(size(Px));
>> zq = zeros(size(Qx)); oq =  ones(size(Qx));
>> zr = zeros(size(Rx)); or =  ones(size(Rx));
>> zs = zeros(size(Sx)); os =  ones(size(Sx));

>> A = [ op zp zp zp Px  Py
>>       zq oq zq zq Qy -Qx
>>       zr zr or zr Rx  Ry
>>       zs zs zs os Sy -Sx]

>> [c, n] = clsq(A,2)

>> % compute the 4 corners of the rectangle
>> B  = [n  [-n(2) n(1)]']
>> X = -B* [c([1 3 3 1])'; c([2 2 4 4])']
>> X = [X X(:,1)]
>> plot(X(1,:), X(2,:))

>> % compute the individual lines, if possible
>> if all([sum(op)>1 sum(oq)>1 sum(or)>1 sum(os)>1]),
>>    [c1, n1] = clsq([op Px Py],2)
>>    [c2, n2] = clsq([oq Qx Qy],2)
>>    [c3, n3] = clsq([or Rx Ry],2)
>>    [c4, n4] = clsq([os Sx Sy],2)

>>    % and their intersection points
>>    aaa = -[n1(1) n1(2); n2(1) n2(2)]\[c1; c2];
>>    bbb = -[n2(1) n2(2); n3(1) n3(2)]\[c2; c3];
>>    ccc = -[n3(1) n3(2); n4(1) n4(2)]\[c3; c4];
>>    ddd = -[n4(1) n4(2); n1(1) n1(2)]\[c4; c1];

>>    plot([aaa(1) bbb(1) ccc(1) ddd(1) aaa(1)], ...
>>          [aaa(2) bbb(2) ccc(2) ddd(2) aaa(2)],':')
>> end
>> hold off;
```

The Program rectangle not only computes the rectangle but also fits the individual lines to the set of points. The result is shown as Figure 6.4. Some

comments may be needed to understand some statements. To find the coordinates of a corner of the rectangle, we compute the point of intersection of the
two lines of the corresponding sides. We have to solve the linear system

$$\begin{array}{rcl} n_1x + n_2y &=& -c_1 \\ -n_2x + n_1y &=& -c_2 \end{array} \iff C\mathbf{x} = -\begin{pmatrix} c_1 \\ c_2 \end{pmatrix}, \quad C = \begin{pmatrix} n_1 & n_2 \\ -n_2 & n_1 \end{pmatrix}$$

Since C is orthogonal, we can simply multiply the right hand side by $B = C^T$
to obtain the solution. By arranging the equations for the 4 corners so that the
matrix C is always the system matrix, we can compute the coordinates of all 4
corners simultaneously with the compact statement

```
>> X = -B* [c([1 3 3 1])'; c([2 2 4 4])'].
```

FIGURE 6.4. *Fitting a Rectangle.*

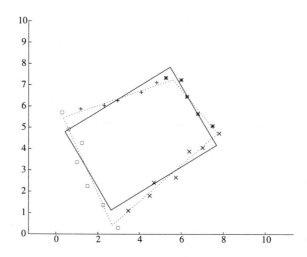

Fitting a Square

If $|d|$ denotes the length of the side of a square, then the four sides, *oriented
counter clock wise*, have the equations

$$\begin{array}{rrcl} \alpha: & c_1 + n_1x + n_2y &=& 0 \\ \beta: & c_2 - n_2x + n_1y &=& 0 \\ \gamma: & d + c_1 + n_1x + n_2y &=& 0 \\ \delta: & d + c_2 - n_2x + n_1y &=& 0 \\ & n_1^2 + n_2^2 &=& 1. \end{array} \tag{6.8}$$

The orientation is important: the parameter d in the third and fourth equation
should have the same sign. If we want to orient the sides clockwise, then the

normal vector in the equations for lines β and δ must have the other sign, i.e. we have to use the equations

$$
\begin{aligned}
\beta: & c_2 + n_2 x - n_1 y = 0, \\
\delta: & d + c_2 + n_2 x - n_1 y = 0.
\end{aligned}
$$

We will assume in the following, that the four sets of points

$$ P_i, i = 1, \ldots, p, \quad Q_j, j = 1, \ldots, q, \quad R_k, k = 1, \ldots, r, \quad S_l, l = 1, \ldots, s $$

belong to the four sides which are oriented counter clockwise. Then our problem becomes

$$ ||\mathbf{r}|| = \sum_{i=1}^{m} r_i^2 = \min $$

subject to

$$
\begin{pmatrix}
0 & 1 & 0 & x_{P_1} & y_{P_1} \\
\vdots & \vdots & \vdots & \vdots & \vdots \\
0 & 1 & 0 & x_{P_p} & y_{P_p} \\
0 & 0 & 1 & y_{Q_1} & -x_{Q_1} \\
\vdots & \vdots & \vdots & \vdots & \vdots \\
0 & 0 & 1 & y_{Q_q} & -x_{Q_q} \\
1 & 1 & 0 & x_{R_1} & y_{R_1} \\
\vdots & \vdots & \vdots & \vdots & \vdots \\
1 & 1 & 0 & x_{R_r} & y_{R_r} \\
1 & 0 & 1 & y_{S_1} & -x_{S_1} \\
\vdots & \vdots & \vdots & \vdots & \vdots \\
1 & 0 & 1 & y_{S_s} & -x_{S_s}
\end{pmatrix}
\begin{pmatrix}
d \\ c_1 \\ c_2 \\ n_1 \\ n_2
\end{pmatrix}
=
\begin{pmatrix}
r_1 \\ r_2 \\ \vdots \\ r_{p+q+r+s}
\end{pmatrix}
\quad \text{and } n_1^2 + n_2^2 = 1. \quad (6.9)
$$

We can reuse a lot of the program `rectangle`. The only change is the new definition of the matrix A and the vector c which is used to plot the square. When entering the points we have to arrange the sides counter clockwise.

```
>> A = [ zp op zp  Px  Py
>>        zq zq oq  Qy -Qx
>>        or or zr  Rx  Ry
>>        os zs os  Sy -Sx]
>> [cc, n] = clsq(A,2)
>> % compute the vector of constants for the four lines
>> c(1) = cc(2); c(2) = cc(3); c(3) = c(1)+cc(1);
>> c(4) = c(2)+cc(1);  c=c(:);
>> % compute the 4 corners of the square
>> B  = [n  [-n(2) n(1)]']
>> X = -B* [c([1 3 3 1])'; c([2 2 4 4])']
>> X = [X X(:,1)]
>> plot(X(1,:), X(2,:))
>> hold off;
```

FIGURE 6.5. *Fitting a Rectangle and a Square.*

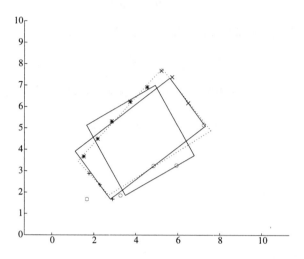

FIGURE 6.6. *"Best" Square for Four Points.*

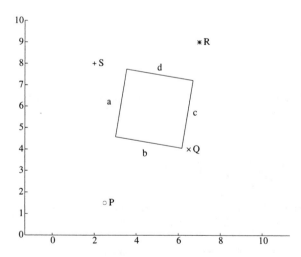

In Figure 6.5 the best rectangle and the best fitted square are shown. Notice that in extreme cases, we may not obtain a result that we may wish. Consider Figure 6.6, where we computed the "best square" for the four points $P = (2.5, 1.5)$, $Q = (6.5, 4)$, $R = (7, 9)$ and $S = (2, 8)$. It seems that the square is not well fitted. However, notice that point P belongs to the first side a, point Q to b, point R to c and S to d. The four points lie perfectly on the lines through the sides so that the sum of squares of their distances is zero!

6.3 Fitting Hyperplanes

Function `clsq` can be used to fit an $(n - 1)$-dimensional hyperplane in R^n to given points. Let the rows of the matrix $X = [\mathbf{x}_1, \mathbf{x}_2, \ldots, \mathbf{x}_m]^T$ contain the coordinates of the given points, i.e. point P_i has the coordinates $\mathbf{x}_i = X(i, :)$, $i = 1, \ldots, m$. Then the call

>> [c, N] = clsq ([ones(m,1) X], n);

determines the hyperplane in normal form $c + N_1 y_1 + N_2 y_2 + \ldots + N_n y_n = 0$.

In this section, we will show how we can compute also best fit hyperplanes of lower dimensions s, where $1 \le s \le n - 1$. We follow the theory developed in [3]. An s-dimensional hyperplane α in R^n can be represented in parameter form:

$$\alpha: \quad \mathbf{y} = \mathbf{p} + \mathbf{a}_1 t_1 + \mathbf{a}_2 t_2 + \cdots \mathbf{a}_s t_s = \mathbf{p} + A\mathbf{t}. \tag{6.10}$$

In this equation $\mathbf{p}$ is a point on the plane and $\mathbf{a}_i$ are linearly independent direction vectors, thus the hyperplane is determined by the data $\mathbf{p}$ and $A = [\mathbf{a}_1, \ldots, \mathbf{a}_s]$.

Without loss of generality, we assume that A is orthogonal, i.e., $A^T A = I_s$. If we now want to fit a hyperplane to the given set of points X, then we have to minimize the distance of the points to the plane. The distance d_i of point $P_i = \mathbf{x}_i$ to the hyperplane is given by

$$d_i = \min_{\mathbf{t}} ||\mathbf{p} - \mathbf{x}_i + A\mathbf{t}||_2.$$

To determine the minimum we solve grad $d_i^2 = 2A^T(\mathbf{p} - \mathbf{x}_i + A\mathbf{t}) = \mathbf{0}$ for $\mathbf{t}$, and, since A is orthogonal, we obtain

$$\mathbf{t} = A^T(\mathbf{x}_i - \mathbf{p}). \tag{6.11}$$

Therefore the distance becomes

$$d_i^2 = ||\mathbf{p} - \mathbf{x}_i + AA^T(\mathbf{x}_i - \mathbf{p})||_2^2 = ||\mathcal{P}(\mathbf{x}_i - \mathbf{p})||_2^2,$$

where we denoted by $\mathcal{P} = I - AA^T$, the projector onto the complement of the range of A, i.e. on the null space of A^T.

Our objective is to minimize the sum of squares of the distances of all points to the hyperplane. We want to minimize the function

$$F(\mathbf{p}, A) = \sum_{i=1}^{m} ||\mathcal{P}(\mathbf{x}_i - \mathbf{p})||_2^2. \tag{6.12}$$

A necessary condition is grad $F = 0$. We first consider the first part of the gradient, the partial derivative

$$\frac{\partial F}{\partial \mathbf{p}} = -\sum_{i=1}^{m} 2\mathcal{P}^T \mathcal{P}(\mathbf{x}_i - \mathbf{p}) = -2\mathcal{P}(\sum_{i=1}^{m} \mathbf{x}_i - m\mathbf{p}) = 0,$$

where we made use of the property of a projector $\mathcal{P}^T \mathcal{P} = \mathcal{P}$. Since $\mathcal{P}$ projects the vector $\sum_{i=1}^{m} \mathbf{x}_i - m\mathbf{p}$ onto 0, this vector must be in the range of A, i.e.

$$\mathbf{p} = \frac{1}{m} \sum_{i=1}^{m} \mathbf{x}_i + A\tau. \tag{6.13}$$

Inserting this expression into Equation (6.12), the objective function to be minimized simplifies to

$$G(A) = \sum_{i=1}^{m} ||\mathcal{P}\hat{\mathbf{x}}_i||_2^2 = ||\mathcal{P}\hat{X}^T||_F^2, \tag{6.14}$$

where we put

$$\hat{\mathbf{x}}_i = \mathbf{x}_i - \frac{1}{m} \sum_{i=1}^{m} \mathbf{x}_i,$$

and where we used the *Frobenius norm* of a matrix $(||A||_F^2 := \sum_{i,j} a_{ij}^2)$. Now since $\mathcal{P}$ is symmetric, we may also write

$$G(A) = ||\hat{X}\mathcal{P}||_F^2 = ||\hat{X}(I - AA^T)||_F^2 = ||\hat{X} - \hat{X}AA^T||_F^2. \tag{6.15}$$

If we define $Y := \hat{X}AA^T$, which is a matrix of rank s, then we can consider the problem of minimizing

$$||\hat{X} - Y||_F^2 = \min, \quad \text{subject to} \quad rank(Y) = s.$$

It is well known, how to approximate in the Frobenius norm a matrix by a matrix of lower rank (cf. e.g. [2]):

1. Compute the singular value decomposition of $\hat{X} = U\Sigma V^T$, with

$$U, V \quad \text{orthogonal and} \quad \Sigma = diag(\sigma_1, \sigma_2 \ldots \sigma_n)$$

   and $\sigma_1 \geq \sigma_2 \geq \cdots \geq \sigma_n$.

2. The minimizing matrix is then given by $Y = U\Sigma_s V^T$, where

$$\Sigma_s = diag(\sigma_1, \sigma_2 \ldots, \sigma_s, 0, 0, \ldots, 0).$$

Now if $Y = U\Sigma_s V^T$, we have to find an orthogonal matrix A such that $\hat{X}AA^T = Y$. It is easy to verify, that if we choose $A = V_1$ where $V_1 = V(:, 1:s)$, then

ALGORITHM 6.3. *Computation of Hyperplanes.*

```
function [V,p ] = hyper(Q);
% Fits a hyperplane of dimension s <n
% to a set of given points Q(i,:) belonging
%  to R^n.
% The hyperplane has the equation
% X = p + V(:,1:s)*tau (Parameter Form) or
% is defined as solution of the linear
% equations  V(:,s+1:n)'*(y - p)=0 (Normal form)
m = max(size(Q));
p = sum(Q)'/m;
Qt = Q - ones(size(Q))*diag(p);
[U S V] = svd(Qt, 0);
```

$\hat{X}AA^T = U\Sigma_s V^T$. Thus the singular value decomposition of $\hat{X}$ gives us all the lower-dimensional hyperplanes that fit best the given points:

$$y = p + V_1 t, \quad \text{with} \quad p = \frac{1}{m}\sum_{i=1}^{m} x_i.$$

Notice that $V_2 = V(:, s+1:n)$ gives us also the normal form of the hyperplane: Here the hyperplane is described as the solution of the linear equations

$$V_2^T y = V_2^T p.$$

In order to compute the hyperplanes, we therefore essentially have to compute one singular value decomposition. This is done by the MATLAB function hyper.m (Algorithm 6.3). The reader should note that the statement [U S V] = svd(Qt, 0) computes the "economy size" singular value decomposition. If Qt is an m-by-n matrix with $m > n$, then only the first n columns of U are computed, and S is an n-by-n matrix.

Fitting a Plane in Space

Suppose that we are given the m points $P_1, \ldots, P_m$ in space, and that we want to fit a plane through these points such that the sum of the squares of the distances from the points to the plane is minimized. We can describe a plane implicitly by the equations

$$c + n_1 x + n_2 y + n_3 z = 0, \quad n_1^2 + n_2^2 + n_3^2 = 1. \tag{6.16}$$

In this case the distance r from a point $P = (x, y, z)$ to the plane is given by

$$r = |c + n_1 x + n_2 y + n_3 z|.$$

Consequently, we want to minimize

$$||r|| = \sum_{i=1}^{m} r_i^2 = \min$$

subject to

$$
\begin{pmatrix} 1 & x_{P_1} & y_{P_1} & z_{P_1} \\ 1 & x_{P_2} & y_{P_2} & z_{P_2} \\ \vdots & \vdots & \vdots & \vdots \\ 1 & x_{P_m} & y_{P_m} & z_{P_m} \end{pmatrix} \begin{pmatrix} c \\ n_1 \\ n_2 \\ n_3 \end{pmatrix} = \begin{pmatrix} r_1 \\ r_2 \\ \vdots \\ r_m \end{pmatrix} \quad \text{and } n_1^2 + n_2^2 + n_3^2 = 1. \quad (6.17)
$$

In MATLAB this problem is easily solved by calling the function clsq:

```
>> [c, n] = clsq ([ones(m,1) P'], 3);
```

Alternatively, a point P on the plane can always be represented explicitly by

$$
P = \mathbf{p} + c_1 \mathbf{v}_1 + c_2 \mathbf{v}_2. \quad (6.18)
$$

If we use the procedure hyper to compute $\mathbf{p}$, $\mathbf{v}_1$, and $\mathbf{v}_2$, then $\mathbf{p}$ denotes the center of gravity of the data points, and $\mathbf{v}_1$ and $\mathbf{v}_2$ denote two orthogonal base vectors of the plane. Additionally, the straight line given by the equation

$$
L = \mathbf{p} + c_1 \mathbf{v}_1 \quad (6.19)
$$

is also the least squares line through the given data points. We can generate a set of sample data points by the MATLAB statements

```
>> % mainhyper.m
>> m = 100;
>> rand ('seed', 10);
>> randn ('seed', 0);
>> P = rand (2, m) - 0.5;
>> P (3, :) = 0.5 * P (1, :) + 0.5 * P (2, :) + ...
>>              0.25 * randn (1, m);
```

These points lie approximately on the plane $x + y - 2z = 0$. The plane, that goes through these points in the least squares sense, is computed by the call

```
>> [V, p] = hyper (P');
```

We can now compute the projections of the points P_j onto this plane by

```
>> proj = p * ones (1, m) + ...
>>          V (:, 1:2) * V (:, 1:2)' * (P - p * ones (1, m));
```

In order to plot a square part of the plane we need to know the coordinates of four corner of the square. They can be computed by the following MATLAB statements:

```
>> corners = [1 -1 -1  1
>>            1  1 -1 -1];
>> corners = p * ones (1, 4) + V (:, 1:2) * corners;
```

In the same way, we can compute two points of the least squares line as follows:

```
>> straight = p * ones (1, 2) + V (:, 1) * [-1 1];
```

We are now ready to plot the least squares line and the least squares plane together with the projections of the data points onto the plane. The resulting static picture, however, would only give an inadequate representation of the three-dimensional situation. It would be hard to tell the relative position of all the different objects.

A moving picture, on the other hand, can create a much more realistic impression. In our example we can have the plane slowly revolve around the z-axis. This will also give us a good opportunity for introducing MATLAB's `movie` command.

A movie in MATLAB consists of a sequence of precomputed frames. The command `frames = moviein (nframes)` allocates the matrix `frames` which will store all the frames of our movie as its columns. By `nframes` we denote the number of single frames that we want to generate. We have to precompute all the frames of the movie and store them in `frames` by calls of `getframe`. Afterwards, we can display our animated pictures by the command `movie`.

In order to rotate the lines and the plane we introduce the rotation matrix

$$R := \begin{pmatrix} \cos\varphi & -\sin\varphi & 0 \\ \sin\varphi & \cos\varphi & 0 \\ 0 & 0 & 1 \end{pmatrix}, \tag{6.20}$$

where $\varphi = 2\pi/n_{\text{frames}}$. If we apply R to a point P we rotate P by the angle φ around the z-axis.

We want to use an orthographic projection to display our three-dimensional pictures on the two-dimensional computer screen, where the observer is located on the y-axis at $-\infty$. This is easily achieved by the MATLAB command `view (viewmtx (0, 0, 0))`.

The projection of the data points onto the plane, the least squares line, and the least squares plane must be plotted in the right sequence in order to achieve the desired effect. First, we plot the projection lines behind the plane. They can be identified by the condition `P (2, j) > proj (2, j)`. Next, we plot the plane by a call of the patch primitive. Such a patch object supports the shading of its surface by prescribing the colors in its corners. If we make use of the gray-colormap we can color the surface of the plane such that it becomes darker along the y-axis. This trick leads to the convincing three-dimensional appearance of the picture. Then, we draw the least squares line by a call of `line`. Finally, the projection lines before the plane are plotted. We can avoid the automatic replotting of the picture elements, which is undesirable in this application, by setting the option `EraseMode` to `none`. In this way the default hidden line removal in MATLAB is disabled.

These considerations lead us to the following MATLAB program, which first precomputes all the frames and then displays the movie infinitely many times:

```
>> nframes = 36;  phi = 2 * pi / nframes;
>> rot = [cos(phi) -sin(phi) 0
>>        sin(phi)  cos(phi) 0
>>        0         0        1];
>> clf;  axis ('off');  colormap (gray);
>> frames = moviein (nframes);
```

```
>> for k = 1:nframes,
>>   clf;
>>   axis ([-1.25 1.25 -1.25 1.25 -1.25 1.25]);  axis ('off');
>>   view (viewmtx (0, 0, 0));  caxis ([-1.3 1.3]);

>>   % plot points behind the plane
>>   for j = 1:m,
>>     if P (2, j) > proj (2, j),
>>       line ([P(1,j) proj(1,j)], ...
>>             [P(2,j) proj(2,j)], ...
>>             [P(3,j) proj(3,j)], ...
>>             'Color', 'yellow', 'EraseMode', 'none');
>>     end
>>   end
>>   drawnow;

>>   % plot plane
>>   patch (corners (1, :), corners (2, :), corners (3, :), ...
>>          -corners (2, :), ...
>>          'EdgeColor', 'none', 'EraseMode', 'none');
>>   drawnow;

>>   % plot least squares line
>>   line (straight (1, :), straight (2, :), straight (3, :), ...
>>         'Color', 'red', 'EraseMode', 'none');
>>   drawnow;

>>   % plot points before the plane
>>   for j = 1:m,
>>     if P (2, j) <= proj (2, j),
>>       line ([P(1,j) proj(1,j)], ...
>>             [P(2,j) proj(2,j)], ...
>>             [P(3,j) proj(3,j)], ...
>>             'Color', 'yellow', 'EraseMode', 'none');
>>     end
>>   end
>>   drawnow;

>>   frames (:, k) = getframe;

>>   % rotate points
>>   P = rot * P;  proj = rot * proj;
>>   corners = rot * corners;  straight = rot * straight;
>> end;
>> clf;  axis ('off');  movie (frames, inf);
```

References

[1] G. GEISE und S. SCHIPKE *Ausgleichsgerade, -kreis, -ebene und -kugel im Raum*, Mathematische Nachrichten, 62, 1974.

[2] G.H. GOLUB and CH. VAN LOAN, *Matrix Computations*. 2nd ed., John Hopkins University Press, Baltimore, 1989.

[3] H. SPÄTH, *Orthogonal least squares fitting with linear manifolds*. Numer. Math., 48, 1986, pp. 441–445.

[4] VDI BERICHTE 761, *Dimensional Metrology in Production and Quality Control*. VDI Verlag, Düsseldorf, 1989.

[5] H.J. WARNECKE und W. DUTSCHKE, *Fertigungsmesstechnik*, Springer, Berlin, 1984.

Chapter 7. The Generalized Billiard Problem

S. Bartoň

7.1 Introduction

We consider the following problem: *Given a billiard table (not necessarily rect-angular) and two balls on it, from which direction should the first ball be struck, so that it rebounds off the rim of the table, and then impacts the second ball?* This problem has previously been solved for a circular table in [1, 3].

Let the billiard table be parametrically described by $X = f(t)$ and $Y = g(t)$. These functions may be arbitrary, with the single requirement being that their first derivatives exist. We aim at the rim points described by the coordi-nates $[X(t_i), Y(t_i)]$. To solve the given problem, we shall use two different methods; *the generalized reflection method* and *the shortest trajectory method*. We use both methods to derive functions of the parameter t, whose roots are the points $t_1, \ldots, t_n$; i.e. the solution to the given problem. We shall first find the analytic solution in the first part of the chapter. In the second part we intuitively check this solution by solving some practical examples. In more complicated cases we shall use numerical methods to solve the final equation.

7.2 The Generalized Reflection Method

The solution of the problem can be divided into two main steps. First, the trajectory of the ball reflected by the general point on the billiard rim is found. Second, the point on the billiard rim is found such that the second ball lies on this trajectory.

The first step can be solved by a generalization of the plane mirror problem, i.e. find the path touching line l from point P to point Q, according to the mirror condition. The mirror condition is satisfied when the impact angle is equal to the reflection angle.

In the second step we calculate the distance between the reflected trajectory and the position of the second ball. This distance is a function of the impact position point. The problem is solved once the point has been found with corresponding distance equal to zero.

7.2.1   Line and Curve Reflection

To begin, we construct point M as a *mirror point* of P, using the line l as an axis of symmetry. The line l' connecting points M and Q intersects line l at point T. Now, the required path passes through point T as shown in Figure 7.1). From point P one must aim at point T in order to hit the ball located at point Q. As the mirror line we use a tangent line of the table boundary. It is possible to

FIGURE 7.1. *The Line Reflection.*

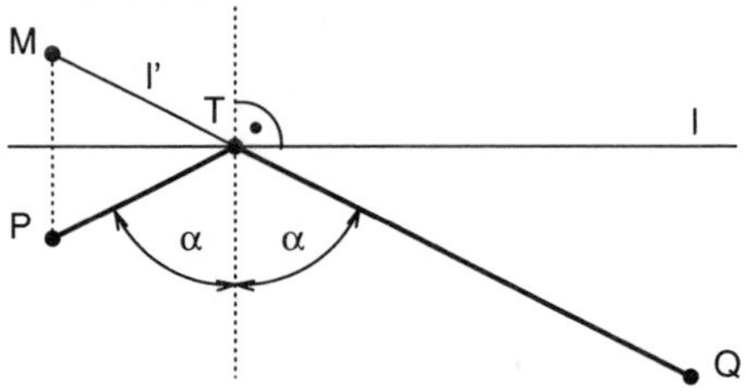

describe this line implicitly:

$$l \equiv l_1 \equiv y - T_y = k(x - T_x), \quad k_1 \equiv k = \left. \frac{\dfrac{d\,X(t)}{d\,t}}{\dfrac{d\,Y(t)}{d\,t}} \right|_T \tag{7.1}$$

where

$$T \equiv [T_x,\ T_y] = [X(t),\ Y(t)],$$

and where T_x, T_y are the coordinates of the common point on the billiard table boundary and k is the derivative of the boundary function at this point. The ball hitting the boundary is reflected just as if it were reflected by this line. To find the mirror image of the position of one ball we use l as a mirror line. The solution of the generalized billiard problem is shown in Figure 7.2. We can aim the ball located at P toward a point T on the billiard boundary. We use the line l_1, tangent to the boundary at point T, to find point M, the mirror image of point P. l_2 is the line segment connecting points P and M. l_1 is perpendicular to l_2, intersecting l_2 in its midpoint C.

If point M is known, then it is possible to draw the line segment l_3 by connecting points M and Q, and the ray l_4 (beginning at point M and going through point T). The ball located at point P will move along this ray after its reflection. Now we can calculate the distance d_2 between point Q and ray l_4. If this distance is equal to zero, i.e., $d_2 = 0$, then the position of the second ball lies on the trajectory of the first ball after its reflection. Hence, the balls must collide. A second possibility is to calculate the distance d_1 between points T and V, the intersection of lines l_1 and l_3. To hit the ball located at point Q, this distance must also be equal to zero, i.e. $d_1 = 0$, since in this case T and V must coincide; i.e., l_4 will then coincide with l_3. The calculation of the distance d_1 is simpler than the calculation of d_2. To determine the distance between two

FIGURE 7.2. The Curve Reflection.

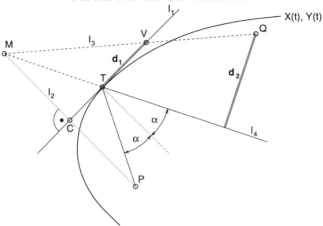

points is simpler than to calculate the distance between a point and a line. The problem is thus reduced to finding point T; for which the corresponding distances d_1, or d_2 respectively, are equal to zero.

7.2.2 Mathematical Description

For the analytic solution, let us again consider Figure 7.2. To simplify the process it is useful to describe all lines using the same type of formula. A line can be described by the coordinates of one point on it and a direction. For the first line l_1 we use Equation (7.1). The other lines are described by

$$l_2 \equiv y - P_y = k_2(x - P_x), \quad \text{where} \quad k_2 = -\frac{1}{k_1}$$
$$l_3 \equiv y - Q_y = k_3(x - Q_x), \quad \text{where} \quad k_3 = \frac{M_y - Q_y}{M_x - Q_x}. \tag{7.2}$$

The simplest way to obtain the coordinates of point M is to find the point C at the intersection of lines l_1 and l_2, and to use the mirror condition

$$C_x = \frac{P_x + M_x}{2}, \quad C_y = \frac{P_y + M_y}{2}. \tag{7.3}$$

Subsequently, it is possible to calculate the coordinates of point V as the intersection of lines l_1 and l_3, and to determine the *aiming error distance* d_1,

$$d_1 = \sqrt{(V_x - T_x)^2 + (V_y - T_y)^2}. \tag{7.4}$$

Using Equations (7.1) through (7.4) we can calculate d_1 as a function of the known coordinates of the positions of the balls, and the billiard cushion shape function and its first derivative

$$d_1(t) = d_1\left(P_x, P_y, Q_x, Q_y, X(t), Y(t), \frac{dX(t)}{dt}, \frac{dY(t)}{dt}\right). \tag{7.5}$$

TABLE 7.1.

List of Main Variables used in the Generalized Billiard Problem.

Variable	Meaning
T_x, T_y	$T_x = X(t)$ and $T_y = Y(t)$, description of the shape of the cushion
k_1	slope of the tangent of the cushion function in point $[T_x, T_y]$ (line l_1)
k_2	slope of line l_2, which is perpendicular to the l_1, $[P_x,\ P_y]$ is a point of line l_2
k_3	slope of the line l_3, connecting the points $[Q_x,\ Q_y]$ and $[M_x, M_y]$
Aimerr	square of the distance between points $[V_x,\ V_y]$ and $[T_x,\ T_y]$, ("aiming error")
DAA, DB	numerator and denominator of the *Aimerr*
P_x, P_y	coordinates of the first ball
Q_x, Q_y	coordinates of the second ball
M_x, M_y	mirror point of the point $[P_x,\ P_y]$, mirror line is l_1
C_x, C_y	coordinates of the intersection point of lines l_1 and l_2
V_x, V_y	coordinates of the intersection point of lines l_1 and l_3

We now wish to solve the equation $d_1(t) = 0$ for t. MAPLE is used to obtain an explicit expression for $d_1(t)$, and its solution.

7.2.3   MAPLE Solution

The program is strictly based on the above mentioned relations. We shall use the following variables (cf. Table 7.1). The generalized billiard problem is solved when $[V_x,\ V_y] \equiv [T_x,\ T_y]$, i.e. if $DAA = 0$.

```
> E1 := solve({V[y] - Q[y] = k[3]*(V[x]-Q[x]),
>         V[y] - T[y] = k[1]*(V[x] - T[x])}, {V[x], V[y]}):
> assign(E1):
> E2 := solve({C[y] - P[y] = k[2]*(C[x] - P[x]),
>         C[y] - T[y] = k[1]*(C[x] - T[x])}, {C[x], C[y]}):
> assign(E2):

> M[x] := P[x] + 2*(C[x] - P[x]): M[y] := P[y] + 2*(C[y] - P[y]):
> k[3] := (M[y] - Q[y])/(M[x] - Q[x]):
> T[y] := Y(t):   T[x] := X(t):
> k[2] := -1/k[1]:
> k[1] := diff(Y(t), t)/diff(X(t),t):
> Aimerr := normal((T[x] - V[x])^2 + (T[y] - V[y])^2):
```

We now have to solve the equation $Aimerr = 0$. If the parametrical description of the billiard rim $X(t)$, $Y(t)$ is used, the expression for $Aimerr$ becomes a fraction. This can be used to simplify the solution, because we only need to find the zeros of the numerator. But to check if such a zero is a correct solution, we must be sure that there are no common zeros of the $Aimerr$ numerator and denominator.

We now attempt to simplify the $Aimerr$ numerator.

```
> DA := op(1, numer(Aimerr)):
> DB := denom(Aimerr):
> DA1 := factor(select(has, DA, {diff(X(t), t)^2,
>           diff(Y(t), t)^2})):
> Da2 := select(has, DA, diff(X(t), t)):
> DA2 := factor(select(has, Da2, diff(Y(t), t))):
> DA - expand(DA2 + DA1);
```

$$0$$

```
> DAA := DA1 + DA2;
```

$$DAA := -2(\frac{\partial}{\partial t} Y(t)) (\frac{\partial}{\partial t} X(t))$$
$$(-Q_y P_y + Q_y Y(t) - X(t) P_x + Y(t) P_y - X(t) Q_x + P_x Q_x - Y(t)^2 + X(t)^2) +$$
$$(\frac{\partial}{\partial t} Y(t)) - (\frac{\partial}{\partial t} X(t))) ((\frac{\partial}{\partial t} Y(t)) + (\frac{\partial}{\partial t} X(t)))$$
$$(-Q_y P_x + Q_y X(t) + X(t) P_y + Y(t) P_x - 2 Y(t) X(t) + Y(t) Q_x - Q_x P_y)$$

The result of the MAPLE code is the variable DAA, which depends on the specific function used for the billiard cushion shape. For any given problem instance these functions must be provided as input.

7.3 The Shortest Trajectory Method

From physics [2] it is known that the trajectory of a light beam from one point to another minimizes the length of the path. In case of a reflection this curve consists of two line segments. The first line segment connects the point of the light source, *(in our case: the position of the first ball)*, with the point of reflection, *(the point on the billiard table boundary)*. The latter point is connected by the second line segment to the point of observation, *(the position of the second ball)*. Using this, we can also solve the generalized billiard problem.

We can construct the first line segment l_P connecting the point P with the reflection point T on the billiard table boundary, and the second line segment l_Q connecting Q with T. Now we must find the smallest sum of the lengths of both line segments S as a function of T.[1] Remember that T is the point on the billiard table boundary.

---

[1]This is a proposal of Josef Bartoñ, (the father of the author)

With $T \equiv [X(t),\, Y(t)]$, our problem is to solve equation

$$\frac{d\left(\sqrt{(T_x(t) - P_x)^2 + (T_y(t) - P_y)^2} + \sqrt{(T_x(t) - Q_x)^2 + (T_y(t) - Q_y)^2}\right)}{dt} = 0$$

for the parameter t.

7.3.1   MAPLE Solution

```
> LP := sqrt((X(t) - P[x])^2 + (Y(t) - P[y])^2):
> LQ := sqrt((X(t) - Q[x])^2 + (Y(t) - Q[y])^2):
> lP := diff(LP,t): lQ := diff(LQ,t):
> dll := numer(lP)*denom(lQ) + numer(lQ)*denom(lP);
```

$$dll := (-(\frac{\partial}{\partial t} X(t))\, P_x + (\frac{\partial}{\partial t} X(t))\, X(t) + (\frac{\partial}{\partial t} Y(t))\, Y(t) - P_y\, (\frac{\partial}{\partial t} Y(t)))$$
$$\sqrt{(X(t) - Q_x)^2 + (Y(t) - Q_y)^2} +$$
$$((\frac{\partial}{\partial t} X(t))\, X(t) - (\frac{\partial}{\partial t} X(t))\, Q_x + (\frac{\partial}{\partial t} Y(t))\, Y(t) - Q_y\, (\frac{\partial}{\partial t} Y(t)))$$
$$\sqrt{(X(t) - P_x)^2 + (Y(t) - P_y)^2}$$

For the final solution of equation $dll = 0$ we first have to specify the billiard table shape functions $X(t)$ and $Y(t)$.

7.4   Examples

We now compare the reflection method with the shortest trajectory method. Both should produce the same result. As a first example we consider the circular billiard table, whose solution is well known [1, 3].

7.4.1   The Circular Billiard Table

To check our program we will recompute the solutions given in [1, 3] with the same assumptions: i.e., $R = 1$ and $Q_y = 0$. We can simply continue the computations with the MAPLE programs presented in Sections 7.2.3 and in 7.3.1 by defining

```
> X(t) := cos(t): Y(t) := sin(t): Q[y]:= 0:
```

Because the shape of the billiard cushion is well defined, the position of the balls is completely general. First, we shall solve $DAA = 0$ (cf. Section 7.2.3).

```
> DA := simplify(DAA);
```

$$DA := \cos(t)\, P_y + 2\sin(t)\cos(t)\, P_x\, Q_x - 2\cos(t)^2\, Q_x\, P_y - \sin(t)\, P_x$$
$$- \sin(t)\, Q_x + Q_x\, P_y$$

```
> SolDA := solve(DA,t);
```

$$SolDA := \arctan(\frac{P_y\,(-\%1 + 2\,\%1^2\,Q_x - Q_x)}{2\,\%1\,P_x\,Q_x - Q_x - P_x},\ \%1)$$
$$\%1 := \mathrm{RootOf}((4\,P_y^2\,Q_x^2 + 4\,P_x^2\,Q_x^2)\,\_Z^4$$
$$+ (-4\,P_y^2\,Q_x - 4\,P_x^2\,Q_x - 4\,P_x\,Q_x^2)\,\_Z^3$$
$$+ (2\,P_x\,Q_x - 4\,P_x^2\,Q_x^2 + Q_x^2 + P_x^2 + P_y^2 - 4\,P_y^2\,Q_x^2)\,\_Z^2$$
$$+ (4\,P_x\,Q_x^2 + 4\,P_x^2\,Q_x + 2\,P_y^2\,Q_x)\,\_Z - P_x^2 + P_y^2\,Q_x^2 - Q_x^2 - 2\,P_x\,Q_x)$$

Using the command `solve` we can calculate an *analytic* value of the parameter t such that $DA = 0$. The solution details are not presented here for the sake of brevity. To compare our result with the results of [1, 3] we shall substitute the general variables P_x, P_y and Q_x by their numerical values. We shall check if the zeros of DA are the correct solutions of the *Aimerr*. If they are substituted into DB its value must not be zero.

```
> P[x] := 1/2: P[y] := 1/2: Q[x] := -3/5:
> DA := simplify(DA):
> SolDA := solve(DA, t);
```

$$SolDA := \arctan(-2\,\%1 - 1 + 6\,\%1^3 + 5\,\%1^2,\ \%1)$$
$$\%1 := \mathrm{RootOf}(-23\,\_Z^2 + 24\,\_Z^3 - 9\,\_Z + 36\,\_Z^4 + 4)$$

```
> FSolDA := evalf(allvalues(SolDA));
```

$$FSolDA := 2.750972642, .9380770508, 2.274796770, -1.251457484$$

```
> for i from 1 by 1 to nops([FSolDA]) do:
>       tA[i] := FSolDA[i];
>       DBi := evalf(subs(t = tA[i], DB));
>       if DBi = 0 then print('Root',i,' unusable!'); fi;
>       print('i =',i, ' tA = ', tA[i],' ==> DB = ', DBi);
> od:
```

$$i =, 1,\ tA =, 2.750972642,\ ==> \ DB =, 2.948610256$$

$$i =, 2,\ tA =, .9380770508,\ ==> \ DB =, 2.742080137$$

$$i =, 3,\ tA =, 2.274796770,\ ==> \ DB =, 2.415361451$$

$$i =, 4,\ tA =, -1.251457484,\ ==> \ DB =, 6.280614819$$

Note that the position angles (*in radians*) of the aiming points are the same as in [1, 3], i.e., `evalf(tA[4]+2*Pi)` = 5.031727826. The coordinates of the four aiming points are computed for $t = t_{A1}, \ldots, t_{A4}$ by $T_x = \cos(t)$ and $T_y = \sin(t)$. We now check the second method $dl = 0$ (cf. Section 7.3.1).

We again use MAPLE to find the general analytic solution:

```
> P[x] := 'P[x]': P[y] := 'P[y]': Q[x] := 'Q[x]': Q[y] := 0:
> dl := simplify(dll);
```

$$dl := \sqrt{-2\cos(t)\,Q_x + Q_x{}^2 + 1}\,\sin(t)\,P_x$$
$$-\sqrt{-2\cos(t)\,Q_x + Q_x{}^2 + 1}\,\cos(t)\,P_y$$
$$+ Q_x\sin(t)\,\sqrt{-2\cos(t)\,P_x + P_x{}^2 + 1 - 2\,P_y\sin(t) + P_y{}^2}$$

```
> # Sold1 := solve(dl,t):
```

The analytic solution exists, but is more complicated than in the preceding example, because the variable dl contains square roots. In such cases it is better to modify the equation first i.e. to remove the square roots before the solution.

```
> eq := op(1, dl) + op(2, dl) = -op(3, dl):
> eq := expand(map(u -> u^2, eq)):
> eq := lhs(eq) - rhs(eq) = 0:
> Sold1 := solve(eq, t):
```

The result is not too complicated, but still is not shown here for the sake of brevity. But we can reassign values for P_x, P_y and Q_x and solve eq again:

```
> P[x] := 1/2: P[y] := 1/2: Q[x] := -3/5:
> eq;
```

$$\frac{33}{50}\sin(t)^2\cos(t) - \frac{1}{5}\sin(t)^2 - \frac{3}{5}\sin(t)\cos(t)^2 - \frac{17}{25}\sin(t)\cos(t) + \frac{3}{10}\cos(t)^3$$
$$+ \frac{17}{50}\cos(t)^2 + \frac{9}{25}\sin(t)^3 = 0$$

```
> Sold1 := solve(eq,t);
```

$$Sold1 := \arctan\left(\frac{\frac{33}{146} + \frac{5}{146}\sqrt{137}}{-\frac{15}{146} + \frac{11}{146}\sqrt{137}}\right),\ \arctan\left(\frac{\frac{33}{146} - \frac{5}{146}\sqrt{137}}{-\frac{15}{146} - \frac{11}{146}\sqrt{137}}\right) - \pi,$$
$$\arctan(-2\,\%1 - 1 + 6\,\%1^3 + 5\,\%1^2,\ \%1)$$
$$\%1 := \mathrm{RootOf}(-23\,\_Z^2 + 24\,\_Z^3 - 9\,\_Z + 36\,\_Z^4 + 4)$$

```
> FSold1 := evalf(allvalues({Sold1}));
```

$$FSold1 := \{.6775335699,\ -2.965871238,\ .9380770491\},$$
$$\{.6775335699,\ -2.965871238,\ 2.750972642\},$$
$$\{.6775335699,\ -2.965871238,\ -1.251457483\},$$
$$\{.6775335699,\ -2.965871238,\ 2.274796770\}$$

As shown, both solutions are the same. Note that the redundant roots of the second solution result from modifying the equation. Thus they are of no practical use for us.

FIGURE 7.3. *Dan and dln Functions.*

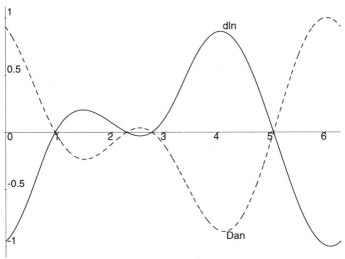

Let us compare the functions $DA(t)$ and $dl(t)$. Because they were significantly simplified they have lost their initial physical meaning. We first normalize the functions by dividing each of them by their largest absolute value. Then we plot both normalized functions on the same axes as shown in Figure 7.3.

```
> DAd := diff(DA, t): dld := diff(dl, t):
> tamax := fsolve(DAd = 0, t, 6..7):
> tlmax := fsolve(dld = 0, t, 6..7):
> Dan := DA/abs(subs(t = tamax, DA)):
> dln := dl/abs(subs(t = tlmax, dl)):
> with(plots):
> p1 := plot(Dan, t = 0..2*Pi, linestyle=3):
> p2 := plot(dln, t = 0..2*Pi, linestyle=0):
> display({p1,p2});
```

Upon initial inspection, it appears that the functions $Dan(t)$ and $dln(t)$ are essentially identical. and that only their signs are different. Let us visualize the difference by plotting the rather unusual curve $PF(x, y) \equiv [Dan(t), \; dln(t)]$, (cf. Figure 7.4). If our conjecture is correct, this graph should be the line segment from point $[-1, 1]$ to point $[1, -1]$. The observed deviation from this line segment, however, indicates that the two functions are in fact different.

```
> plot([Dan, dln, t = 0..2*Pi]);
```

It is interesting to create the following new curve:

$$\tilde{X}(t) = X(t)(1 + Dan), \quad \tilde{Y}(t) = Y(t)(1 + Dan). \qquad (7.6)$$

This curve is constructed by superimposing the function Dan (or dln) as a "perturbation" to the table shape function. Since at the aim points the functions DA, Dan are equal to zero, the coordinates of these points are on the

FIGURE 7.4. *Graphical Comparison of Dan and dln.*

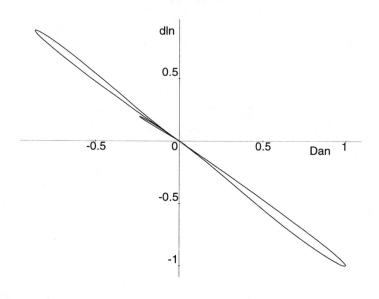

FIGURE 7.5. *Graphical Solution for Circle.*

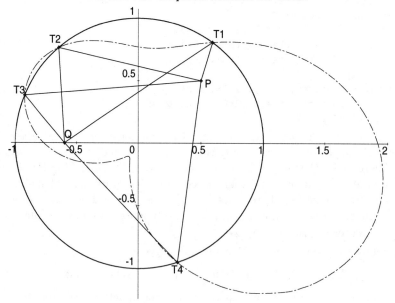

ALGORITHM 7.1. *Procedure* ResultPlot.

```
ResultPlot := proc(P1x, P1y, P2x, P2y, Xborder, Yborder, Perturb,
                Start, End, Zerosvec, Nzeros, Xrange, Yrange)
    local i, n, Trajectories, AimPointX, AimPointY,
            PlotTraject, PlotPoints, PlotBorder, PlotPert;
    for i from 1 by 1 to Nzeros do
        AimPointX[i] := evalf(subs(t = Zerosvec[i], Xborder));
        AimPointY[i] := evalf(subs(t = Zerosvec[i], Yborder));
        Trajectories[i] := [[P1x, P1y], [AimPointX[i],
                            AimPointY[i]], [P2x, P2y]];
    od;
    PlotTraject := plot({seq(Trajectories[i], i = 1..Nzeros)},
                    style = LINE, thickness=1):
    PlotPoints := plot({[P1x, P1y], [P2x, P2y],
                    seq([AimPointX[i], AimPointY[i]],
                    i = 1..Nzeros)}, style=POINT):
    PlotBorder := plot([Xborder, Yborder, t = Start..End],
                    thickness=2):
    PlotPert := plot([Xborder*(1 + Perturb), Yborder*(1 + Perturb),
                    t = Start..End], linestyle=3):
    display({PlotTraject, PlotPoints, PlotBorder, PlotPert},
        scaling=constrained, view=[Xrange, Yrange]);
    end;
```

cushion (cf. Figure 7.5). The aim points, therefore, can be found graphically as the points of intersection of the cushion shape function and the disturbed shape function (7.6).

Because graphical output will be used several times, we shall create a MAPLE procedure ResultPlot (see Algorithm 7.1), which will be used for plotting the graphical solution. Procedure ResultPlot is easy to use. For example, Figure 7.5 was created via

```
> ResultPlot(P[x], P[y], Q[x], Q[y], X(t), Y(t),
>           Dan, 0, 2*Pi, tA, 4, -1..2, -1.25..1);
```

7.4.2   The Elliptical Billiard Table

An ellipse with the semi axes a and b is described in parametric form by:

$$X(t) = a \cos(t), \quad Y(t) = b \sin(t). \tag{7.7}$$

We consider the ellipse with the major semi axis $a = 1$ and with the minor semi axis $b = 0.8$. Let $P \equiv [-0.6, -0.3]$, and $Q \equiv [0.5, 0.5]$ be the coordinates of the positions of the balls. The previously derived equations DA (cf. Section 7.2.3) and dl (cf. Section 7.3.1) are used.

```
> X(t) :=  cos(t):  Y(t) := 4/5*sin(t):
> P[x] := -3/5: P[y] := -3/10: Q[x] := 1/2: Q[y] := 1/2:
> DA := simplify(DAA);
```

$$DA := \frac{42}{125}\sin(t)\cos(t) + \frac{18}{625}\sin(t)\cos(t)^2 + \frac{369}{500}\cos(t)^2 + \frac{9}{125}\cos(t)^3$$
$$+ \frac{2}{25}\sin(t) + \frac{7}{125}\cos(t) - \frac{9}{20}$$

```
> SolDA := solve(DA,t);
```

$$SolDA := \arctan(\frac{32992953}{45907900}\%1^2 - \frac{2162333043}{459079000}\%1^3 - \frac{586629}{1836316}$$
$$- \frac{2457054}{57384875}\%1^5 - \frac{10344267}{11476975}\%1^4 + \frac{60534443}{18363160}\%1, \%1)$$
$$\%1 := \text{RootOf}(4183641\_Z^4 + 785160\_Z^5 + 326640\_Z^3 - 4826050\_Z^2$$
$$+ 37584\_Z^6 - 651000\_Z + 1225625)$$

```
> irreduc(op(1,op(2,SolDA)));
```

$$true$$

As shown, an analytic solution to the given cannot be obtained. Thus, numerical methods are used instead.

```
> tA[1] := fsolve(DA, t, -3..-2);
> tA[2] := fsolve(DA, t, -1..0);
> tA[3] := fsolve(DA, t, 1..2);
> tA[4] := fsolve(DA, t, 2..3);

> tl[1] := fsolve(dll, t, -3..-2);
> tl[2] := fsolve(dll, t, -1..0);
> tl[3] := fsolve(dll, t, 1..2);
> tl[4] := fsolve(dll, t, 2..3);
```

$$tA_1 := -2.425591133$$
$$tA_2 := -.5260896824$$
$$tA_3 := 1.038696884$$
$$tA_4 := 2.761693514$$
$$tl_1 := -2.425591133$$
$$tl_2 := -.5260896824$$
$$tl_3 := 1.038696884$$
$$tl_4 := 2.761693514$$

```
> Dan := DA/abs(evalf(subs( t = fsolve(diff(DA, t),
>               t, -2..-1), DA))):
> dln := dll/abs(evalf(subs( t = fsolve(diff(dll, t),
>               t, -2..-1), dll))):
> p1:=plot(Dan, t = 0..2*Pi,linestyle=0):
> p2:=plot(dln, t = 0..2*Pi,linestyle=3):
> display({p1,p2});
> ResultPlot(P[x], P[y], Q[x], Q[y], X(t), Y(t),
>               Dan, -Pi, Pi, tA, 4, -1.5..2, -0.8..0.8);
```

FIGURE 7.6. *Solution for Ellipse – Dan and dln Functions.*

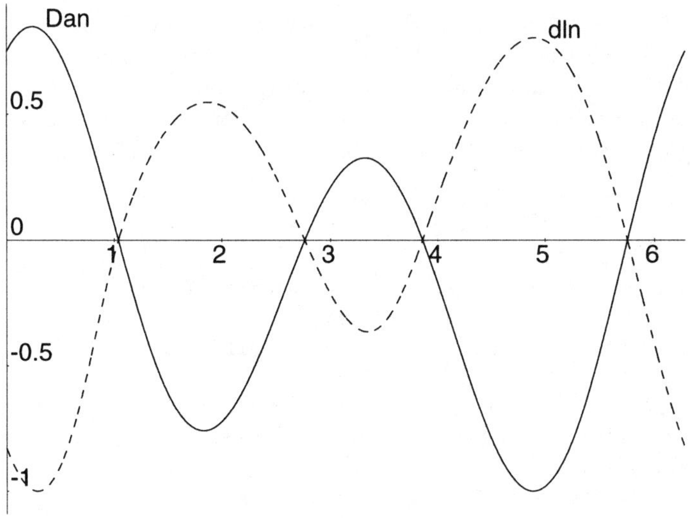

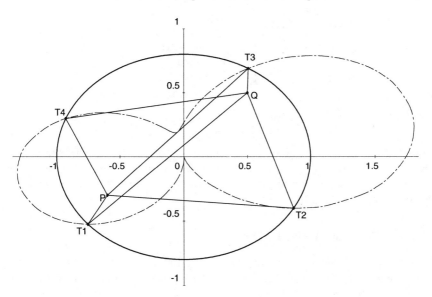

FIGURE 7.7. *Graphical Solution for Ellipse.*

7.4.3   The Snail Billiard Table

A snail (or helix) shape of the cushion of a billiard table is rather unusual, but it can be used to further demonstrate the capabilities of MAPLE. The "snail shape curve", or *the spiral of Archimedes*, is described best in the polar coordinate system by the function $\varrho = a\,\varphi$. The transformation the rectangular coordinates system gives us the parametric description of the spiral of Archimedes: $X(t) = t\cos(t)$, $Y(t) = t\sin(t)$, where $a = 1$. These equations are used to solve the snail billiard problem. Let the balls have coordinates $P \equiv [-6,\ -12]$, and $Q \equiv [7,\ 5]$.

```
> X(t) := t*cos(t): Y(t) := t*sin(t):
> P[x] := -6: P[y] := -12:  Q[x] := 7:  Q[y] := 5:
> DA := simplify(DAA):
> Dan := DA/abs(evalf(subs(t = fsolve(diff(DA, t),
>                 t, 15..17), DA))):
> dln := dll/abs(evalf(subs(t = fsolve(diff(dll, t),
>                 t, 15..17), dll))):
> p1:=plot(Dan, t = 4*Pi..6*Pi,linestyle=0):
> p2:=plot(dln, t = 4*Pi..6*Pi,linestyle=3):
> display({p1,p2});
> p1:=plot(Dan, t = 12.5..12.85,linestyle=0):
> p2:=plot(dln, t = 12.5..12.85,linestyle=3):
> display({p1,p2});
> tA[1] := fsolve(DA, t, 12.5..12.6);
> tA[2] := fsolve(DA, t, 12.7..12.8);
> tA[3] := fsolve(DA, t, 14..15);
> tA[4] := fsolve(DA, t, 16..17);
```

$$tA_1 := 12.56987562$$
$$tA_2 := 12.75992609$$
$$tA_3 := 14.95803948$$
$$tA_4 := 16.82749442$$

```
> ResultPlot(P[x], P[y], Q[x], Q[y], X(t), Y(t),
>            dln, 4*Pi, 6*Pi, tA, 4, -20..30, -31..26);
```

The results are presented in the same way as in the preceding subsections. Figure 7.8 shows the functions *Dan* and *dln*, with an enlarged detail. Note there are two nearly identical zeros of the function. The plots are necessary to help find starting values for `fsolve`.

7.4.4   The Star Billiard Table

Finally we consider a rather eccentric billiard shape, a five pointed star (very suitable for corpulent players). This shape can be successfully approximated by parametric equations

$$X(t) = \cos(t)\left(1 + \frac{\sin(5\,t)}{5}\right), \qquad Y(t) = \sin(t)\left(1 + \frac{\sin(5\,t)}{5}\right).$$

FIGURE 7.8.
Snail Billiard
Functions Dan
and dln.

FIGURE 7.9.
Dan and dln
Enlarged
Detail.

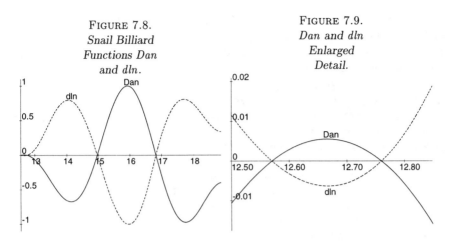

FIGURE 7.10. Graphical Solution for Snail Billiard.

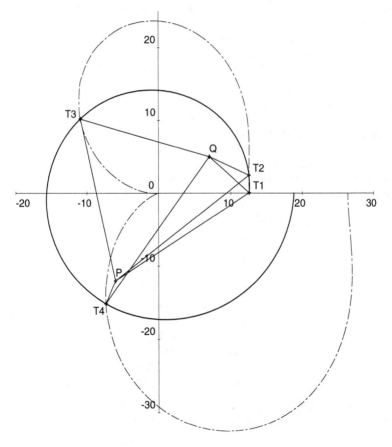

Let the ball coordinates be $P \equiv [-0.8, 0.2]$ and $Q \equiv [0.6, -0.8]$. Thus the balls
are located close to the tops of the star corners (see Figure 7.13). We shall solve
this example as shown previously. But we shall see, that there are many zeros
of the functions DA and dll.

```
> X(t) := cos(t)*(1 + sin(5*t)/5):
> Y(t) := sin(t)*(1 + sin(5*t)/5):
> P[x] := -4/5: P[y] := 1/5: Q[x] := 3/5: Q[y] := -4/5:
> plot({DAA, dll}, t = 0..2*Pi);
> Dan := DAA/evalf(abs(subs(t = fsolve(diff(DAA, t),
>               t, 1..1.5), DAA))):
> dln := dll/evalf(abs(subs(t = fsolve(diff(dll, t),
>               t, 1.7..1.9), dll))):
> p1:=plot(Dan,t = 0..2*Pi,linestyle=0):
> p2:=plot(dln,t = 0..2*Pi,linestyle=3):
> display({p1,p2});
> plot([Dan, dln, t = 0..2*Pi]);
```

Two final plots (cf. Figures 7.11 and 7.12) can be used to show that there are
two different functions DAA and dll describing positions of the aiming points.
As shown there are 10 aiming points but some of them may not be acceptable,
because the trajectory of the ball is intersecting the billiard cushion. We can
only find correct aiming points by examining the graphical solution.

```
> tA[1]  := fsolve(DAA = 0, t, 0.3..0.4);
> tA[2]  := fsolve(DAA = 0, t, 0.9..1.0);
> tA[3]  := fsolve(DAA = 0, t, 1.5..1.6);
> tA[4]  := fsolve(DAA = 0, t, 2.3..2.4);
> tA[5]  := fsolve(DAA = 0, t, 2.7..2.8);
> tA[6]  := fsolve(DAA = 0, t, 3.3..3.5);
> tA[7]  := fsolve(DAA = 0, t, 4.0..4.2);
> tA[8]  := fsolve(DAA = 0, t, 4.7..4.9);
> tA[9]  := fsolve(DAA = 0, t, 5.3..5.5);
> tA[10] := fsolve(DAA = 0, t, 5.5..5.7);
```

$$tA_1 := .3824290257$$
$$tA_2 := .9324776977$$
$$tA_3 := 1.527744928$$
$$tA_4 := 2.341434311$$
$$tA_5 := 2.760135295$$
$$tA_6 := 3.396070545$$
$$tA_7 := 4.072484425$$
$$tA_8 := 4.807203452$$
$$tA_9 := 5.401272765$$
$$tA_{10} := 5.638437107$$

```
> ResultPlot(P[x], P[y], Q[x], Q[y], X(t), Y(t),
>            Dan, 0, 2*Pi, tA, 10, -1.4..1.8, -1.6..2.2);
```

A second possibility to determine the correct aiming points employs the parametric description of the ball trajectory. This trajectory contains two line seg-

ments, l_1 from the position of the first ball P to the aiming (*i.e.* *reflection*) point T_i on the billiard cushion and l_2 from T_i to the second ball position Q. These line segments may be described parametrically using the parameter u.

$$l_1: \begin{array}{l} x = P_x + u\left(T_{i_x} - P_x\right) \\ y = P_y + u\left(T_{i_y} - P_y\right) \end{array} \qquad l_2: \begin{array}{l} x = Q_x + u\left(T_{i_x} - Q_x\right) \\ y = Q_y + u\left(T_{i_y} - Q_y\right) \end{array}$$

Now we can find points of intersection of lines l_1 and l_2 with the billiard cushion. We simply calculate the corresponding parameters $u_1, \ldots, u_2$. If their values satisfy $0 < u_i < 1$, it indicates that there is at least one other point of intersection on the line segment. So the trajectory which is located between the position of the ball and the reflection point cannot be used. Figure 7.13 shows that there are two unusable reflection points T_1 and T_7. We can use this to show that line segment QT_1 intersects with the billiard cushion inside:

```
> T1[x] := evalf(subs(t = tA[1], X(t))):
> T1[y] := evalf(subs(t = tA[1], Y(t))):
> eq1 := X(t) = Q[x] + u*(T1[x] - Q[x]):
> eq2 := Y(t) = Q[y] + u*(T1[y] - Q[y]):
> fsolve({eq1, eq2}, {u, t}, {u = 0..0.5, t=3*Pi/2..2*Pi});
```

$$\{\, t = 5.810766136, u = .3288529247 \,\}$$

```
> fsolve({eq1, eq2}, {u, t}, {u = 0.33..0.8, t = 5.82..6.2});
```

$$\{\, t = 6.162750379, u = .5576715283 \,\}$$

We see that line segment QT_1 intersects the billiard cushion twice between the points Q and T_1, corresponding to values of the parameter $u_1 = 0.3288529247$ and $u_2 = 0.5576715283$.

7.5 Conclusions

Using MAPLE, we have solved the generalized billiard problem by two methods. The pictures and numbers indicate that the results of both methods are the same. It is more difficult to derive the final equation with the generalized reflection method than with the shortest trajectory method. But when all variables are substituted, the final equation DA for the first method is simpler than dl, the result of the second method. Using the second method, we must calculate square roots of various functions and their first derivatives. Therefore the final equation of the second method is more complicated. To what degree depends on the shape of the boundary cushion. For some boundary functions it may happen that the result of the second method is simpler.

The first method can be generalized to an N-reflection trajectory. By this we mean that the first ball must first hit the cushion N times before colliding with the other ball. However the second method does not permit such a generalization. The property of shortest trajectory does not fulfill the reflection conditions in this case.

FIGURE 7.11.
Star Billiard
Functions *Dan*
and *dln*.

FIGURE 7.12.
Dan and *dln*
Parametric
Plot.

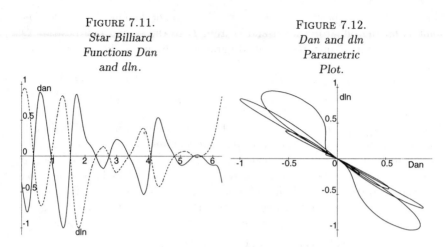

FIGURE 7.13. Graphical Solution for Star Billiard.

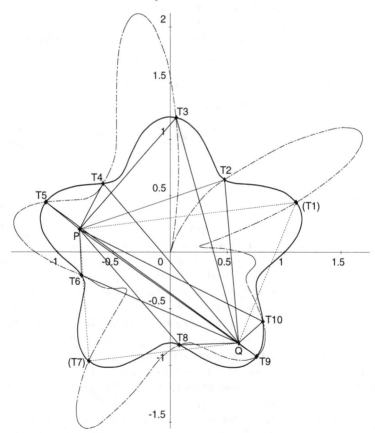

References

[1] W. GANDER AND D. GRUNTZ, *The Billiard Problem*, Int. J. Educ. Sci. Technol., 23, 1992, pp. 825 – 830.

[2] R. G. LERNER AND G. L. TRIGG, *Encyclopedia of Physics*, VCH Publishers, New York, 1991.

[3] J. WALDVOGEL, *The Problem of the Circular Billiard*, El. Math., 47, 1992, pp. 108 – 113.

Chapter 8. Mirror Curves

S. Bartoň

8.1 The Interesting Waste

To solve the generalized billiard problem we used the generalized reflection method. This method is based on the calculation of M, the mirror image point of the position of the first ball. Point M moves as we move point T along the boundary of the billiard cushion shape, (see Chapter 7, Figure 7.2). M traces a *mirror curve*, which depends on the change of the tangent line at point T. This mirror curve is dependent on the position of point P and the shape of the billiard cushion.

In solving the generalized billiard problem it is not necessary to determine the shapes of the mirror curves. However, these curves are very interesting. It would be a pity not to examine them further. We may do so using MAPLE's graphical facilities.

8.2 The Mirror Curves Created by MAPLE

Because the position of the mirror point is given by two coordinates $M \equiv [M_x, M_y]$, both dependent on t, we obtain the mirror curve as a parametric function, see [1]. Its shape can be easily derived using part of the calculations performed to obtain equation DA, (see Chapter 7, Section 7.2.3 and Table 7.1). We show here only the necessary steps for the calculation of M_x, M_y. The result is

```
> E2 := solve({C[y] - P[y] = k[2]*(C[x] - P[x]),
>              C[y] - Y(t) = k[1]*(C[x] - X(t))}, {C[x], C[y]}):
> assign(E2):
> k[1] := diff(Y(t), t)/diff(X(t),t):
> k[2] := -1/k[1]:
> M[x] := normal(P[x] + 2*(C[x] - P[x]));
> M[y] := normal(P[y] + 2*(C[y] - P[y]));
```

$$M_x := -\left(-(\frac{\partial}{\partial t}X(t))^2\, P_x + P_x\,(\frac{\partial}{\partial t}Y(t))^2 - 2\,P_y\,(\frac{\partial}{\partial t}Y(t))\,(\frac{\partial}{\partial t}X(t))\right.$$
$$\left. + 2\,Y(t)\,(\frac{\partial}{\partial t}Y(t))\,(\frac{\partial}{\partial t}X(t)) - 2\,(\frac{\partial}{\partial t}Y(t))^2\,X(t)\right) \Big/ \qquad (8.1)$$
$$\left((\frac{\partial}{\partial t}X(t))^2 + (\frac{\partial}{\partial t}Y(t))^2\right)$$

$$M_y := (-P_y\,(\frac{\partial}{\partial t}\,X(t))^2 + P_y\,(\frac{\partial}{\partial t}\,Y(t))^2 + 2\,(\frac{\partial}{\partial t}\,X(t))^2\,Y(t)$$
$$- 2\,X(t)\,(\frac{\partial}{\partial t}\,Y(t))\,(\frac{\partial}{\partial t}\,X(t)) + 2\,P_x\,(\frac{\partial}{\partial t}\,Y(t))\,(\frac{\partial}{\partial t}\,X(t)))\Big/ \qquad (8.2)$$
$$((\frac{\partial}{\partial t}\,X(t))^2 + (\frac{\partial}{\partial t}\,Y(t))^2)$$

Now we can create the mirror curves for a given pattern curve and a mirror point P. For example consider the parabola curve. Let the point P move along the y axis from the initial position $P_y = -3$, to the final position $P_y = 3$, with step $\Delta y = 1$, see Figure 8.1. We see the continuous deformation of the mirror curve as a function of P_y by looking at the 3d-plot, see Figure 8.2.

```
> X(t) := t:   Y(t) := t^2/2:
> P[x]:= 0:
> m[x] := simplify(M[x]):  m[y] := simplify(M[y]):
> SPy := [seq(i/2, i = -6..6)]:
> p1 := plot({seq([m[x], m[y], t=-4..4], P[y] = SPy)},
>            color = black, thickness = 1):
> p2 := plot([X(t), Y(t), t = -4..4], view = [-4 .. 4, -4 .. 4],
>            thickness = 2):
> with(plots):
> display({p1, p2}, scaling = constrained);
> P[y] := 'P[y]':
> plot3d(subs(P[y] = Py,[P[y], m[x], m[y]]), Py = -3..3,
>        t = -3..3, axes = boxed, orientation = [15,30],
>        labels = ['Py', 'Mx', 'My'], grid = [40, 40] );
```

FIGURE 8.1. *The Parabola's Mirror Curves.*

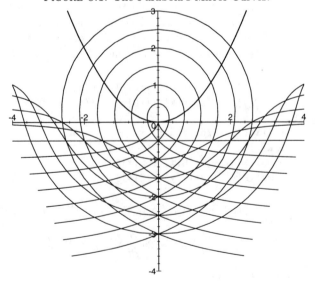

FIGURE 8.2.
Continuous Deformation of the Parabola's Mirror Curve.

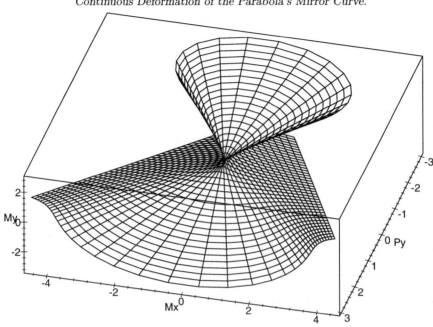

8.3   The Inverse Problem

The inverse problem – to find the original pattern curve for a known mirror curve and a fixed point, is very complicated. In this case $M_x = M_x(t)$, $M_y = M_y(t)$, P_x, P_y are known, and we must solve the following system of differential equations for the functions $X(t)$, $Y(t)$ Equations (8.1) and (8.2). In general, this system of differential equations can not be solved analytically using MAPLE's dsolve function, so numeric methods must be used.

8.3.1   Outflanking Manoeuvre

However, it is very interesting that we can obtain an explicit analytic solution of the system of differential Equations (8.1) and (8.2) by using geometrical arguments! To solve the problem geometrically we consider Figure 8.3. It shows how to find the pattern curve for the given mirror curve and the mirror point. Let M_c be the given mirror curve, P_c be the pattern curve (*The original curve*), P be the given mirror point, M be a point on the mirror curve, M_1 and M_2 be points on the mirror curve to the left and right of M respectively, T be a point on the pattern curve and l_t be the line tangent to P_c at point T.

First we must remember how to create the mirror curve M_c. We select a point T of the pattern curve P_c. We create the tangent line l_t to the pattern curve at this point. This line is used as the mirror line for the construction

FIGURE 8.3.
Construction of the Mirror Curve M_c
and the Pattern Curve P_c.

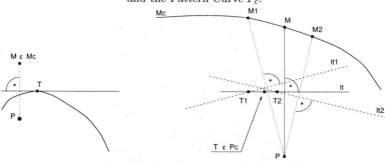

of point M, (l_t is the axis of symmetry of P, and M. As the point T moves along P_c, M traces the mirror curve M_c.

8.3.2 Geometrical Construction of a Point on the Pattern Curve

Given the fixed point P and the mirror curve, our problem is to determine the pattern curve. It is enough to find only one point. If it is possible to find a point on the pattern curve, then it is possible to construct the whole curve, point by point. It is obvious from the preceding section that tangent lines to the pattern curve are axes of symmetry between the fixed point P and a general point M at the mirror curve M_c. This will be useful in solving the inverse problem.

We can select a general point M, the point M_1, and the point M_2 on the mirror curve M_c. Let the point M_1 be at a "small" distance to the left of point M, and M_2 to the right of point M. We then construct line segments l_c, l_{c1} and l_{c2} connecting the point P with the corresponding points on the mirror curve. Let l_t, l_{t1} and l_{t2} be the corresponding axes of symmetry. Let T_1 be the intersection of lines l_t and l_{t1}, and let T_2 the intersection of lines l_t and l_{t2}. The line segment traced by the points T_1 and T_2 is a secant of the constructed pattern curve. This line segment becomes shorter as the points M_1 and M_2 approach the center point M. In the limit

$$\lim_{\substack{M_1 \to M^- \\ M_2 \to M^+}} \overleftrightarrow{T_1 T_2} = 0$$

this line segment is converted to a point. This point is our point P_c on the pattern curve. Now we repeat the same process for the construction of the next point.

TABLE 8.1.
List of Main Variables Used in the Inverse Problem.

Variable	Meaning
$X(t)$, $Y(t)$	$X(t) = f(t)$ and $Y(t) = g(t)$: description of the mirror curve
Ξ, Θ	$\Xi = \varphi(t)$ and $\Theta = \psi(t)$: calculated pattern curve
P_x, P_y	coordinates of the fixed point
dt	Δt, a "small" step of parameter t,
$dX(t)$	$X(t \pm \Delta t)/\Delta t$
$dY(t)$	$Y(t \pm \Delta t)/\Delta t$

8.3.3 MAPLE Solution

The analytic solution of the inverse mirror problem is based on the previous geometrical construction. We can use Taylor's series to express the functions $M_{1_x}(t)$, $M_{1_y}(t)$, $M_{2_x}(t)$, $M_{2_x}(t)$, because the points M_1 and M_2 are at a "small" distances from the central point M.

To solve the inverse mirror problem we use the functions and variables from the Table 8.1.

```
> X(t) := 'X(t)': Y(t) := 'Y(t)':
> P[x] := 'P[x]': P[y] := 'P[y]':
> C1[x] := (X1(t) + P[x])/2:
> C1[y] := (Y1(t) + P[y])/2:
> C2[x] := (X2(t) + P[x])/2:
> C2[y] := (Y2(t) + P[y])/2:
> k[1] := -1*(P[x] - X1(t))/(P[y] - Y1(t)):
> k[2] := -1*(P[x] - X2(t))/(P[y] - Y2(t)):
> E1 := solve({Theta - C1[y] = k[1]*(Xi - C1[x]),
>         Theta - C2[y] = k[2]*(Xi - C2[x])}, {Xi, Theta}):
> assign(E1):
> X1(t) := X(t) - dX(t)*dt: Y1(t) := Y(t) - dY(t)*dt:
> X2(t) := X(t) + dX(t)*dt: Y2(t) := Y(t) + dY(t)*dt:
> Xi := normal(Xi); Theta := normal(Theta);
```

$$\Xi := -\frac{1}{2}(2\,dt^3\,dY(t)\,dX(t)^2 + (4\,P_y\,X(t)\,dt - 4\,Y(t)\,X(t)\,dt)\,dX(t)$$
$$+ 2\,dt^3\,dY(t)^3$$
$$+ (2\,dt\,X(t)^2 - 2\,dt\,P_x{}^2 + 4\,P_y\,Y(t)\,dt - 2\,Y(t)^2\,dt - 2\,dt\,P_y{}^2)\,dY(t))\Big/($$
$$(-2\,P_y\,dt + 2\,Y(t)\,dt)\,dX(t) + (2\,P_x\,dt - 2\,X(t)\,dt)\,dY(t))$$

$$\Theta := \frac{1}{2}(2\,dt^3\,dX(t)^3 +$$
$$(2\,dt^3\,dY(t)^2 + 4\,X(t)\,dt\,P_x - 2\,dt\,X(t)^2 + 2\,Y(t)^2\,dt - 2\,dt\,P_y{}^2 - 2\,dt\,P_x{}^2)$$
$$dX(t) + (-4\,Y(t)\,X(t)\,dt + 4\,P_x\,Y(t)\,dt)\,dY(t))\Big/($$
$$(-2\,P_y\,dt + 2\,Y(t)\,dt)\,dX(t) + (2\,P_x\,dt - 2\,X(t)\,dt)\,dY(t))$$

8.3.4   Analytic Solution

The MAPLE expressions for Ξ and Θ contain only terms dt^2, but no dt terms. If the expressions $\lim_{dt\to 0} \Xi$ and $\lim_{dt\to 0} \Theta$ are evaluated, the dt^2 terms approach zero, and the $dX(t)$ and $dY(t)$ terms become $X_t(t)$ and $Y_t(t)$.

```
> Xi := limit(Xi, dt = 0):   Theta := limit(Theta, dt = 0):
> dX(t) := diff(X(t), t):   dY(t) := diff(Y(t), t):
> Xi := collect(Xi, [diff(X(t), t), diff(Y(t) ,t)]);
> Theta := collect(Theta, [diff(X(t), t), diff(Y(t), t)]);
```

$$
\begin{aligned}
\Xi := \frac{1}{2}\Big(& (2\,P_y\,X(t) - 2\,Y(t)\,X(t))\,(\frac{\partial}{\partial t}X(t)) \\
& + (2\,P_y\,Y(t) - Y(t)^2 + X(t)^2 - P_x{}^2 - P_y{}^2)\,(\frac{\partial}{\partial t}Y(t))\Big)\Big/ \Big(\\
& (P_y - Y(t))\,(\frac{\partial}{\partial t}X(t)) + (-P_x + X(t))\,(\frac{\partial}{\partial t}Y(t))\Big)
\end{aligned}
$$

$$
\begin{aligned}
\Theta := \frac{1}{2}\Big(& (-P_y{}^2 + 2\,X(t)\,P_x - X(t)^2 + Y(t)^2 - P_x{}^2)\,(\frac{\partial}{\partial t}X(t)) \\
& + (-2\,Y(t)\,X(t) + 2\,P_x\,Y(t))\,(\frac{\partial}{\partial t}Y(t))\Big)\Big/ \Big(\\
& (-P_y + Y(t))\,(\frac{\partial}{\partial t}X(t)) + (P_x - X(t))\,(\frac{\partial}{\partial t}Y(t))\Big)
\end{aligned}
\tag{8.3}
$$

The system of Equations (8.1) and (8.2) was in general too complicated for MAPLE to be solved by `dsolve` for $X(t)$ and $Y(t)$. But via our outflanking manoeuvre we used MAPLE to derive an explicit analytic solution (8.3) for the general case.

8.4   Examples

Now we can check our solution using Equations (8.3) for the pattern curve computation. The mirror curve for the computed pattern curve must be the same as the given mirror curve, which was used as input for the pattern curve computation.

8.4.1   The Circle as the Mirror Curve

We can find the pattern curve for the circle and a fixed point. Without loss of generality, we may assume that the mirror circle radius is 1, i.e. we are using a unit circle. Now for every position of the fixed point P we can rotate the coordinate system so that $P_y = 0$. The P_x coordinate can be variable. If P is located inside the circle, then $P_x < 1$. If $P_x > 1$, then P is placed outside the circle. When $P_x = 1$, the point P is on the circle.

```
> X(t) := cos(t):  Y(t) := sin(t):  P[y] := 0:
> xi := simplify(Xi);  theta := simplify(Theta);
```

$$\xi := \frac{1}{2} \frac{\cos(t)\,(-1 + P_x{}^2)}{-1 + \cos(t)\,P_x}$$

$$\theta := \frac{1}{2} \frac{\sin(t)\,(-1 + P_x{}^2)}{-1 + \cos(t)\,P_x}$$

```
> X(t) := xi:  Y(t) := theta:
> m[x] := simplify(M[x]);  m[y] := simplify(M[y]);
```

$$m_x := \cos(t)$$

$$m_y := \sin(t)$$

When the pattern curve ξ, θ is substituted into the equations to calculate the mirror curve $X(t)$, $Y(t)$, the input and the output are equal. We can use this as a demonstration that the derivation of both systems of curves is correct.

Now we must find the equation of the pattern curve. Its parametrical description is:

$$x = \frac{\cos(t)\,\left(P_x^2 - 1\right)}{2\,(\,P_x\,\cos(t)\, - 1)}, \qquad y = \frac{\sin(t)\,\left(P_x^2 - 1\right)}{2\,(\,P_x\,\cos(t)\, - 1)}.$$

It looks like a conic section. To be sure we have to derive an analytic description of this curve.

```
> E1 := simplify(subs(t = solve(xi = X, t),
>                     Y^2  = theta^2), symbolic):
> E2 := collect(4*E1, [X^2, X ,P[x]]):
> E3 := factor((E2)):
> E4 := normal(map(t->t/(P[x]^2-1), subs(X = (Xs+P[x])/2, E3))):
> E5 := subs(Xs = 2*X-P[x] ,map(t->t+Xs^2, -expand(E4)));
```

$$E5 := -4\,\frac{Y^2}{-1 + P_x{}^2} + (2\,X - P_x)^2 = 1$$

As we see from the last equation $E5$, the pattern curve for the unit circle is always a conic section. If the fixed point is inside the circle the pattern curve is an ellipse. The hyperbola results for a point outside the circle. The center C of the conic always has the coordinate $C_x = P_x/2$, $C_y = 0$. The major semi axis of the conic is $a = 1/2$ and the minor semi axis is $b = \sqrt{|P_x{}^2 - 1|}/2$. With $e = \sqrt{a^2 - b^2}$ for the ellipse and $e = \sqrt{a^2 + b^2}$ for the hyperbola, e is the linear eccentricity. We can calculate the distance between the focuses which is equal to $2\,e$ and to P_x. Because the center of the conic has coordinate $C_x = P_x/2$, the foci coordinates are always equal to $F_1 \equiv [0,\,0]$ and $F_2 \equiv [P_x,\,0]$. We

plot some examples. To avoid discontinuities it is necessary to split the p2 plot
into two parts, p21 and p2r.

```
> X(t) := cos(t):  Y(t) := sin(t):
> P[y] := 0:   P[x] := 'P[x]':
> SPx:=[seq(i/10,i=0..9)]:
> p1 := plot({seq([xi, theta, t = 0..2*Pi], P[x] = SPx)},
>            color = black, thickness = 0):
> P[x] := 'P[x]':
> SPx := [seq(1+i/10, i = 1..10)]:
> PxR := arccos(1/P[x])*99/100:
> p2r := plot({seq([xi, theta, t = -PxR..PxR], P[x]=SPx)},
>            color = black, linestyle = 3):
> P[x] := 'P[x]': PxL := arccos(1/P[x])*101/100:
> p21 := plot({seq([xi, theta, t = PxL..2*Pi - PxL],
>            P[x]=SPx)}, color = black, linestyle = 3):
> p3 := plot([X(t), Y(t), t = 0..2*Pi],
>            view = [-1.5 ..2, -1.5.. 1.5], thickness = 2):
> display({p1, p2r, p21, p3}, scaling = constrained);
```

FIGURE 8.4. *The Circle as Mirror Curve.*

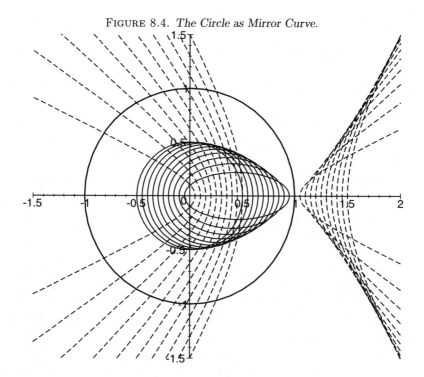

8.4.2   The Line as the Mirror Curve

Now let us consider the line as the mirror curve. Because this problem has translation symmetry, we can put the mirror line on the X axis of the coordinate system, and the fixed point P can be put on the Y axis, thus $P_x = 0$ and P_y is variable. Therefore we can compute the pattern curves as a function of P_y.

```
> X(t) := t:  Y(t) := 0:  P[x] := 0:  P[y] := 'P[y]':
> xi := simplify(Xi);  theta := simplify(Theta);
```

$$\xi := t$$

$$\theta := \frac{1}{2} \frac{P_y{}^2 + t^2}{P_y}$$

```
> X(t) := xi: Y(t) := theta:
> m[x] := simplify(M[x]);  m[y] := simplify(M[y]);
```

$$m_x := t$$

$$m_x := 0$$

```
> E1 := subs(xi = X, Y = theta);
```

$$E1 := Y = \frac{1}{2} \frac{P_y{}^2 + X^2}{P_y}$$

We obtain a family of parabolas depending on the parameter P_y. Let us try to find the envelope and plot some examples.

```
> E2 := Y = X:  E3 := Y =- X:
> solve({E1, E2}, {X,Y});  solve({E1, E3}, {X,Y});
```

$$\{X = P_y, Y = P_y\}, \{X = P_y, Y = P_y\}$$

$$\{Y = P_y, X = -P_y\}, \{Y = P_y, X = -P_y\}$$

We see that the pattern curve for the line is always a parabola equation. The top of the parabola is at the mid point of the line segment beginning at the point P and perpendicular to the mirror line. The focus point of the parabola coincides with the fixed point P. As the coordinate P_y changes its value, the parabolas fill the area defined by conditions $y \geq x$ and $y \geq -x$. These two lines create the envelope for the parabolas. Figure 8.5 shows this bounding of the parabolas.

```
> SPy := [seq(i/4, i = 1..16)]:
> p1 := plot({seq([xi, theta, t = -6..6], P[y] = SPy)},
             color = black, thickness = 1):
> p2 := plot([[-6, 6], [0, 0], [6, 6]], thickness = 2):
> display({p1, p2}, view = [-6..6, 0..8], scaling = constrained);
```

FIGURE 8.5. *The Line as the Mirror Curve.*

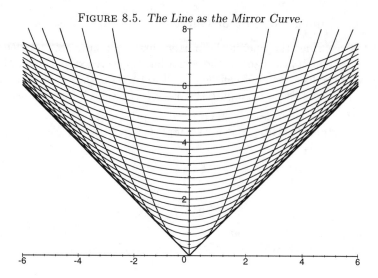

8.5 Conclusions

The inverse mirror problem was successfully solved using MAPLE. It was possible to find a general shape of the pattern curve for both the circle and the line, as a function of the fixed point P. To do this it was necessary to solve a large system of equations, i.e. (8.1), (8.2) and (8.3) but we found the solution without difficulty.

We can demonstrate the pattern curve's continuous deformation as the function of the fixed (*mirror*) point, see Fig. 8.6 and 8.7. If we move with this point along any suitable curve, and the mirror curves corresponding to the fixed point's position in the plane perpendicular to the given curve. The point of intersection of this plane and the given curve coincides with the fixed point's instantaneous position. Using the displacement length as the third coordinate we produced the 3d plot as shown in Figure 8.8.

```
> a:=1: b:=5/8:
> rho := sqrt(a^2*cos(t)^2 + b^2*sin(t)^2):
> X(t) := rho*cos(t):  Y(t) := rho*sin(t):
> p1 := plot([X(t), Y(t), t = -Pi..Pi], thickness = 3):
> P[x] := 'P[x]':  P[y] := 0:
> xi := simplify(Xi,symbolic):  theta := simplify(Theta,symbolic):
> SPx := [seq(i/10, i = -9..9)]:
> p2 := plot({seq([xi, theta, t = -Pi..Pi], P[x] = SPx)},
>              thickness = 1, color = black):
> display({p1, p2}, axes = boxed, scaling = constrained);
> P[x] := 0:  P[y] := 'P[y]':
> xi := simplify(Xi,symbolic): theta := simplify(Theta, symbolic):
> SPy := [seq(i/10, i =-6..6)]:
```

```
> p2 := plot({seq([xi, theta, t= -Pi..Pi], P[y] = SPy)},
>           thickness = 1, color = black):
> display({p1, p2}, axes = boxed, scaling = constrained);
> P[x] := cos(tau)/2: P[y]:=sin(tau)/2:
> xi := simplify(Xi, symbolic): theta := simplify(Theta, symbolic):
> plot3d([xi, theta, tau] ,t = -Pi..Pi, tau = -Pi..Pi ,
>        grid = [90, 90], axes = none, orientation = [-110, 90]);
```

FIGURE 8.6.

Pattern Curves,

$-0.9 \leq P_x \leq$

$0.9, P[y] = 0$

FIGURE 8.7.

Pattern Curves,

$P_x = 0, -0.6 \leq$

$P_y \leq 0.6$

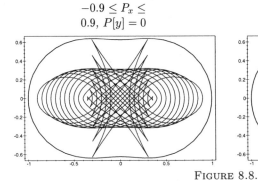
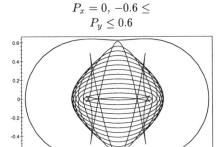

FIGURE 8.8.

The Continuous Deformation of The Pattern Curve,

$P_x = \frac{\cos(\tau)}{2}, P_y = \frac{\sin(\tau)}{2}, -\pi \leq \tau \leq \pi$

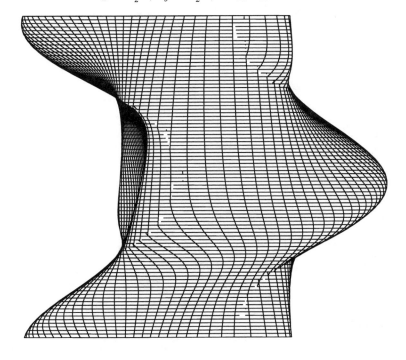

We do not know yet if the mirror curves will be useful in real-world applications. It seems that the analytic solution obtained by our geometric argument is new. We can imagine some applications in physics, especially in optics. E.g. for given optical properties of an optical instrument *described by the pattern curve* one can find technical parameters *described by the mirror curve*.

References

[1] H. J. BARTSCH, *Taschenbuch Mathematischer Formeln*, Fachbuchverlag, BRD, Leipzig, 1991.

Chapter 9. Smoothing Filters

W. Gander and U. von Matt

9.1 Introduction

In many applications one is measuring a variable that is both slowly varying and also corrupted by random noise. Then it is often desirable to apply a smoothing filter to the measured data in order to reconstruct the underlying smooth function. We may assume that the noise is independent of the observed variable. Furthermore we assume that the noise obeys a normal distribution with mean zero and standard deviation δ.

In this chapter we will discuss two different approaches to this smoothing problem, the Savitzky-Golay filter and the least squares filter. We will analyze the properties of these two filters with the help of the test function

$$F(x) := e^{-100(x-1/5)^2} + e^{-500(x-2/5)^2} + e^{-2500(x-3/5)^2} + e^{-12500(x-4/5)^2}. \quad (9.1)$$

This function has four bumps of varying widths (cf. Figure 9.1).

In MATLAB we can generate a vector $\mathbf{f}$ of length $n = 1000$ consisting of measured data corrupted by random noise of standard deviation $\delta = 0.1$ by the statements of Algorithm 9.1. We get the sample data shown as Figure 9.2.

In the following sections we will denote the measured data by f_i, $i = 1, \ldots, n$, and the smoothed data by g_i, $i = 1, \ldots, n$.

9.2 Savitzky-Golay Filter

This approach to smoothing has been introduced by A. Savitzky and M.J.E. Golay in 1964 [10]. The original paper contains some errors that are corrected in [11]. The reader can also find another introduction to this subject in [8].

ALGORITHM 9.1. *Generation of Noisy Data.*

```
n = 1000;
delta = 0.1;
x = [0:n-1]'/(n-1);
F = exp (- 100*(x - 1/5).^2) + exp (-  500*(x - 2/5).^2) + ...
    exp (-2500*(x - 3/5).^2) + exp (-12500*(x - 4/5).^2);
randn ('seed', 0);
f = F + delta * randn (size (x));
```

FIGURE 9.1. *Smooth Function $F(x)$.*

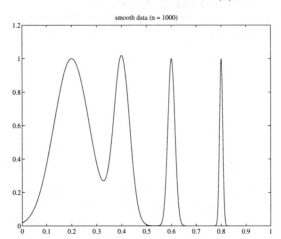

The key idea is that the smoothed value g_i in the point x_i is obtained by taking an average of the neighboring data. The simplest method consists in computing a moving average of a fixed number of f_i's.

More generally we can also fit a polynomial through a fixed number of points. Then the value of the polynomial at x_i gives the smoothed value g_i. This idea is also shown as Figure 9.3, where n_L denotes the number of points to the left of x_i, and n_R denotes the number of points to the right of x_i. By $p_i(x)$ we denote a polynomial of degree M which is fit in the least squares sense through the $n_L + n_R + 1$ points. Then, we have $g_i = p_i(x_i)$.

9.2.1   Filter Coefficients

The polynomial $p_i(x)$ of degree M, which is to be fitted through the data f_i, can be written as

$$p_i(x) := \sum_{k=0}^{M} b_k \left(\frac{x - x_i}{\Delta x}\right)^k. \tag{9.2}$$

We assume that the abscissas x_i are uniformly spaced with $x_{i+1} - x_i \equiv \Delta x$. In order to fit $p_i(x)$ in the least squares sense through the measured data we have to determine the coefficients b_k such that

$$\sum_{j=i-n_L}^{i+n_R} (p_i(x_j) - f_j)^2 = \min. \tag{9.3}$$

FIGURE 9.2. *Noisy Function* $f(x)$.

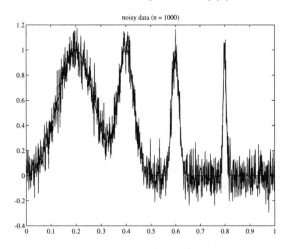

Let us define the matrix

$$A := \begin{bmatrix} (-n_L)^M & \cdots & -n_L & 1 \\ \vdots & & \vdots & \vdots \\ 0 & \cdots & 0 & 1 \\ \vdots & & \vdots & \vdots \\ n_R^M & \cdots & n_R & 1 \end{bmatrix} \in \mathbf{R}^{(n_L+n_R+1)\times(M+1)} \tag{9.4}$$

and the two vectors

$$\mathbf{b} := \begin{bmatrix} b_M \\ \vdots \\ b_1 \\ b_0 \end{bmatrix} \in \mathbf{R}^{M+1} \tag{9.5}$$

and

$$\mathbf{f} := \begin{bmatrix} f_{i-n_L} \\ \vdots \\ f_i \\ \vdots \\ f_{i+n_R} \end{bmatrix} \in \mathbf{R}^{n_L+n_R+1}. \tag{9.6}$$

It should be noted that the matrix A neither depends on the abscissa x_i nor on the spacing Δx.

Using these definitions we can restate the least squares problem (9.3) in matrix terms as

$$\|A\mathbf{b} - \mathbf{f}\|_2 = \min. \tag{9.7}$$

FIGURE 9.3. *Least Squares Polynomial* $p_i(x)$.

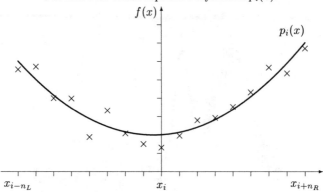

It would now be possible to solve (9.7) for **b** by means of the QR-decomposition of A (cf. [5, Chapter 5]). For our filtering purpose, however, we only need to know $g_i = p_i(x_i) = b_0$. The solution **b** of (9.7) can also be expressed as the solution of the normal equations

$$A^T A \mathbf{b} = A^T \mathbf{f}. \tag{9.8}$$

Thus, we get

$$g_i = \mathbf{e}_{M+1}^T (A^T A)^{-1} A^T \mathbf{f}, \tag{9.9}$$

where $\mathbf{e}_{M+1}$ denotes the $(M + 1)$st unit vector.

Obviously, we can represent g_i as a linear combination of the f_i's. We define the vector

$$\mathbf{c} := A(A^T A)^{-1} \mathbf{e}_{M+1} \tag{9.10}$$

containing the filter coefficients $c_{-n_L}, \ldots, c_{n_R}$. Since **c** does not depend on x_i and Δx, it only needs to be evaluated once. Then, all the smoothed values g_i can be computed by the simple scalar product

$$g_i = \mathbf{c}^T \mathbf{f} = \sum_{j=i-n_L}^{i+n_R} c_{j-i} f_j. \tag{9.11}$$

The calculation of the vector **c** according to Equation (9.10) may be inaccurate for large values of M. This loss of accuracy can be attributed to the fact that the condition number of A is squared in forming $A^T A$. On the other hand it is well known that the least squares system (9.7) can be stably solved by means of the QR-decomposition

$$A = QR. \tag{9.12}$$

By Q we denote an orthogonal $(n_L + n_R + 1)$-by-$(M + 1)$ matrix, and by R we denote an upper triangular $(M + 1)$-by-$(M + 1)$ matrix. If we substitute decomposition (9.12) into (9.10) we get for the vector **c** the expression

$$\mathbf{c} = \frac{1}{r_{M+1,M+1}} Q \mathbf{e}_{M+1}. \tag{9.13}$$

ALGORITHM 9.2. *Savitzky-Golay Smoothing Filter.*

```
function g = SavGol (f, nl, nr, M)

A = ones (nl+nr+1, M+1);
for j = M:-1:1,
  A (:, j) = [-nl:nr]' .* A (:, j+1);
end
[Q, R] = qr (A);
c = Q (:, M+1) / R (M+1, M+1);

n = length (f);
g = filter (c (nl+nr+1:-1:1), 1, f);
g (1:nl) = f (1:nl);
g (nl+1:n-nr) = g (nl+nr+1:n);
g (n-nr+1:n) = f (n-nr+1:n);
```

This is the numerically preferred way of computing **c**. Amazingly enough, this has been pointed out neither in [8] nor in [10].

9.2.2 Results

We present the Savitzky-Golay smoothing filter as Algorithm 9.2. If we apply this algorithm to our initial smoothing problem from Figure 9.2, we get the smoothed curve in Figure 9.4. For easy reference we have also superimposed the graph of the function F from Equation (9.1). We have chosen the parameters $n_L = n_R = 16$ and $M = 4$ which seem to be optimal for this test case. The execution of Algorithm 9.2 needs about 6.41 milliseconds of CPU-time on a PC with an Intel Pentium Pro running at 200 MHz.

The main advantage of a Savitzky-Golay filter is its speed. For given values of n_L, n_R, and M, the filter parameters **c** need to be evaluated only once. Then each filtered value g_i can be computed by the simple scalar product (9.11) of length $n_L + n_R + 1$. It is conceivable that this operation could even be implemented in hardware for special purpose applications.

It is a disadvantage that it is not obvious how to choose the filter parameters n_L, n_R, and M. In [1, 8, 13] some practical hints are given. But in many cases some visual optimization is needed in order to obtain the best results.

Finally, the boundary points represent yet another problem. They cannot be smoothed by a filter with $n_L > 0$ or $n_R > 0$. Either these boundary values are just dropped, as it is done in the case of Figure 9.4, or one constructs a special Savitzky-Golay filter with $n_L = 0$ or $n_R = 0$ for these special cases.

9.3 Least Squares Filter

Another approach to our filtering problem consists in requiring that the filtered curve $g(x)$ be as smooth as possible. In the continuous case we would require

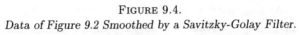

FIGURE 9.4.

Data of Figure 9.2 Smoothed by a Savitzky-Golay Filter.

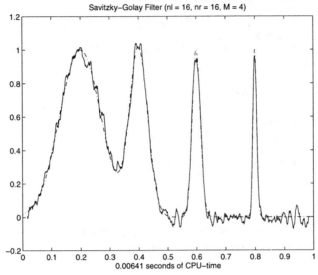

that

$$\int_{x_{\min}}^{x_{\max}} g''(x)^2\, dx = \min. \tag{9.14}$$

Since the function $f(x)$ is only available as measured data f_i at discrete points x_i we will only compute the smoothed values g_i at the same points. Therefore, we must express the second derivative of $g(x)$ by a finite difference scheme. A popular choice is

$$g''(x_i) \approx \frac{g_{i+1} - 2g_i + g_{i-1}}{\Delta x^2}. \tag{9.15}$$

We could give similar stencils for non-equally spaced abscissas x_i. Consequently, condition (9.14) is replaced by the condition

$$\sum_{i=2}^{n-1} (g_{i+1} - 2g_i + g_{i-1})^2 = \min. \tag{9.16}$$

Besides this smoothness condition we also require that the function $g(x)$ approximates $f(x)$ within the limits of the superimposed noise. If we assume the f_i's to be corrupted by random noise of mean zero and standard deviation δ we require that

$$|g_i - f_i| \leq \delta \tag{9.17}$$

on the average. For n samples this condition can also be written as

$$\sum_{i=1}^{n} (g_i - f_i)^2 \leq n\delta^2. \tag{9.18}$$

Let us now define the matrix

$$A := \begin{bmatrix} 1 & -2 & 1 & & & \\ & 1 & -2 & 1 & & \\ & & \ddots & \ddots & \ddots & \\ & & & 1 & -2 & 1 \end{bmatrix} \in \mathbf{R}^{(n-2) \times n}. \qquad (9.19)$$

Then, we can restate the optimization problem (9.16,9.18) in matrix notation as the minimization of

$$\|A\mathbf{g}\|_2 = \min \qquad (9.20)$$

subject to the quadratic inequality constraint

$$\|\mathbf{g} - \mathbf{f}\|_2^2 \le \alpha^2 := n\delta^2. \qquad (9.21)$$

9.3.1 Lagrange Equations

Let us now study the solution of the optimization problem (9.20,9.21). First we assume that there are elements from the null space $\mathcal{N}(A)$ of the matrix A satisfying the constraint (9.21). In this case we want to determine the vector $\mathbf{g}$ which satisfies the constraint (9.21) best, i.e. we consider the minimization problem

$$\|\mathbf{g} - \mathbf{f}\|_2 = \min \qquad (9.22)$$

subject to the linear equality constraint

$$A\mathbf{g} = \mathbf{0}. \qquad (9.23)$$

It is not hard to show that the unique optimal solution in this case is given by

$$\mathbf{g} = \mathbf{f} - A^{\mathrm{T}}\mathbf{z}, \qquad (9.24)$$

where $\mathbf{z}$ can be computed as the unique solution of the linear least squares problem

$$\|A^{\mathrm{T}}\mathbf{z} - \mathbf{f}\|_2 = \min. \qquad (9.25)$$

The vector $\mathbf{z}$ can also be characterized as the solution of the normal equations

$$AA^{\mathrm{T}}\mathbf{z} = A\mathbf{f}, \qquad (9.26)$$

which correspond to (9.25).

Now, let us assume that there is no element from the null space $\mathcal{N}(A)$ which is compatible with the constraint (9.21). In particular, this means that the vector $\mathbf{g} \in \mathcal{N}(A)$ defined by the Equations (9.24) and (9.25) satisfies

$$\|\mathbf{g} - \mathbf{f}\|_2 > \alpha. \qquad (9.27)$$

We can study the least squares problem (9.20,9.21) by introducing the Lagrange principal function

$$\Phi(\mathbf{g}, \lambda, \mu) := \mathbf{g}^{\mathrm{T}}A^{\mathrm{T}}A\mathbf{g} + \lambda(\|\mathbf{g} - \mathbf{f}\|_2^2 + \mu^2 - \alpha^2), \qquad (9.28)$$

where μ denotes a so-called slack variable. By differentiating Φ with respect to $\mathbf{g}$, λ, and μ we get the Lagrange equations

$$(A^{\mathrm{T}}A + \lambda I)\mathbf{g} = \lambda \mathbf{f}, \tag{9.29}$$

$$\|\mathbf{g} - \mathbf{f}\|_2^2 + \mu^2 = \alpha^2, \tag{9.30}$$

$$\lambda \mu = 0. \tag{9.31}$$

For $\lambda = 0$, we would have $A^{\mathrm{T}}A\mathbf{g} = \mathbf{0}$ or $A\mathbf{g} = \mathbf{0}$, since the matrix A has full rank. But according to our above assumption, the vector $\mathbf{g}$ cannot be an element of the null space $\mathcal{N}(A)$. Therefore we will assume $\lambda \neq 0$ from now on.

Because of $\lambda \neq 0$, we have $\mu = 0$ from Equation (9.31). Therefore, we can simplify the Lagrange Equations (9.29,9.30,9.31) to

$$(A^{\mathrm{T}}A + \lambda I)\mathbf{g} = \lambda \mathbf{f}, \tag{9.32}$$

$$\|\mathbf{g} - \mathbf{f}\|_2 = \alpha. \tag{9.33}$$

Since $\lambda \neq 0$, we can make use of Equation (9.32) to express $\mathbf{g}$ by

$$\mathbf{g} = \mathbf{f} - A^{\mathrm{T}}\mathbf{z}, \tag{9.34}$$

where

$$\mathbf{z} = \frac{1}{\lambda}A\mathbf{g}. \tag{9.35}$$

By substituting the expression (9.34) for $\mathbf{g}$ into (9.32,9.33), we get the dual Lagrange equations

$$(AA^{\mathrm{T}} + \lambda I)\mathbf{z} = A\mathbf{f}, \tag{9.36}$$

$$\|A^{\mathrm{T}}\mathbf{z}\|_2 = \alpha. \tag{9.37}$$

As soon as λ and $\mathbf{z}$ have been determined, we can compute $\mathbf{g}$ according to Equation (9.34).

In [3, 4] it is shown that there is a unique Lagrange multiplier $\lambda > 0$ and an associated vector $\mathbf{z}$ which solve the dual Lagrange Equations (9.36,9.37). Furthermore, the vector $\mathbf{g}$ from (9.34) will then solve the least squares problem (9.20,9.21).

The dual Lagrange Equations (9.36,9.37) have the advantage that the matrix $AA^{\mathrm{T}} + \lambda I$ is nonsingular for $\lambda \geq 0$. The Lagrange multiplier λ can be found by solving the nonlinear secular equation

$$s(\lambda) := \|A^{\mathrm{T}}(AA^{\mathrm{T}} + \lambda I)^{-1}A\mathbf{f}\|_2^2 = \alpha^2. \tag{9.38}$$

This will be discussed in more detail in Section 9.3.2.

As the result of our analysis we can present Algorithm 9.3 for the solution of the constrained least squares problem (9.20,9.21). The next sections will concentrate on the practical issues of the implementation in MATLAB.

ALGORITHM 9.3.

Solution of the Constrained Least Squares
Problem (9.20,9.21).

Solve the linear least squares problem (9.25) for **z**.

if $\|A^{\mathrm{T}}\mathbf{z}\|_2 > \alpha$ **then**

Solve the secular equation (9.38) for the unique zero $\lambda > 0$.

Solve the linear system (9.36) for **z**.

end

$\mathbf{g} := \mathbf{f} - A^{\mathrm{T}}\mathbf{z}$

9.3.2 Zero Finder

The solution of the secular Equation (9.38) for its unique zero $\lambda > 0$ represents the major effort of Algorithm 9.3. We show the graph of the secular function $s(\lambda)$ as Figure 9.5. Since $s(\lambda)$ is a nonlinear function an iterative method is needed to solve the secular Equation (9.38). A good choice would be Newton's method which is defined by the iteration

$$\lambda_{k+1} := \lambda_k - \frac{s(\lambda_k) - \alpha^2}{s'(\lambda_k)}. \tag{9.39}$$

However, for our particular equation, Reinsch [9] has proposed the accelerated Newton iteration

$$\lambda_{k+1} := \lambda_k - 2\frac{s(\lambda_k)}{s'(\lambda_k)}\Big(\frac{\sqrt{s(\lambda_k)}}{\alpha} - 1\Big). \tag{9.40}$$

If this iteration is started with $\lambda_0 = 0$ we get a strictly increasing sequence of λ_k's. Proofs of this key property can be found in [9] and in [12, pp. 65–66].

FIGURE 9.5. *Secular Function $s(\lambda)$.*

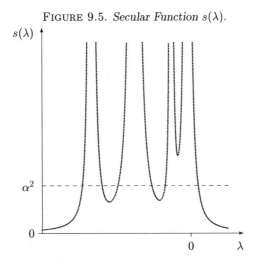

However, this mathematical property cannot be preserved in floating-point arithmetic. This observation leads to the numerical termination criterion

$$\lambda_{k+1} \leq \lambda_k. \tag{9.41}$$

9.3.3   Evaluation of the Secular Function

In order to use the iteration (9.40) to solve the secular Equation (9.38), we need a way to evaluate numerically the secular function $s(\lambda)$ and its derivative $s'(\lambda)$. The values of $s(\lambda)$ and $s'(\lambda)$ can be expressed by

$$s(\lambda) = \|A^{\mathrm{T}}\mathbf{z}\|_2^2, \tag{9.42}$$
$$s'(\lambda) = -2\mathbf{z}^{\mathrm{T}}(\mathbf{z} + \lambda\mathbf{z}'), \tag{9.43}$$

where $\mathbf{z}$ and $\mathbf{z}'$ satisfy the equations

$$(AA^{\mathrm{T}} + \lambda I)\mathbf{z} = A\mathbf{f} \tag{9.44}$$

and

$$(AA^{\mathrm{T}} + \lambda I)\mathbf{z}' = -\mathbf{z}. \tag{9.45}$$

The last two Equations (9.44) and (9.45) suggest that we have to solve linear systems involving the matrix $AA^{\mathrm{T}} + \lambda I$. However, we can also read Equation (9.44) as the normal equations corresponding to the linear least squares problem

$$\left\|\begin{bmatrix} A^{\mathrm{T}} \\ \sqrt{\lambda}I \end{bmatrix}\mathbf{z} - \begin{bmatrix} \mathbf{f} \\ \mathbf{0} \end{bmatrix}\right\|_2 = \min. \tag{9.46}$$

Similarly, we can compute $\mathbf{z}'$ as the solution of the least squares problem

$$\left\|\begin{bmatrix} A^{\mathrm{T}} \\ \sqrt{\lambda}I \end{bmatrix}\mathbf{z}' - \begin{bmatrix} \mathbf{0} \\ -\mathbf{z}/\sqrt{\lambda} \end{bmatrix}\right\|_2 = \min, \tag{9.47}$$

provided that $\lambda > 0$. It should be noted that for $\lambda = 0$ the vector $\mathbf{z}'$ is not needed to compute $s'(\lambda)$ in Equation (9.43). Numerically, we prefer computing $\mathbf{z}$ and $\mathbf{z}'$ from the two linear least squares problems (9.46,9.47) over solving the two linear systems (9.44,9.45).

We can solve the two least squares problems (9.46,9.47) with the help of the QR-decomposition

$$\begin{bmatrix} A^{\mathrm{T}} \\ \sqrt{\lambda}I \end{bmatrix} = Q\begin{bmatrix} R \\ 0 \end{bmatrix}, \tag{9.48}$$

where Q and R denote an orthogonal $(2n-2)$-by-$(2n-2)$ matrix, and an upper triangular $(n-2)$-by-$(n-2)$ matrix, respectively. In MATLAB we could compute this decomposition by the statement

```
>> [Q, R] = qr ([A'; sqrt(lambda)*eye(n-2)]);
```

FIGURE 9.6. *First Stage of the QR-decomposition (9.48).*

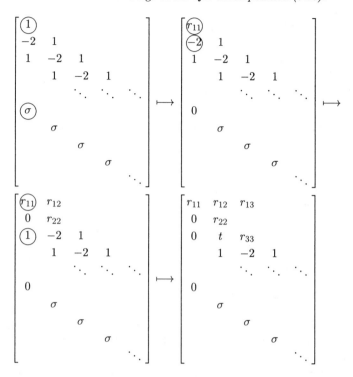

However, this command does not take advantage of the special structure of the matrix, and its numerical complexity increases cubically with the size of A. Fortunately, we can devise an algorithm to compute the QR-decomposition (9.48) whose numerical complexity increases only linearly with the size of the matrix A. We can achieve this improvement by an appropriate sequence of Givens transformations. In Figure 9.6 we show the application of the first three Givens rotations when processing the first column. The quantity σ is an abbreviation for $\sqrt{\lambda}$. If we apply this step $n - 2$ times we get the desired QR-decomposition (9.48). We show the same procedure in pseudo-code as Algorithm 9.6. The construction and application of Givens rotations is described by calls of the BLAS routines rotg and rot. Their precise definition is given in [2, 6]. The actual implementation in MATLAB or in a conventional programming language like C or Fortran is now straightforward.

We stress that the matrix $\bar{Q}$ computed by Algorithm 9.6 is not identical to the matrix Q from the QR-decomposition (9.48). Rather the matrix $\bar{Q}$ contains the information on the Givens rotations needed to apply the matrix Q to a vector $\mathbf{x}$. We present the evaluation of the products $\mathbf{y} := Q\mathbf{x}$ and $\mathbf{y} := Q^{\mathrm{T}}\mathbf{x}$ as Algorithms 9.4 and 9.5.

ALGORITHM 9.4. *Evaluation of the Product* $\mathbf{y} := Q\mathbf{x}$.

$\mathbf{y} := \mathbf{x}$
for $k := n - 2$ **to** 1 **by** -1 **do**
    rot $(y_k, y_{k+2}, \bar{Q}_{k5}, -\bar{Q}_{k6})$
    rot $(y_k, y_{k+1}, \bar{Q}_{k3}, -\bar{Q}_{k4})$
    rot $(y_k, y_{n+k}, \bar{Q}_{k1}, -\bar{Q}_{k2})$
end

ALGORITHM 9.5. *Evaluation of the Product* $\mathbf{y} := Q^{\mathrm{T}}\mathbf{x}$.

$\mathbf{y} := \mathbf{x}$
for $k := 1$ **to** $n - 2$ **do**
    rot $(y_k, y_{n+k}, \bar{Q}_{k1}, \bar{Q}_{k2})$
    rot $(y_k, y_{k+1}, \bar{Q}_{k3}, \bar{Q}_{k4})$
    rot $(y_k, y_{k+2}, \bar{Q}_{k5}, \bar{Q}_{k6})$
end

9.3.4 MEX-Files

Since MATLAB is an interpretative language it is not well suited to run Algorithms 9.4, 9.5, and 9.6. The overhead introduced by its interpreter would be considerable. Rather our code performs much better when implemented in a conventional programming language like C or Fortran. Fortunately, MATLAB provides a means to execute separately compiled code—the so-called MEX-files.

Let us assume we have implemented Algorithms 9.4, 9.5, and 9.6 in the C programming language. If we want to execute this code from inside MATLAB we need an interface procedure for each of these algorithms. In MATLAB a function call has the general syntax

$$[l_1, l_2, \ldots, l_{\mathrm{nlhs}}] = \mathtt{fct} \ (r_1, r_2, \ldots, r_{\mathrm{nrhs}}) ;$$

The l_i's are called output parameters, and the r_i's are called input parameters. If this function is to be implemented as an external C subroutine we need the following interface procedure:

```
#include "mex.h"

void mexFunction (int nlhs, mxArray *plhs[],
                  int nrhs, mxArray *prhs[])
{
   ...
}
```

The two parameters nlhs and nrhs give the number of left-hand side arguments and the number of right-hand side arguments with which fct has been called in MATLAB. The parameter plhs is a pointer to an array of length nlhs where we must put pointers for the returned left-hand side matrices. Likewise, the

ALGORITHM 9.6.

Calculation of the QR-decomposition (9.48).

Allocate the $(n-2)$-by-6 matrix $\bar{Q}$.

$r_{11} := 1$

$t := -2$

if $n > 3$ **then**

    $r_{22} := 1$

end

for $k := 1$ **to** $n - 2$ **do**

    $\text{tmp} := \sqrt{\lambda}$

    rotg $(r_{kk}, \text{tmp}, \bar{Q}_{k1}, \bar{Q}_{k2})$

    rotg $(r_{kk}, t, \bar{Q}_{k3}, \bar{Q}_{k4})$

    if $k < n - 2$ **then**

        $r_{k,k+1} := 0$

        rot $(r_{k,k+1}, r_{k+1,k+1}, \bar{Q}_{k3}, \bar{Q}_{k4})$

    end

    $\text{tmp} := 1$

    rotg $(r_{kk}, \text{tmp}, \bar{Q}_{k5}, \bar{Q}_{k6})$

    if $k < n - 2$ **then**

        $t := -2$

        rot $(r_{k,k+1}, t, \bar{Q}_{k5}, \bar{Q}_{k6})$

        if $k < n - 3$ **then**

            $r_{k,k+2} := 0$

            $r_{k+2,k+2} := 1$

            rot $(r_{k,k+2}, r_{k+2,k+2}, \bar{Q}_{k5}, \bar{Q}_{k6})$

        end

    end

end

parameter `prhs` is a pointer to an array of length `nrhs`, whose entries point to the right-hand side matrices.

This interface routine should perform the following tasks:

1. It checks whether the proper number of input and output arguments has been supplied.

2. It makes sure that the dimensions of the input matrices meet their specification.

3. It allocates the storage for the output matrices.

4. It calls another subroutine to perform the actual calculation.

The included file `mex.h` contains the MEX-file declarations and prototypes. There are a number of auxiliary subroutines available that can be called by the interface routine. For details the reader is referred to the External Interface Guide of MATLAB.

ALGORITHM 9.7. *MEX-File for Algorithm 9.6.*

```
#include "mex.h"

#define max(A, B)   ((A) > (B) ? (A) : (B))
#define min(A, B)   ((A) < (B) ? (A) : (B))

#define n     prhs[0]
#define sigma prhs[1]

#define Qbar plhs[0]
#define R    plhs[1]

void mexFunction (int nlhs, mxArray *plhs[],
                  int nrhs, mxArray *prhs[])
{ int size, nnz;

  if (nrhs != 2) {
    mexErrMsgTxt ("spqr requires two input arguments.");
  } else if (nlhs != 2) {
    mexErrMsgTxt ("spqr requires two output arguments.");
  }

  if ((mxGetM (n) != 1) || (mxGetN (n) != 1) ||
      (*mxGetPr (n) < 3.0)) {
    mexErrMsgTxt ("n must be a scalar greater or equal 3.");
  }

  if ((mxGetM (sigma) != 1) || (mxGetN (sigma) != 1)) {
    mexErrMsgTxt ("sigma must be a scalar.");
  }

  size = (int) *mxGetPr (n);
  Qbar = mxCreateDoubleMatrix (size-2, 6, mxREAL);
  if (size == 3) {nnz = 1;} else {nnz = 3*size - 9;}
  R = mxCreateSparse (size-2, size-2, nnz, mxREAL);

  QR (size, *mxGetPr (sigma), mxGetPr (Qbar), R);
}
```

ALGORITHM 9.8.

Calculation of the QR-decomposition (9.48) in C.

```c
void QR (int n, double sigma, double *Qbar, mxArray *R)
{ int    i, j, k, nnz, n2, n3, n4, *ir, *jc;
  double co, *pr, si, t, tmp;

  nnz = mxGetNzmax (R);  n2 = n-2;  n3 = n-3;  n4 = n-4;
  ir = mxGetIr (R);  jc = mxGetJc (R);  pr = mxGetPr (R);
  /* diagonal of R */
  ir [0] = 0;  for (i = 1; i < n2; i++) {ir [3*i - 1] = i;}
  /* first upper off-diagonal of R */
  for (i = 0; i < n3; i++) {ir [3*i + 1] = i;}
  /* second upper off-diagonal of R */
  for (i = 0; i < n4; i++) {ir [3*i + 3] = i;}
  /* columns of R */
  jc [0] = 0;  jc [1] = 1;
  for (j=2; j < n2; j++) {jc [j] = 3*j - 3;}
  jc [n2] = nnz;

#define r(i, j) pr [k = jc [j], k + i - ir [k]]
  r (0, 0) = 1.0;  t = -2.0;  if (n > 3) {r (1, 1) = 1.0;}
  for (j = 0; j < n2; j++) {
    tmp = sigma;
    rotg (&r (j, j), &tmp, &Qbar [j], &Qbar [n2 + j]);
    rotg (&r (j, j), &t, &Qbar [2*n2 + j], &Qbar [3*n2 + j]);
    if (j < n3) {
      r (j, j+1) = 0.0;
      rot (&r (j, j+1), &r (j+1, j+1),
           Qbar [2*n2 + j], Qbar [3*n2 + j]);
    }
    tmp = 1.0;
    rotg (&r (j, j), &tmp, &Qbar [4*n2 + j], &Qbar [5*n2 + j]);
    if (j < n3) {
      t = -2.0;
      rot (&r (j, j+1), &t, Qbar [4*n2 + j], Qbar [5*n2 + j]);
      if (j < n4) {
        r (j, j+2) = 0.0;  r (j+2, j+2) = 1.0;
        rot (&r (j, j+2), &r (j+2, j+2),
             Qbar [4*n2 + j], Qbar [5*n2 + j]);
      }
    }
  }
#undef r
}
```

ALGORITHM 9.9. *Least Squares Smoothing Filter.*

```
function g = lsq (f, delta)

n = length (f);
alpha = sqrt (n) * delta;
e = ones (n, 1);
A = spdiags ([e -2*e e], 0:2, n-2, n);

lambda = 0;
while 1,
  [Qbar, R] = spqr (n, sqrt (lambda));
  z = QTx (Qbar, [f; zeros(n-2, 1)]);
  z = R \ z (1:n-2);

  F = norm (A' * z)^2;
  if (F <= alpha^2), break; end;

  if (lambda > 0),
    zp = QTx (Qbar, [zeros(n, 1); -z/sqrt(lambda)]);
    zp = R \ zp (1:n-2);
    Fp = -2 * z' * (z + lambda * zp);
  else
    Fp = -2 * z' * z;
  end;

  lambdaold = lambda;
  lambda = lambda - 2 * (F / Fp) * (sqrt (F) / alpha - 1);
  if (lambda <= lambdaold), break; end;
end;
g = f - A' * z;
```

As an example we present as Algorithm 9.7 the C code which serves as an interface to the QR-decomposition of Algorithm 9.6. A translation of the pseudo-
code of Algorithm 9.6 into C is shown as Algorithm 9.8. We would also like
to remind the reader that all the code is available in machine-readable form
(see the preface for more information). In MATLAB we can now execute the
statement

```
>> [Qbar, R] = spqr (n, sqrt (lambda));
```

to compute the sparse QR-decomposition (9.48).

   In the same way we can implement Algorithms 9.4 and 9.5 by MEX-files.
We can call them in MATLAB by the statements y = Qx (Qbar, x) and y =
QTx (Qbar, x), respectively.

9.3.5   Results

We are now ready to present the implementation of the least squares filter as
Algorithm 9.9. The matrix A from equation (9.19) is created as a sparse matrix.

FIGURE 9.7.

Data of Figure 9.2 Smoothed by a Least Squares Filter.

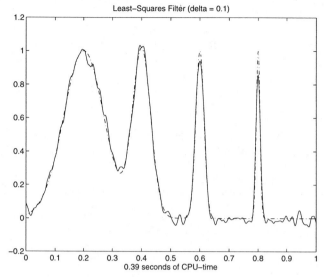

Least–Squares Filter (delta = 0.1)

TABLE 9.1.

CPU-Times Corresponding to Different Data Sizes.

n	Savitzky-Golay Filter	Least Squares Filter
10^3	0.006 sec	0.39 sec
10^4	0.036 sec	4.55 sec
10^5	0.344 sec	77.93 sec

In this way only the nonzero entries of A need to be stored. This is accomplished by the function `spdiags` which defines a sparse matrix by its diagonals.

The QR-decomposition (9.48) is computed by a call of `spqr` which implements Algorithm 9.6 as a MEX-file. Similarly, a call of the function `QTx` executes the MEX-file corresponding to Algorithm 9.5.

If we apply Algorithm 9.9 to our test data represented in Figure 9.2 we get the smoothed curve from Figure 9.7. We have set $\delta = 0.1$, which corresponds to the standard deviation of the noise in the function $f(x)$.

From a visual point of view the least squares filter returns a smoother result than the Savitzky-Golay filter whose output has been presented as Figure 9.4. The results get even better when the number of points n is increased. On the other hand the CPU-times required to execute Algorithm 9.9 are substantially higher than those needed to execute the Savitzky-Golay filter from Algorithm 9.2. We have summarized these CPU-times for a number of different data sizes n as Table 9.1.

We conclude that a least squares filter can produce a smoother curve $g(x)$ than a Savitzky-Golay filter. However, we need to pay for this advantage with a more complex algorithm and with an increased CPU-time. A least squares filter has also the advantage that only one parameter, the standard deviation δ of the noise, has to be estimated. This quantity is more directly related to the physical data than the parameters n_L, n_R, and M from the Savitzky-Golay filter.

References

[1] M. U. A. BROMBA AND H. ZIEGLER, *Application Hints for Savitzky-Golay Digital Smoothing Filters*, Analytical Chemistry, 53 (1981), pp. 1583–1586.

[2] J. J. DONGARRA, C. B. MOLER, J. R. BUNCH AND G. W. STEWART, *LINPACK Users' Guide*, SIAM Publications, Philadelphia, 1979.

[3] W. GANDER, *On the Linear Least Squares Problem with a Quadratic Constraint*, Habilitationsschrift ETH Zürich, STAN-CS-78-697, Stanford University, 1978.

[4] W. GANDER, *Least Squares with a Quadratic Constraint*, Numer. Math., 36 (1981), pp. 291–307.

[5] G. H. GOLUB AND C. F. VAN LOAN, *Matrix Computations*, Second Edition, The Johns Hopkins University Press, Baltimore, 1989.

[6] C. L. LAWSON, R. J. HANSON, D. R. KINCAID AND F. T. KROGH, *Basic Linear Algebra Subprograms for Fortran Usage*, ACM Trans. Math. Softw., 5 (1979), pp. 308–325.

[7] W. H. PRESS, B. P. FLANNERY, S. A. TEUKOLSKY AND W. T. VETTERLING, *Numerical Recipes*, Cambridge University Press, Cambridge, 1986.

[8] W. H. PRESS AND S. A. TEUKOLSKY, *Savitzky-Golay Smoothing Filters*, Computers in Physics, 4 (1990), pp. 669–672.

[9] C. H. REINSCH, *Smoothing by Spline Functions. II*, Numer. Math., 16 (1971), pp. 451–454.

[10] A. SAVITZKY AND M. J. E. GOLAY, *Smoothing and Differentiation of Data by Simplified Least Squares Procedures*, Analytical Chemistry, 36 (1964), pp. 1627–1639.

[11] J. STEINIER, Y. TERMONIA AND J. DELTOUR, *Comments on Smoothing and Differentiation of Data by Simplified Least Square Procedure*, Analytical Chemistry, 44 (1972), pp. 1906–1909.

[12] U. VON MATT, *Large Constrained Quadratic Problems*, Verlag der Fachvereine, Zürich, 1993.

[13] H. ZIEGLER, *Properties of Digital Smoothing Polynomial (DISPO) Filters*, Applied Spectroscopy, 35 (1981), pp. 88–92.

Chapter 10. The Radar Problem

S. Bartoň and I. Daler

10.1 Introduction

The controlling system for a multiradar display, in an air traffic long-distance control center, receives different information from different kinds of radars on the globe. The information coming from each radar contains among others the coordinates x, y and z (with respect to the Cartesian coordinate system of the given radar) of an airplane which is "seen" by the radar. The two–dimensional

FIGURE 10.1. *Airplane in two Radar Systems in Plane.*

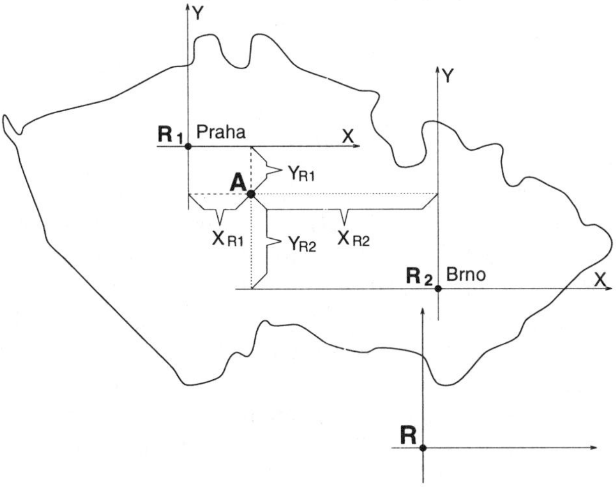

situation in the case of two radars R_1, R_2 is shown on Figure 10.1. We can see that the airplane is described by the coordinates $[x_{R_1}, y_{R_1}]$ of the radar R_1 and by the coordinates $[x_{R_2}, y_{R_2}]$ of the radar R_2. For the controlling system of the multiradar display it is necessary to work with the "absolute" coordinate system (with the origin R). The data from all the radars concerning the position of the airplane A are transformed into this system. That means that in this coordinate system the airplane A is described by several points which can be processed using appropriate criteria. *(For example, the airplane will be represented by the point which corresponds to its most probable position)*. In the coordinate system with the origin R, the long-distance control is able to follow the relative position of airplanes from the visual information on the display. The controlling program for a multiradar display is then able to watch the whole region of operation of the airplane using chosen windows. For simplicity let the absolute coordinate system with the origin R be identical to the coordinate system of the radar R_2. For a multiradar display it is necessary to solve the following basic problem:

Let the radars R_1 and R_2 be situated at points P_1 and P_2 on the globe. An airplane A is "seen" by the radar R_1 in its local Cartesian coordinate system $\mathbf{P_1}$ with the coordinates $[x_1, y_1, z_1] \equiv A(\mathbf{P_1})$, and it is necessary to find the coordinates $[x_2, y_2, z_2] \equiv A(\mathbf{P_2})$ of the airplane A in the Cartesian coordinate system $\mathbf{P_2}$ of the radar R_2 (cf. Figure 10.2). The points P_1, P_2 on the

FIGURE 10.2. *Airplane in two Radar Systems.*

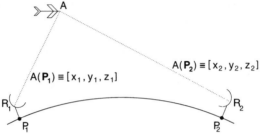

globe are given by the geographic coordinates $[f_1, l_1]$ and $[f_2, l_2]$, where f_1, f_2 are the geographic latitudes and l_1, l_2 are the geographic longitudes. These geographic coordinates $\mathbf{G} \equiv [f, l]$ are usually given in degrees, minutes, and seconds ($-90° \leq f \leq 90°$, $-180° \leq l \leq 180°$). Also, let the Cartesian coordinate systems $(P_1; +x_1, +y_1, +z_1) \equiv \mathbf{P_1}$, $(P_2; +x_2, +y_2, +z_2) \equiv \mathbf{P_2}$ have their origins in the points P_1, P_2 (cf. Figure 10.5). The $x_1 y_1$ plane is a tangent plane to the ellipsoid in the point P_1. The x_1−axis lies in the plane of the latitude f_1, the y_1−axis lies in the plane of the meridian l_1. The z_1−axis is positively oriented towards the earth's center. The x_1−axis and the y_1−axis, are oriented in the direction of increasing longitude and latitude, respectively. The axes of the coordinate system $\mathbf{P_2}$ are similarly oriented.

10.2   Converting Degrees into Radians

The geographic coordinates $[f, l]$ are usually given in degrees, minutes, seconds, and their decimal parts. For further calculations it is necessary to transform the angles f and l into radians. Since we will use this transformation several times, we design a MATLAB function $\mathbf{deg2rad}(\alpha°, ['], ["])$. The input variable α of this function will be a vector of 1 to 4 components (cf. Algorithm 10.1, Table 10.1 and Example 1).

ALGORITHM 10.1. *Function* deg2rad.

```
function rad = deg2rad(alpha)
%----------------------------
% conversion: alpha deg., min., sec.  ->  radians
%
d = length(alpha);
if d > 4
   disp(alpha), error('invalid input list in function deg2rad')
end
alpha = [alpha(:); zeros(4-d,1)];
alpha(3) = alpha(3) + alpha(4)/100;
rad = pi/180*((alpha(3)/60 + alpha(2))/60 + alpha(1));
```

TABLE 10.1. *Data Organization of Angle* α.

Number of Components	Input Angle α
1	integer.decfrac$^\circ$
2	integer$^\circ$ integer.decfrac$'$
3	integer$^\circ$ integer$'$ integer.decfrac$''$
4	integer$^\circ$ integer$'$ integer$''$ integer(decfrac)$''$

Example 1:

We test all the four possibilities for the input angle α as shown in Table 10.1. For the given input vectors we should always obtain the same value in radians:

```
>> format long
>> a = [16.641038888888888];       % 1 component
>> b = [16 38.462333333333333];    % 2 components
>> c = [16 38 27.74];              % 3 components
>> d = [16 38 27 74];              % 4 components
>> radians = [deg2rad(a) deg2rad(b) deg2rad(c) deg2rad(d)]

radians =

0.29044091956353 0.29044091956353 0.29044091956353 0.29044091956353
```

10.3   Transformation of Geographic into Geocentric Coordinates

In this section we develop a function to transform the geographic coordinates $[f, l]$ of a point P into geocentric Cartesian coordinates: $(C; +x_c, +y_c, +z_c) \equiv \mathbf{C}$, with origin C in the earth's center. The z_c−axis goes through the earth's poles, with the positive orientation to the North Pole. The x_c−axis passes through the null and 180th meridian, the positive orientation to the null meridian. The

y_c−axis is perpendicular to the x_c−axis and to the z_c−axis, with positive orientation to the 90th meridian (cf. Figure 10.5). For this transformation from the geographic coordinates $[f, l]$ into the geocentric Cartesian coordinates $[x_c, z_c, y_c]$, (according to Figure 10.3), we will write the MATLAB function gg2gc(f, l). The

FIGURE 10.3.
Transformation from Geographic to Geocentric Coordinates.

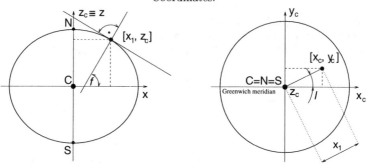

shape of the Earth is a *geoid*, but for our purposes we will use the approximation described by a circular ellipsoid. The axes of revolutions of both bodies are the same. The cross section of an ellipsoid is the (Krasovsky) ellipse, with the major semiaxis $A = 6378.245\ [km]$ and the minor semiaxis $B = 6356.863\ [km]$. To derive the transformation equations it is convenient to use the parametric description of the ellipse (cf. Figure 10.4).

FIGURE 10.4.
An Ellipse Description by the Slope of the Normal Vector.

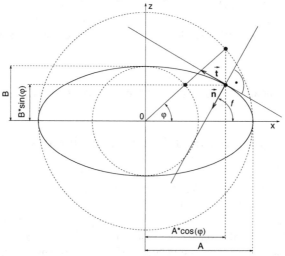

It is important to remember that the geographic coordinates are not the spherical coordinates! The geographic latitude is determined by measuring the angle of an object *(sun, moon, known star)* in the sky above the horizon, i.e. the angle between a tangent plane to the earth's surface and the object. The usual parametric representation of the ellipse is

$$x(\varphi) = A\cos(\varphi), \qquad z(\varphi) = B\sin(\varphi).$$

In order to relate the angle f to x and z we first compute the normal vector $\vec{n}$ at $[x, z]$

$$\vec{t} \equiv \begin{pmatrix} \dot{x} \\ \dot{z} \end{pmatrix} = \begin{pmatrix} -A\sin(\varphi) \\ B\cos(\varphi) \end{pmatrix} \qquad \Longrightarrow \qquad \vec{n} \equiv \begin{pmatrix} -\dot{z} \\ \dot{x} \end{pmatrix} = \begin{pmatrix} -B\cos(\varphi) \\ -A\sin(\varphi) \end{pmatrix}.$$

Therefore,

$$\tan(f) = -\frac{\dot{x}}{\dot{z}} = \frac{A}{B}\tan(\varphi).$$

We are interested in $\sin(\varphi)$ and $\cos(\varphi)$. From $\tan(\varphi) = \frac{B}{A}\tan(f)$ we obtain:

$$\cos(\varphi) = \frac{1}{\sqrt{1 + \tan^2(\varphi)}}, \qquad \sin(\varphi) = \tan(\varphi)\cos(\varphi).$$

Now we can present the corresponding MATLAB function as Algorithm 10.2.

ALGORITHM 10.2. *Function* gg2gc.

```
function [P] = gg2gc(f, l)
% transformation geographic. -> geocentric. coordinates
% f, l in radians
%
A = 6378.245; B = 6356.863;   % Krasovsky ellipse
tanfi = B/A*tan(f);
cosfi = 1/sqrt(1 + tanfi^2);
sinfi = tanfi*cosfi;
P = [ A*cosfi*cos(l); A*cosfi*sin(l); B*sinfi ];
```

Example 2:

Let us check both functions by computing the geocentric coordinates of Brno.

```
>> f = deg2rad([49 3 37 2]);
>> l = deg2rad([16 38 27 74]);
>> BRNO = gg2gc(f,l)

BRNO =

   1.0e+003 *

    4.01206298678276
    1.19917767600837
    4.79503887137397
```

10.4   The Transformations

We can now perform the transformation from the coordinate system $\mathbf{P_1}$ into
the coordinate system $\mathbf{P_2}$. For more information on coordinate transformations,
see e. g. [1, 2]. This complicated transformation consists of four partial transformations, more precisely, of one translation and three rotations. Since the
coordinate system $\mathbf{P_1}$ changes after each transformation, we must distinguish
the coordinate systems using a superscript:

$$\mathbf{P_1^i}, \quad i = 0, \cdots, 4, \quad \text{where } \mathbf{P_1^0} \equiv \mathbf{P_1}, \quad \mathbf{P_1^4} \equiv \mathbf{P_2}.$$

1. The Translation

The values of the displacement $[\Delta x, \Delta y, \Delta z] = P_2(\mathbf{P_1}) - P_1(\mathbf{P_1}) \equiv \vec{\Delta}_P$ are the
coordinates of the point P_2 in the first coordinate system, i. e. we denote it by
$P_2(\mathbf{P_1})$. The $\vec{\Delta}_P$ is known in the $\mathbf{C}$ system: $\vec{\Delta}_P(\mathbf{C}) = P_2(\mathbf{C}) - P_1(\mathbf{C})$. We must
transform $\vec{\Delta}_P(\mathbf{C})$ into $\vec{\Delta}_P(\mathbf{P_1})$. This transformation is shown as Figure 10.5
and can be computed by the following steps:

FIGURE 10.5. $\vec{\Delta}_P(\mathbf{C}) \Longrightarrow \vec{\Delta}_P(\mathbf{P_1})$ Transformation
C, $\mathbf{P_1}$ and $\mathbf{P_2}$ Definitions
Basic Transformations.

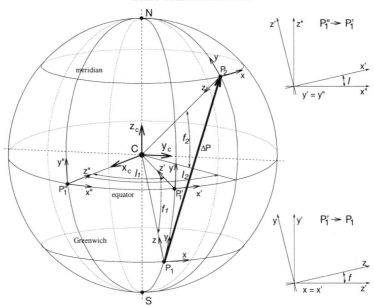

1. Let us put the origin P_1'' of the coordinate system $\mathbf{P_1''}$ at the intersection
   of the equator and the Greenwich meridian. From Figure 10.5 we can see

that the transformation is simply

$$
\begin{pmatrix} \Delta x \\ \Delta y \\ \Delta z \end{pmatrix}_{\mathbf{P}_1''} = \begin{pmatrix} \Delta y \\ \Delta z \\ -\Delta x \end{pmatrix}_{\mathbf{C}} = M \times \begin{pmatrix} P_{2x} - P_{1x} \\ P_{2y} - P_{1y} \\ P_{2z} - P_{1z} \end{pmatrix}_{\mathbf{C}}, \quad \text{where } M = \begin{pmatrix} 0 & 1 & 0 \\ 0 & 0 & 1 \\ -1 & 0 & 0 \end{pmatrix}.
$$

2. To move $\mathbf{P}_1''$ into $\mathbf{P}_1'$, we perform a rotation around the y''-axis with the angle l_1. This transformation can be described by the rotation matrix R_1:

$$
\begin{pmatrix} \Delta x \\ \Delta y \\ \Delta z \end{pmatrix}_{\mathbf{P}_1'} = R_1 \times \begin{pmatrix} \Delta x \\ \Delta y \\ \Delta z \end{pmatrix}_{\mathbf{P}_1''}, \quad \text{where } R_1 = \begin{pmatrix} \cos(l_1) & 0 & \sin(l_1) \\ 0 & 1 & 0 \\ -\sin(l_1) & 0 & \cos(l_1) \end{pmatrix}.
$$

3. To move $\mathbf{P}_1'$ into $\mathbf{P}_1$, we use a rotation around the x'-axis with the angle f_1. This transformation is described by the matrix R_2:

$$
\begin{pmatrix} \Delta x \\ \Delta y \\ \Delta z \end{pmatrix}_{\mathbf{P}_1} = R_2 \times \begin{pmatrix} \Delta x \\ \Delta y \\ \Delta z \end{pmatrix}_{\mathbf{P}_1'}, \quad \text{where } R_2 \equiv \begin{pmatrix} 1 & 0 & 0 \\ 0 & \cos(f_1) & \sin(f_1) \\ 0 & -\sin(f_1) & \cos(f_1) \end{pmatrix}.
$$

4. Now the coordinate system $\mathbf{P}_1$ is in a general position. Summarizing steps $1-4$, the translation vector $\vec{\Delta}_P(\mathbf{P}_1)$ can be expressed by

$$
\begin{pmatrix} \Delta x \\ \Delta y \\ \Delta z \end{pmatrix}_{\mathbf{P}_1} = R_2 \times R_1 \times M \times \begin{pmatrix} P_{2x} - P_{1x} \\ P_{2y} - P_{1y} \\ P_{2z} - P_{1z} \end{pmatrix}_{\mathbf{C}}.
$$

We obtain the new coordinates $A(\mathbf{P}_1^1)$ of the airplane from the previous coordinates $A(\mathbf{P}_1)$ by using the relation

$$
A(\mathbf{P}_1^1) \equiv \begin{pmatrix} x \\ y \\ z \end{pmatrix}_{\mathbf{P}_1^1} = A(\mathbf{P}_1) - \begin{pmatrix} \Delta x \\ \Delta y \\ \Delta z \end{pmatrix}_{\mathbf{P}_1}.
$$

2. First Rotation

We rotate the $\mathbf{P}_1^1$ coordinate system around the x-axes by the angle f_1 in order to make the y-axis parallel to the earth axis z_c. Then, the new coordinates of the airplane $A(\mathbf{P}_1^2)$ are obtained from the previous coordinates $A(\mathbf{P}_1^1)$ using

$$
A(\mathbf{P}_1^2) \equiv \begin{pmatrix} x \\ y \\ z \end{pmatrix}_{\mathbf{P}_1^2} = R_3 \times \begin{pmatrix} x \\ y \\ z \end{pmatrix}_{\mathbf{P}_1^1}, \quad \text{where } R_3 = R_2^T \equiv \begin{pmatrix} 1 & 0 & 0 \\ 0 & \cos(f_1) & -\sin(f_1) \\ 0 & \sin(f_1) & \cos(f_1) \end{pmatrix}.
$$

3. Second Rotation

We rotate the $\mathbf{P_1^2}$ coordinate system by the angle $\Delta l = l_2 - l_1$ around the $y-$axis in order to let the two x-axes of $\mathbf{P_1^2}$ and $\mathbf{P_2}$ coincide. The new coordinates $A(\mathbf{P_1^3})$ of the airplane are then computed from the previous coordinates $A(\mathbf{P_1^2})$ by

$$A(\mathbf{P_1^3}) \equiv \begin{pmatrix} x \\ y \\ z \end{pmatrix}_{\mathbf{P_1^3}} = R_4 \times \begin{pmatrix} x \\ y \\ z \end{pmatrix}_{\mathbf{P_1^2}} , \text{where } R_4 \equiv \begin{pmatrix} \cos(\Delta l) & 0 & \sin(\Delta l) \\ 0 & 1 & 0 \\ -\sin(\Delta l) & 0 & \cos(\Delta l) \end{pmatrix}.$$

4. Third Rotation

We rotate the system $\mathbf{P_1^3}$ by the angle f_2 around the $x-$axis to let it coincide with the system $\mathbf{P_2}$. Thus, the final coordinates $A(\mathbf{P_2}) \equiv A(\mathbf{P_1^4})$ of the airplane A are obtained from the previous coordinates $A(\mathbf{P_1^3})$ by

$$A(\mathbf{P_2}) \equiv \begin{pmatrix} x \\ y \\ z \end{pmatrix}_{\mathbf{P_1^4}} = R_5 \times \begin{pmatrix} x \\ y \\ z \end{pmatrix}_{\mathbf{P_1^3}} , \text{where } R_5 \equiv \begin{pmatrix} 1 & 0 & 0 \\ 0 & \cos(f_2) & \sin(f_2) \\ 0 & -\sin(f_2) & \cos(f_2) \end{pmatrix}.$$

10.5   Final Algorithm

The above steps can be concatenated. By the using matrices $R_1 - R_5$ we get the overall transformation

$$A(\mathbf{P_2}) \equiv \begin{pmatrix} x \\ y \\ z \end{pmatrix}_{\mathbf{P_2}} = R_5 \times R_4 \times R_2^T \times \left(A(\mathbf{P_1}) - R_2 \times R_1 \times M \times \begin{pmatrix} P_{2x} - P_{1x} \\ P_{2y} - P_{1y} \\ P_{2z} - P_{1z} \end{pmatrix}_{\mathbf{C}} \right).$$

This algorithm can be used as the basis for the MATLAB function transf, (see Algorithm 10.3).

$$A(\mathbf{P_2}) = \text{transf}(f_1, l_1, A(\mathbf{P_1}), f_2, l_2).$$

10.6   Practical Example

Our problem is now completely solved. We can check our solution with a practical example:

```
>> format long
>> f1 = deg2rad([49 45 18]);          % Praha
>> l1 = deg2rad([14 7 58]);           %    position
>> f2 = deg2rad([49 3 37 2]);         % Brno
>> l2 = deg2rad([16 38 27 74]);       %    position
>> A1praha = [200; 140; -5.790];      % 1. A. seen from Praha
>> A2praha = [157; 397; -9.870];      % 2. A. seen from Praha
```

ALGORITHM 10.3. *Function* transf.

```
function [A2] = transf(f1, l1, A1, f2, l2)
%------------------------------------------
% the airplane radar position recalculation
% f, l = geographic radar positions in radians
% A1 = [x;  y; z](P1) = position of airplane seen by radar at P1
%
M  = [    0        1         0
          0        0         1       % M = almost permutations
         -1        0         0   ];  % matrix
%
R1 = [ cos(l1)     0       sin(l1)
          0        1         0       % R1 = Displacement
       -sin(l1)    0       cos(l1)]; %      1. rotation matrix
%
R2 = [    1        0         0
          0      cos(f1)   sin(f1)   % R2 = Displacement
          0     -sin(f1)   cos(f1)]; %      2. rotation matrix
%
P1c = gg2gc(f1, l1);                 % 1. radar    geocentric
P2c = gg2gc(f2, l2);                 % 2. radar    coordinates
%
DPc = P2c - P1c;                     % DPc - radar's displacement
%                                    %    (1. radar pos. [f=0, l=0])
T = R2*R1*M*DPc;                     % - radar's displacement
%                                    %    (1. radar pos. [f1, l1])
a = A1 - T;                          % A. pos. after translation
%
Dl = l2 - l1;                        % longitude difference
%
R4 = [ cos(Dl)     0       sin(Dl)
          0        1         0       % R4 = rotation matrix,
       -sin(Dl)    0       cos(Dl)]; %      rot. around y by l2-l1
%
R5 = [    1        0         0
          0      cos(f2)   sin(f2)   % R5 = rotation matrix,
          0     -sin(f2)   cos(f2)]; %      rot.  around x by f2
%
A2 = R5*R4*R2'*a;                    % A2 = [x; y; z](P2)
%                                    % position of airplane
%                                    % seen by radar at P2
% end of function
```

```
>> A1brno = transf(f1, l1, A1praha, f2, l2) % 1. A. seen from Brno

A1brno =

   1.0e+02 *

   0.23632088742730
   2.13625791235094
  -0.06831714145899

>> a1praha = transf(f2, l2, A1brno, f1, l1) % inv. transformation

a1praha =

   1.0e+02 *

   2.00000000000000
   1.40000000000000
  -0.05790000000000

>> A2brno = transf(f1, l1, A2praha, f2, l2) % 2. A. seen from Brno

A2brno =

   1.0e+02 *

  -0.10857206384904
   4.71937837075423
  -0.06683743260798

>> a2praha = transf(f2, l2, A2brno, f1, l1) % inv. transformation

a2praha =

   1.0e+02 *

   1.57000000000000
   3.97000000000000
  -0.09870000000000
```

Acknowledgments

The authors thank J. Hřebíček and W. Gander for their suggestions and help to improve this chapter.

References

[1] H. J. BARTSCH, *Taschenbuch Mathematischer Formeln*, Fachbuchverlag, Leipzig, 1991.

[2] D. G. ZILL AND M. R. CULLEN, *Advanced Engineering Mathematics*, PWS-KENT, Boston, 1992.

Chapter 11. Conformal Mapping of a Circle

H.J. Halin, L. Jaschke

11.1 Introduction

Mapping techniques are mathematical methods which are frequently applied for solving fluid flow problems in the interior involving bodies of nonregular shape. Since the advent of supercomputers such techniques have become quite important in the context of numerical grid generation [1]. In introductory courses in fluid dynamics students learn how to calculate the circulation of an incompressible potential flow about a so-called "Joukowski airfoil" [3] which represent the simplest airfoils of any technical relevance. The physical plane where flow about the airfoil takes place is in a complex $p = u + iv$ plane where $i = \sqrt{-1}$. The advantage of a Joukowski transform consists in providing a conformal mapping of the p-plane on a $z = x + iy$ plane such that calculating the flow about the airfoil gets reduced to the much simpler problem of calculating the flow about a displaced circular cylinder. A special form of the mapping function $p = f(z) = u(z) + iv(z)$ of the Joukowski transform reads

$$p = \frac{1}{2}(z + \frac{a^2}{z}). \tag{11.1}$$

In this chapter we shall demonstrate how the mathematical transformations required in applying mapping methods can be handled elegantly by means of a language for symbolic computation and computer algebra. Rather than choosing a large physical problem that would be beyond the scope of this book, we select a very simple application of conformal mapping to illustrate the essential steps involved.

11.2 Problem Outline

It is suitable to express both x and y in terms of some parameter t such that $z(t) = x(t) + iy(t)$. Consequently also $p(z(t)) = f(z(t))$ holds. Inserting expression $z(t)$ into $p(z) = f(z(t))$ leads to

$$P(t) = U(t) + iV(t) = p(z(t)) \tag{11.2}$$

Twofold differentiation on either side of the equal sign with respect to t yields at next

$$\dot{P}(t) = p'(z(t))\dot{z}(t), \tag{11.3}$$

and subsequently

$$\ddot{P}(t) = p''(z(t))(\dot{z}(t))^2 + p'(z(t))\ddot{z}(t) \tag{11.4}$$

where dots and primes denote derivatives with respect to t and z, respectively.

The relations outlined so far are needed for handling the following problem. We assume a mapping function p which results from solving the second order differential equation presented in [2]

$$z^2 p'' + z p' + (\alpha z^2 + \beta z + \gamma)p = 0 \tag{11.5}$$

subject to the initial conditions $p'(0) = p'_0$ and $p(0) = p_0$.

Obviously the problem of mapping $z(t)$ on the domain $Y(t) = U(t) + iV(t)$ requires the solution of the complex second order differential equation specified by (11.5). This can be done by proceeding as follows: For the considered problem $z(t)$ is chosen as a circle of radius r, i.e. $z = re^{it}$. Hence, derivatives of $z(t)$ with respect to t can be readily evaluated. Inserting the complex expressions for $z(t)$, $\dot{z}(t)$, and $\ddot{z}(t)$, respectively, into Equations (11.2)-(11.5) permits one to completely eliminate any explicit appearance of z and its t-derivatives from all these equations. After this p' and p'' in (11.3) and (11.4) can be expressed only in terms of $\dot{P}(t)$ and $\ddot{P}(t)$. Finally, when using these expressions together with (11.2), our second order differential equation can be modified such that it only entails $P(t)$, $\dot{P}(t)$, and $\ddot{P}(t)$. Considering (11.2) we can simplify this complex differential equation by collecting the real and imaginary components. This yields a system of two coupled second order differential equations. The solution $U(t)$ and $V(t)$ can now be obtained by application of a standard numerical integration algorithm.

For the following two sets of initial conditions and values of parameter r illustrative solutions can be found:

r	$=$	1	r	$=$	0.6
$U(0)$	$=$	-0.563	$U(0)$	$=$	-0.944
$\dot{U}(t)$	$=$	0	$\dot{U}(t)$	$=$	0
$V(0)$	$=$	0	$V(0)$	$=$	0
$\dot{V}(t)$	$=$	0.869	$\dot{V}(t)$	$=$	0.658

In each of the two runs integration is done over the interval $(0 \le t \le 6\pi)$. Moreover, the following parameter values will be used $\alpha = 1$, $\beta = 0.5$, and $\gamma = -4/9$. Subsequently the latter three parameters will be referred to as a, b, and c.

11.3   MAPLE Solution

The solution steps outlined above are not very difficult to perform. Anyhow, since there are several manipulations required, deriving the formulas manually and programming them in any of the well known higher computer languages is error prone and time consuming. It will be shown in the following that an advanced mathematical computation tool such as MAPLE with its numerous symbolic, numeric, and graphical features, can be very helpful in making the solution process or at least parts of it more elaborate and elegant.

The first major task in dealing with the sample problem is to derive the two second order differential equations for $U(t)$ and $V(t)$, respectively. To do this we start with the definition of the parametric representation of the transformed circle $P(t) = p(z(t))$ in the complex $P = U + iV$ plane:

```
> p(z(t))=P(t):
```

Now we would like to express the first derivative of the as yet unknown function $p(z(t))$ with respect to t. For doing this we apply MAPLE's differentiation operator diff and solve for the first derivative with respect to z, i.e. $p'(t) = D(p)(z(t))$, which we denote as pprime.

```
> pprime:=solve(diff(",t),D(p)(z(t)));
```

$$pprime := \frac{\frac{\partial}{\partial t} P(t)}{\frac{\partial}{\partial t} z(t)}$$

Similarly, by a second application of the operator D to p we obtain an expression that can be solved for $p''(t)$ which we denote as p2prime.

```
> p2prime:=solve(diff("",t,t),D(D(p))(z(t)));
```

$$p2prime := -\frac{D(p)(z(t))\left(\frac{\partial^2}{\partial t^2} z(t)\right) - \left(\frac{\partial^2}{\partial t^2} P(t)\right)}{\left(\frac{\partial}{\partial t} z(t)\right)^2}$$

Since $D(p)(z(t))$ is by definition identically to $p'(t)$ we can make this replacement in the expression that yielded p2prime.

```
> D(p)(z(t)):=pprime;
```

$$D(p)(z(t)) := \frac{\frac{\partial}{\partial t} P(t)}{\frac{\partial}{\partial t} z(t)}$$

The new form of p2prime is

```
> p2prime;
```

$$-\frac{\frac{\left(\frac{\partial}{\partial t} P(t)\right)\left(\frac{\partial^2}{\partial t^2} z(t)\right)}{\frac{\partial}{\partial t} z(t)} - \left(\frac{\partial^2}{\partial t^2} P(t)\right)}{\left(\frac{\partial}{\partial t} z(t)\right)^2}$$

Now everything is available for insertion into the complex differential Equation (11.5) which will be named ode.

```
> ode:=z(t)^2*p2prime+z(t)*pprime+(a*z(t)^2+b*z(t)+c)*P(t);
```

$$ode := -\frac{z(t)^2 \left(\frac{(\frac{\partial}{\partial t} P(t))(\frac{\partial^2}{\partial t^2} z(t))}{\frac{\partial}{\partial t} z(t)} - (\frac{\partial^2}{\partial t^2} P(t))\right)}{(\frac{\partial}{\partial t} z(t))^2} + \frac{z(t)(\frac{\partial}{\partial t} P(t))}{\frac{\partial}{\partial t} z(t)} \quad (11.6)$$
$$+ (a z(t)^2 + b z(t) + c) P(t)$$

So far the process of handling the problem has been strictly formal without taking note of the fact that $z(t)$ is a circle. Next this information is provided

```
> z(t):=r*exp(I*t):
```

so that we can specify $P(t)$ as a function depending only on t.

```
> P(t):=U(t)+I*V(t):
```

Our complex differential equation ode will now be split into two coupled ones re and im of the same order. All real components will be collected and allocated to re. Likewise all imaginary components will contribute to im. In either case we replace the complex exponential function of the circle by its well known trigonometric representation involving a sine and cosine function. This is accomplished by the argument trig in MAPLE's combine command combine(...,trig). The commands for doing the decomposition read in detail

```
> re:=combine(evalc(Re(ode)),trig);
```

$$re := -(\frac{\partial^2}{\partial t^2} U(t)) + U(t) a r^2 \cos(2 t) + U(t) b r \cos(t) + U(t) c$$
$$- r^2 V(t) a \sin(2 t) - r V(t) b \sin(t)$$

```
> im:=combine(evalc(Im(ode)),trig);
```

$$im := -(\frac{\partial^2}{\partial t^2} V(t)) + r^2 U(t) a \sin(2 t) + r U(t) b \sin(t) + V(t) a r^2 \cos(2 t)$$
$$+ V(t) b r \cos(t) + V(t) c$$

As can be seen from the above lines the terms are not yet grouped optimally. Therefore we collect in re all coefficients of $U(t)$ and $V(t)$, respectively. This yields

```
> re:=collect(collect(re,U(t)),V(t))=0:
> im:=collect(collect(im,U(t)),V(t))=0:
```

For integration we will combine the differential equations for the real and imaginary components of $P(t)$ to form the overall system odes.

```
> odes:= re,im;
```

$$odes := (-b r \sin(t) - a r^2 \sin(2 t)) V(t)$$
$$+ (a r^2 \cos(2 t) + b r \cos(t) + c) U(t) - (\frac{\partial^2}{\partial t^2} U(t)) = 0,$$
$$(a r^2 \cos(2 t) + b r \cos(t) + c) V(t) + (a r^2 \sin(2 t) + b r \sin(t)) U(t)$$
$$- (\frac{\partial^2}{\partial t^2} V(t)) = 0$$

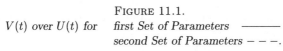

FIGURE 11.1.
$V(t)$ over $U(t)$ for   first Set of Parameters   ———
                         second Set of Parameters – – –.

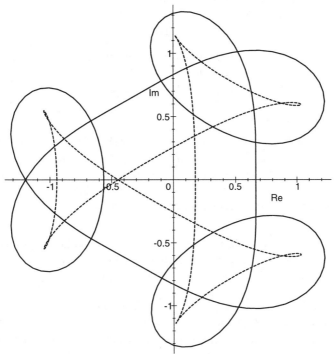

A numerical solution of such a set of differential equations is feasible and thus potentially very time consuming especially since MAPLE's mode of operation is interpretative rather than doing number crunching executions on compiled programs. Moreover, at the present MAPLE offers just a single fourth-fifth order Runge-Kutta-Fehlberg algorithm. Since MAPLE permits output also in Fortran and C notation it would already be a tremendous saving to derive the above equations and get them automatically coded for further usage in a Fortran or C environment.

MAPLE is capable of offering a numeric solution to our problem by means of its 4-5th order Runge-Kutta-Fehlberg algorithm. Prior to applying this algorithm we specify the initial conditions for our first parameter set

```
> init:=V(0)=0,U(0)=-.563,D(V)(0)=.869,D(U)(0)=0:
```

With these initial conditions MAPLE's routine dsolve will give as a solution which we can look at as a procedure F that can be evaluated subsequently to give the numerical values of $U(t)$ and $V(t)$ for various arguments t. Of course we still need to specify the remaining parameters, i.e. a, b, c, and r.

```
> a:=1: b:=0.5: c:=-0.444444443: r:=1:
> F1:=dsolve({odes, init}, {U(t),V(t)},numeric);
```

$$F1 := \mathbf{proc}(rkf45\_x) \dots \mathbf{end}$$

Since we want to illustrate the solution by means of graphics we will have to evaluate our procedure F1 at for instance 201 equally spaced points in the interval $(0 \le t \le 6\pi)$ and temporarily store the values of $U(t)$ and $V(t)$ in a list.

```
> numpts:= 200: Tend:= 6*Pi:
> L1:= [seq(subs(F1(j*Tend/numpts),[U(t),V(t)]),j=0..numpts)]:
```

where numpts is the number of points. L1 is a list of lists. Each time when invoking procedure F1 a new set of values of t, $U(t)$ and $V(t)$ will be added to the list. At the very end we use a MAPLE command to produce a plot, Figure 11.3, where $V(t)$ is plotted over $U(t)$, *(line ———)*

```
> plot(L1);
```

By using the second set of initial conditions and a new value of parameter r we will get another solution which is also shown in Figure 11.3. *(line − − −)*.

```
> init:=V(0)=0,U(0)=-.944,D(V)(0)=.658,D(U)(0)=0:
> r:=0.6:
> F2:=dsolve({odes, init}, {U(t),V(t)},numeric):
> L2:=[seq(subs(F2(j*Tend/numpts),[U(t),V(t)]),j=0..numpts)]:
```

The plot shown in Figure 11.3 can now be produced by the following command:

```
> plot({L1,L2},labels=[Re,Im],linestyle=[0,3],color=black,
        font=[HELVETICA,12]);
```

11.4 MATLAB Solution

The numerical integration of the differential equation developed in the previous section can also be done using MATLAB. The most recent version MATLAB 5.0 provides several functions for integrating complex ODEs.

Let us start with equation (11.6) from the previous section. Letting

$$z(t) := re^{-it}$$

turns it into the following differential equation

```
> ode = 0;
```

$$-(\frac{\partial^2}{\partial t^2} P(t)) + (a r^2 (e^{(It)})^2 + b r e^{(It)} + c) P(t) = 0 \qquad (11.7)$$

For solving this ODE by MATLAB we will convert this second order ODE into a System of two differential equation of first order. Using the substitutions

$$\mathbf{y}(t) := \begin{bmatrix} p(t) \\ \dot{p}(t) \end{bmatrix} \quad,$$

Equation (11.7) can be rewritten as

$$\dot{\mathbf{y}} = \begin{bmatrix} \dot{p}(t) \\ \ddot{p}(t) \end{bmatrix} = \begin{bmatrix} 0 & 1 \\ ar^2 e^{2it} + bre^{it} + c & 0 \end{bmatrix} \underbrace{\begin{bmatrix} p(t) \\ \dot{p}(t) \end{bmatrix}}_{=\,\mathbf{y}}$$

This linear differential equation can easily be coded in MATLAB as a separate M-file ode.m as shown in Algorithm 11.1. Such a so called ode function takes at least two input parameters t and $\mathbf{y}$ and returns $\dot{\mathbf{y}}$, referred to in the function as yd.

ALGORITHM 11.1. *Differential Equation* ode.m

```
function yd= ode(t,y);
global a b c r;

yd= [y(2); (a*r^2*exp(2*i*t)+b*r*exp(i*t)+c)*y(1)];
```

Prior to solving the differential equation we have to initialize the global variables a, b and c. Since the global variable r is being used as a parameter, it will be initialized later on for each separate solution.

```
>> global a b c r
>> a= 1; b= 0.5; c= -4/9;
```

For integration we use the function ode45, the MATLAB implementation of the 4/5th order Runge-Kutta-Fehlberg algorithm, which had previously been applied for getting the solution by means of MAPLE.

```
>> options= odeset('RelTol',1e-8,'AbsTol',1e-8);
>>
>> % first set of parameters
>> r= 1;
>> [t1,p1]= ode45('ode',[0 6*pi],[-0.563 0.869*i],options);
>>
>> % second set of parameters
>> r= 0.6;
>> [t2,p2]= ode45('ode',[0 6*pi],[-0.944 0.658*i],options);
```

For obtaining a good printing quality the relative and the absolute tolerance are set to 10^{-8}. The reference of function ode45 is performed using four arguments in total: 1st the name 'ode' of the function providing the vector $\dot{\mathbf{y}}$ of derivatives; 2nd the range of the integration; 3rd the vector $\mathbf{y_0} = [U(0) + iV(0), \dot{U}(0) + i\dot{V}(0)]^T$; 4th options, e.g. the tolerance in the present case.

Plots of the solutions will be generated by the following commands:

```
>> clf;
>> plot(p1(:,1),'b-');
>> hold on;
>> plot(p2(:,1),'r--');
>> hold off;
```

As expected, these commands produce a figure almost identical to Fig. 11.3 obtained by MAPLE.

References

[1] J. HÄUSER and C. TAYLOR, *Numerical Grid Generation in Computational Fluid Dynamics*, Pineridge Press, Swansea, U.K., 1986.

[2] J. HEINHOLD and U. KULISCH, *Analogrechnen*, BI-Hochschultaschenbücher Reihe Informatik, Bibliographisches Institut Mannheim / Zürich, 168/168a, 1968.

[3] W.F. HUGHES and J.A. BRIGHTON, *Fluid Dynamics*, Schaum's Outline Series, McGraw-Hill, USA, 1967.

Chapter 12. The Spinning Top

F. Klvaňa

12.1 Introduction

In this chapter we will study the motion of a spinning top—the well known children's toy. From the physical point of view we can represent it as a symmetric rigid rotor in a homogeneous gravitational field. Let O be the point at the tip of the top.

FIGURE 12.1. *Coordinate Systems.*

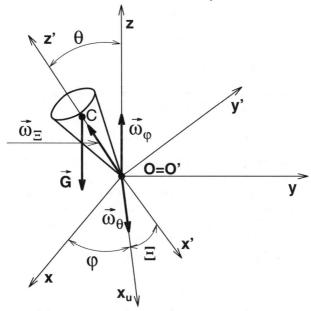

Let (x, y, z) be an inertial coordinate system having its origin at the point O of the rotor where the z axis is oriented vertically (see Figure 12.1). Let $\vec{G} = (0, 0, -mg)$ be the weight of the rotor (acting at the center of mass), where m is the mass of the rotor and g is the gravitational acceleration. Then the kinetic energy of this rotor with angular velocity $\vec{\omega} = (\omega_1, \omega_2, \omega_3)$ is

$$T = \frac{1}{2} \sum_{i,j=1}^{3} I_{ij} \omega_i \omega_j.$$

Notice that the vector $\vec{\omega}$ has the direction of the axis of rotation. I_{ij} is the tensor of inertia of the rotor (see [1, 2]). There exists at least one body-fixed coordinate system (x', y', z') with origin $O' = O$, in which the tensor of inertia is diagonal. In such a system the coordinate axes are called principal axes of inertia. For a symmetric rotor the axis of symmetry is one of the principal axes which we denote by z'. The two other principal axes are in the (x', y') plane orthogonal to z'. The kinetic energy has the form

$$T = \frac{1}{2} I_1 (\omega_{x'}^2 + \omega_{y'}^2) + \frac{1}{2} I_3 \omega_{z'}^2, \tag{12.1}$$

where I_1 and I_3 are the corresponding principal moments of inertia.

It is useful to describe the rotation of a body about the fixed point O by the Euler angles (Φ, Θ, Ξ). Using them we express the transformation from the coordinate system (x, y, x) to (x', y', z') by successive rotations (see Figure 12.1):

1. about the z-axis by the angle Φ $(x \to x_u)$,

2. about the x_u-axis (called the line of nodes) by the angle Θ $(z \to z')$,

3. about the z'-axis by the angle Ξ $(x_u \to x')$.

The transformation matrix $\hat{A}(\Phi, \Theta, \Xi)$ from the (x, y, z) system to the (x', y', z') system is expressed as the product of matrices of rotation,

$$\hat{A}(\Phi, \Theta, \Xi) = \hat{R}_z(\Xi) \cdot \hat{R}_x(\Theta) \cdot \hat{R}_z(\Phi), \tag{12.2}$$

where $\hat{R}_x$, and $\hat{R}_z$, respectively, represents the matrices of rotations about the x-axis or the z-axis, respectively.

The Euler angles $(\Phi(t), \Theta(t), \Xi(t))$ may be used as generalized coordinates of the rotor. Let the rotor rotate in the time interval dt about the instantaneous axis by the angle $|\vec{\omega}| \, dt$. This rotation can be expressed as the succession of rotations about the z-, x_u-axis and z'- axis by the angles $|\vec{\omega}_\Phi| \, dt = \dot{\Phi} \, dt$, $|\vec{\omega}_\Theta| \, dt = \dot{\Theta} \, dt$, and $|\vec{\omega}_\Xi| \, dt = \dot{\Xi} \, dt$, respectively. We can write

$$\vec{\omega} = \vec{\omega}_\Phi + \vec{\omega}_\Theta + \vec{\omega}_\Xi. \tag{12.3}$$

The angular velocities $\vec{\omega}_\Phi$, $\vec{\omega}_\Theta$, $\vec{\omega}_\Xi$ have the following components in the rotating coordinate system $(x'y'z')$:

$$
\begin{aligned}
((\omega_\Phi)_{x'}, (\omega_\Phi)_{y'}, (\omega_\Phi)_{z'}) &= \hat{A}(\Phi, \Theta, \Xi) \cdot (0, 0, \dot{\Phi}) \\
((\omega_\Theta)_{x'}, (\omega_\Theta)_{y'}, (\omega_\Theta)_{z'}) &= \hat{R}_z(\Xi) \cdot (\dot{\Theta}, 0, 0) \\
((\omega_\Xi)_{x'}, (\omega_\Xi)_{y'}, (\omega_\Xi)_{z'}) &= (0, 0, \dot{\Xi}).
\end{aligned}
$$

Using (12.1) and (12.4) we express the kinetic energy T as a function of the generalized coordinates $(\Phi(t), \Theta(t), \Xi(t))$ and the velocities $(\dot{\Phi}(t), \dot{\Theta}(t), \dot{\Xi}(t))$

$$T = \frac{1}{2}I_1(\dot{\Theta}^2 + \dot{\Phi}^2 sin^2\Theta) + \frac{1}{2}I_3(\dot{\Xi} + \dot{\Phi}cos\Theta)^2.$$

For the potential energy of the rotor we find

$$V = mgl \cdot cos\Theta \qquad (12.4)$$

where l is the distance between the center of mass C of the rotor and the fixed point O.

Our problem is well suited to be solved by the Lagrange formalism. Let $L = T - V$ be the Lagrangian and let $(q_1, ..., q_n)$ and $(\dot{q}_1, ..., \dot{q}_n)$ be the generalized coordinates and velocities of the system with n degrees of freedom. Then, the equations of motion are the Lagrange equations

$$\frac{d}{dt}\left(\frac{\partial L}{\partial \dot{q}_i}\right) - \frac{\partial L}{\partial q_i} = 0 \qquad i = 1, ..., n.$$

If L doesn't depend explicitly on some coordinate q_k (such a coordinate is called a cyclic coordinate) then the corresponding Lagrange equation is

$$\frac{d}{dt}\left(\frac{\partial L}{\partial \dot{q}_k}\right) = 0 \quad \Longrightarrow \quad \frac{\partial L}{\partial \dot{q}_k} = p_k = const, \qquad (12.5)$$

which means, that the conjugate momentum p_k is a constant of motion.

Because of the relatively complicated nature of the system of the three Lagrange equations we will use MAPLE for its derivation and study.

12.2 Formulation and Basic Analysis of the Solution

In the MAPLE solution of our problem we use the vector formalism of the package `linalg` together with the following notation for the variables:

$$(\Phi, \Theta, \Xi) \quad \longrightarrow \quad Euler := [\Phi, \Theta, \Xi]$$
$$(\dot{\Phi}, \dot{\Theta}, \dot{\Xi}) \quad \longrightarrow \quad dEuler := [\phi, \theta, \xi]$$

```
> with(linalg):
> Euler  := vector([Phi, Theta, Xi]):
> dEuler := vector([phi, theta, xi]):
```

As the first step of our solution we will express the Lagrangian L in terms of Euler angles. To express the matrices R_x, R_z of rotation about the coordinate axes x and z, we define the matrix operators

```
> Rz := matrix([[cos, sin, 0], [-sin, cos, 0], [0, 0, 1]]);
> Rx := matrix([[1, 0, 0], [0, cos, sin], [0, -sin, cos]]);
```

$$Rz := \begin{bmatrix} \cos & \sin & 0 \\ -\sin & \cos & 0 \\ 0 & 0 & 1 \end{bmatrix}$$

$$Rx := \begin{bmatrix} 1 & 0 & 0 \\ 0 & \cos & \sin \\ 0 & -\sin & \cos \end{bmatrix}$$

The transformation matrix A is then given by Equation (12.2),

```
> A   := evalm(Rz(Xi(t)) &* Rx(Theta(t)) &* Rz(Phi(t))):
```

Using Equations (12.3) and (12.4), the angular velocity $\vec{\omega}$ in the coordinate system $(x'y'z')$ is given by

$$(\omega_{x'}, \omega_{y'}, \omega_{z'}) = A \cdot (0, 0, \dot{\Phi}) + R_z(\Xi) \cdot (\dot{\Theta}, 0, 0) + (0, 0, \dot{\Xi}).$$

```
>                 # vectors of angular velocities about Euler axis
> omega[Phi]   := vector([0,0,phi(t)]):
>                 #rotation about axis z - phi
> omega[Theta] := vector([theta(t),0,0]):
>                 #rotation about the 1. xu  - theta
>                 # vector of the resulting angular velocity
>                 # Omega in (x',y',z')
> Omega := evalm(A &* omega[Phi] + Rz(Xi(t)) &* omega[Theta]
>             + vector([0,0,xi(t)]));
```

$$\Omega := [\sin(\Xi(t)) \sin(\Theta(t)) \phi(t) + \cos(\Xi(t)) \theta(t),$$
$$\cos(\Xi(t)) \sin(\Theta(t)) \phi(t) - \sin(\Xi(t)) \theta(t), \cos(\Theta(t)) \phi(t) + \xi(t)]$$

Then, using Equations (12.1) and (12.4) we can write the Lagrangian as

$$L = T - V = \frac{1}{2} I_1 (\omega_{x'}^2 + \omega_{y'}^2) + \frac{1}{2} I_3 \omega_{z'}^2 - mgl \cos \Theta.$$

```
> T := 1/2*I1*(Omega[1]^2 + Omega[2]^2) + 1/2*I3*Omega[3]^2;
>                           #kinetic energy
```

$$T := \frac{1}{2} I1 \left((\sin(\Xi(t)) \sin(\Theta(t)) \phi(t) + \cos(\Xi(t)) \theta(t))^2 \right.$$
$$+ (\cos(\Xi(t)) \sin(\Theta(t)) \phi(t) - \sin(\Xi(t)) \theta(t))^2 \right) + \frac{1}{2} I3 \left(\cos(\Theta(t)) \phi(t) + \xi(t) \right)^2$$

```
> V := M*g*L*cos(Theta(t)):    #potential energy
> LF := T - V:                 # Lagrangian
```

The second step is the development of the Lagrange equations for the rotor. For their derivation it will be useful to define the MAPLE function

```
LEq2( LagrFce, var, dvar, indep)
```

(see Algorithm 12.1), where `LagrFce` is the Lagrangian, `var` and `dvar` are vectors of the generalized variables and velocities, and `indep` is the independent variable (time t). In the case of a cyclic coordinate, this function returns the Equation (see 12.5)

$$\frac{\partial L}{\partial \dot{q}_k} = IMq_k,$$

where IMq_k (conjugate momentum to q_k) is a constant of motion. The function returns the equations as a set.

Because the standard function `diff(g,u)` differentiates only if `u` is a name, we will define a generalized differentiation function `sdiff(g,u)`, where `u` can be an expression.

To concatenate the string IM and the name of the cyclic variable q_k (by our definition a function $q_k(t)$), we should define a function `cname(name,fce)` which concatenates the name `name` and the name of function `fce`. The definitions of these functions are in Algorithm (12.1).

Since the coordinates Φ and Ξ are cyclic, Function `LEq2` returns two equations for the constants of motion p_Φ (with name IM_Φ), and p_Ξ (IM_Ξ) and the Lagrange equation corresponding to variable Θ. For easier access to the right-hand sides of these equations we will transfer this set of equations into a table where the left-hand sides ($IM_\Phi, IM_\Xi, 0$) will be the indices for the corresponding right-hand sides.

```
>               # generation of the Lagrange equation, or integrals
>               # of motion (for Phi and Xi)
> LEqn := LEq2(LF,Euler(t),dEuler(t),t):
> LEqn := table([op(LEqn)]);
>               #transform the result from set to table
```

$LEqn := \text{table}([$
$\quad IM_\Phi = I1\,\phi(t) - I1\,\phi(t)\cos(\Theta(t))^2 + I3\cos(\Theta(t))^2\,\phi(t) + I3\cos(\Theta(t))\,\xi(t)$
$\quad IM_\Xi = I3\cos(\Theta(t))\,\phi(t) + I3\,\xi(t)$
$\quad 0 = I1\,(\frac{\partial}{\partial t}\,\theta(t)) - I1\cos(\Theta(t))\,\phi(t)^2\sin(\Theta(t))$
$\quad + I3\sin(\Theta(t))\,\phi(t)^2\cos(\Theta(t)) + I3\sin(\Theta(t))\,\phi(t)\,\xi(t) - M\,g\,L\sin(\Theta(t))$
$])$

The most important constant of motion is the energy $E = T + V$. Because $\omega_{z'}$ is proportional to the constant of motion p_Ξ the quantity $E_c = E - 1/2 I_3 \omega_{z'}^2$ is also a constant of motion. We will use it instead of E for further study.

```
>               # Integrals of motion: IM[Phi], IM[Xi] and energy Ec
> Ec := simplify(T + V- 1/2*I3*Omega[3]^2);
```

$$Ec := \frac{1}{2}\,I1\,\phi(t)^2 - \frac{1}{2}\,I1\,\phi(t)^2\cos(\Theta(t))^2 + \frac{1}{2}\,I1\,\theta(t)^2 + M\,g\,L\cos(\Theta(t))$$

ALGORITHM 12.1.
Functions for Generating a Lagrange Equations.

Function sdiff.

```
sdiff := proc(u,svar) local p;
    #differentiation of u with respect to expression svar
    subs(p = svar,diff(subs(svar = p,u),p))
end;
```

Function cname.

```
cname:= proc (nam,fce) local p;
    #add name of function  fce  to the name  nam
    if type(fce,function) then
        p:=op(0,fce); nam[p]
    else
        nam.fce
    fi
end;
```

Function LEq2.

```
LEq2 := proc (LagrFce, var::vector,dvar::vector,indep::name)
    local  i,j,N,res;
    #the generation of the Lagrange equations or constants
    # of motion IM (for cyclic variables) for the Lagrange
    # function LagrFce and vector of generalized coordinates
    # var and velocities dvar and independent variable indep
    N:=vectdim(var);
    for i to N do
        res[i]:=simplify(sdiff(LagrFce,var[i]));
        if res[i]=0 then
            res[i] := cname('IM',var[i])
                        = sdiff(LagrFce, dvar[i])
        else
            res[i] := 0 = diff(sdiff(LagrFce, dvar[i]), indep)
                            - res[i]
        fi;
    od;
    {seq(simplify(res[i]), i=1..N)}
end;
```

For simplification we will carry out the following steps:

1. Solving the equations for p_Φ and p_Ξ for the unknowns $\{\dot{\Phi}, \dot{\Xi}\}$ and substituting them into E_c.

2. Introducing the parameters

$$b = \frac{p_\Phi}{I_1}, \quad a = \frac{p_\Xi}{I_1}, \quad \beta = \frac{1}{2}\frac{I_1}{mgl}, \quad \alpha = \frac{2E_c}{I_1}$$

and solving the equation for E_c for the unknown $\dot{\Theta}^2$.

3. Substitution of $u = cos\Theta$ in the resulting equation.

As result we obtain the following differential equation for $u(t)$, which describes the so-called nutation of the rotor

$$\dot{u}^2 = y(u) := \beta u^3 - (\alpha + a^2)u^2 + (2ab - \beta)u + \alpha - b^2. \qquad (12.6)$$

From the fact that $y(u) > 0$ it follows:

1. Because $\beta > 0$ we have $y(\pm\infty) = \pm\infty$ and $y(\pm 1) \leq 0$; so at least one root of $y(u)$ exists with $u > 1$.

2. Because $y(\pm 1) < 0$, two or zero roots have to be in the interval $[-1, 1]$. For physically realistic conditions two roots should exist: $u_1 \leq u_2 \in [-1, 1]$. Because $\dot{\Theta} = 0$ for $\cos\Theta_1 = u_1$ and $\cos\Theta_2 = u_2$ we have $\Theta(t) \in [\Theta_2, \Theta_1]$, so that the nutation angle of the top oscillates between Θ_1 and Θ_2 .

```
>               # find phi, xi with subs. IMXi=a*I1, IMPhi=b*I1
> phixi := solve({LEqn[IM[Phi]] = b*I1, LEqn[IM[Xi]] =a *I1},
>               {phi(t),xi(t)});
```

$$phixi := \{\xi(t) = -\frac{I3\cos(\Theta(t))^2 a - I3\cos(\Theta(t))b + aI1 - aI1\cos(\Theta(t))^2}{(-1 + \cos(\Theta(t))^2)I3},$$

$$\phi(t) = \frac{\cos(\Theta(t))a - b}{-1 + \cos(\Theta(t))^2}\}$$

```
> Econst := subs(phixi,Ec):
>               # substitution for  xi and  phi in  Ec
> theta2 := solve(subs(M = I1/(2*g*L)*beta,
>               Econst = I1*alpha/2),theta(t)^2):
>               # substitution
>               # cos(Theta) = u -> -sin(Theta)*theta = du
>               # then du2 = diff(u(t),t)^2
> du2 := simplify(subs(cos(Theta(t)) = u,
>               theta2*(1-cos(Theta(t))^2)));
```

$$du2 := \beta u^3 - u^2 a^2 - \alpha u^2 + 2uab - \beta u - b^2 + \alpha$$

```
>                   # analysis of the solution due to the form of du2
> collect(du2,u);
```

$$\beta u^3 + (-a^2 - \alpha) u^2 + (2ab - \beta) u - b^2 + \alpha$$

```
> seq(factor(subs(u = i, du2)), i = {-1,1});
```

$$-(a + b)^2, \; -(a - b)^2$$

```
>                   # so du2 (+-1) < 0 for b <> 0
```

The rotation of the top about the vertical z-axis (the so-called precession) is described by $\Phi(t)$ which is the solution of the differential equation (see the value of the variable phixi[1] above)

$$\dot{\Phi} = \frac{b - acos\Theta}{sin^2\Theta} \tag{12.7}$$

Then we can classify the type of motion of the rotor by the sign of $\dot{\Phi}$ at the points Θ_1, Θ_2, where $\dot{\Theta} = 0$. Let $\Theta_1 \geq \Theta_2 > 0$, so $u_2 = cos\Theta_2 > u_1$:

1. if $b/a > u_2$ or $b/a < u_1$, then $\dot{\Phi}$ doesn't change its sign (see Figure 12.2);

2. if $b - au_2$ and $b - au_1$ have opposite signs, then $\dot{\Phi}$ is periodically changing its sign (see Figure 12.3).

The differential Equation (12.6) may be solved in terms of elliptic functions. However, we will solve our problem numerically. The above form of the differential Equation (12.6) is not appropriate for the numerical solution, though.

12.3   The Numerical Solution

We will start from the Lagrange equations which were generated by MAPLE and saved under the variable LEq2 . The most important characteristics of the motion of the rotor are given by the time dependence of $\Phi(t)$ and $\Theta(t)$ (the precession and nutation angles of the rotor), so we will restrict ourselves to find these variables.

Because we found the solution for $\dot{\Phi}$ and $\dot{\Xi}$ (which are dependent on Θ only and which are saved in the MAPLE variable phixi), we can use them to substitute $\dot{\Phi}$ and $\dot{\Xi}$ in the last Lagrange equation connected with the variable Θ. This equation together with the trivial equation diff(Theta(t),t) = theta(t), and with the equation for $\dot{\Phi}$ (see 12.7) from phixi, form a system of three differential equations of first order for the variables Phi(t), Theta(t) and theta(t). We will now solve this system numerically.

FIGURE 12.2. *Motion of the Top for* $\lambda = 1.1$.

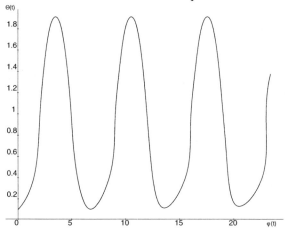

```
>                  # we substitute for xi and phi
>                  # in LEqn[0]  (DE for Theta)
> eq[Theta] := simplify(subs(phixi, LEqn[0])):
> eq[Theta] := simplify(eq[Theta]*I1*(cos(Theta(t))^2 - 1)^2):
>                  # simplify constant parameters in DE
> eq[Theta] := simplify(subs(M = I1/(2*g*L)*beta, eq[Theta]/I1^2));
```

$$
\begin{aligned}
eq_\Theta := {}& (\frac{\partial}{\partial t}\,\theta(t)) - 2\,(\frac{\partial}{\partial t}\,\theta(t))\cos(\Theta(t))^2 + (\frac{\partial}{\partial t}\,\theta(t))\cos(\Theta(t))^4 \\
& + \cos(\Theta(t))^2\sin(\Theta(t))\,a\,b - \cos(\Theta(t))\sin(\Theta(t))\,b^2 \\
& - \sin(\Theta(t))\cos(\Theta(t))\,a^2 + \sin(\Theta(t))\,b\,a - \frac{1}{2}\,\beta\sin(\Theta(t)) \\
& + \beta\sin(\Theta(t))\cos(\Theta(t))^2 - \frac{1}{2}\,\beta\sin(\Theta(t))\cos(\Theta(t))^4
\end{aligned}
$$

```
>                  # DE for Phi
> eq[Phi] := diff(Phi(t),t) = subs(phixi, phi(t));
```

$$
eq_\Phi := \frac{\partial}{\partial t}\,\Phi(t) = \frac{\cos(\Theta(t))\,a - b}{-1 + \cos(\Theta(t))^2}
$$

The system to be solved depends on the three parameters a, b and β, and on the three initial values $\Phi(0)$, $\Theta(0)$ and $\dot{\Theta}(0)$, respectively. Note that $\beta > 0$ while a and b can be chosen positive. Let us choose the initial conditions as

$$
\Phi(0) = 0, \quad \Theta(0) = \Theta_0, \quad \dot{\Theta}(0) = 0, \qquad \text{where} \quad \Theta_0 = 0.1,
$$

and the parameters as

$$
a = 1, \quad b = \lambda a \cos(\Theta_0), \quad \beta = 1, \qquad \text{where} \quad \lambda = 1.1.
$$

FIGURE 12.3. *Motion of the Top for* $\lambda = 0.96$.

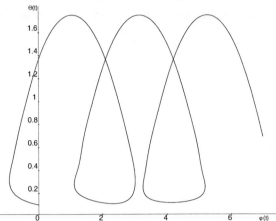

As follows from the discussion of Equation (12.7) two types of precession exist. If the constant $\lambda > 1$, as above, we obtain a solution with constant sign of $\dot{\Phi}$ giving the monotonic precession. For $\lambda < 1$ we obtain a solution with $\dot{\Phi}$ changing its sign corresponding to oscillatory precession.

The solution is illustrated using a parametric plot of $\Phi(t)$ over $\Theta(t)$. Results are shown in Figure 12.2 ($\lambda = 1.1$) and Figure 12.3 ($\lambda = 0.96$), respectively.

```
>                #d definition of initial conditions
> Theta0 := 0.1: #Theta(0)
> initc := Theta(0) = Theta0, theta(0) = 0, Phi(0) = 0:
>                # definition of parameters
> a := 1: beta := 1:
> lambda := 1.1: # 0.96 second plot
> b := a*cos(Theta0)*lambda:
>                # dependent variables of the system DE
> var := {Phi(t),Theta(t),theta(t)}:
>                # solving DE
> res := dsolve({eq[Theta] = 0, diff(Theta(t), t) = theta(t),
>          eq[Phi], initc}, var, numeric);
```

$$res := \mathbf{proc}(rkf45\_x) \ldots \mathbf{end}$$

```
>                # resulting plot
> plot([seq([subs(res(i*0.05), Phi(t)),
>            subs(res(i*0.05), Theta(t))], i = 0..300)]);
```

References

[1] HERBERT GOLDSTEIN, *Classical Mechanics*, Addison-Wesley, 1980.

[2] M. R. SPIEGEL, *Theoretical Mechanics*, McGraw-Hill, 1980.

Chapter 13. The Calibration Problem

J. Buchar and J. Hřebíček

13.1 Introduction

When measuring gas pressure changes in a container, for example an engine cylinder or a gun, by means of a piezoelectric pressure transducer, highly relatively accurate values must be made available in order to obtain the specified absolute accuracy. For this, special measuring and calibrating techniques are necessary, which allow the quantitative determination of the dynamic measuring properties of the transducer. The output from the transducer is in electric voltage. Therefore we must perform the calibration of the transducer so as to finally get the pressure. This is not difficult when we are working with a static pressure. The design of the equipment which enables well defined dynamic pressure measurement is much more complicated. This problem was solved by different authors by using a hydraulic pressure chamber, see [3]. For such a system we developed in our recent research project an experimental method for the dynamic pressure calibration. The essential problem connected with this method consists in the development of a physical model which allows a mathematical description of the hydraulic pressure pulses. This model enables us to calibrate a dynamic pulse pressure transducer in absolute pressure units. The schema of the experimental method is given in Figure 13.1. In the next chapter we will define a physical model using the results contained in [1].

13.2 The Physical Model Description

The pressure transducer is calibrated using hydrostatic pressure in a suitable chamber. For some applications; e.g. for the measurement of pressures in automobile engines or gun chambers, etc.; it is necessary to perform a dynamic calibration. For this purpose we suggest a simple experimental arrangement schematically drawn in Figure 13.1. A cylindrical chamber is filled with a liquid of a given compressibility. The initial volume of this liquid is V_0. The piston has the same cross section S as the chamber and is loaded by the impact of the body of total mass M accelerated to the velocity $v_0 > 0$. The movement $x(t)$ of the piston leads to the volume change $\Delta V = Sx(t)$. This volume change corresponds to a pressure $p(t)$, which (for almost all compressible liquids) can

FIGURE 13.1. *Cross Section of Chamber.*

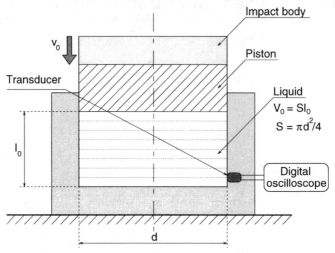

be obtained from the equation

$$p(t) = \alpha \frac{\Delta V}{V_0} + \beta \left(\frac{\Delta V}{V_0}\right)^2, \tag{13.1}$$

where $\alpha > 0$ and $\beta > 0$ are parameters of the given liquid.

Let us suppose that the impact velocity v_0 is limited to some reasonable value. We may then neglect the wave motion processes. The movement of the piston can be described by the differential equation

$$M\ddot{x} = -p(t)S, \tag{13.2}$$

with initial conditions

$$\begin{array}{rcl} x(0) & = & 0, \\ v = \dot{x}(0) & = & v_0 \end{array}. \tag{13.3}$$

Substituting Equation (13.1) for the pressure $p(t)$ and $\Delta V = Sx(t)$ in Equation (13.2) results in:

```
> M*diff(x(t), t, t) = - (alpha*dV/V0 + beta*(dV/V0)^2)*S;
```

$$M\left(\frac{\partial^2}{\partial t^2}\,\mathrm{x}(t)\right) = -\left(\frac{\alpha\,dV}{V0} + \frac{\beta\,dV^2}{V0^2}\right)S$$

```
> subs(dV = S*x(t), ")/S;
```

$$\frac{M\left(\frac{\partial^2}{\partial t^2}\,\mathrm{x}(t)\right)}{S} = -\frac{\alpha\,S\,\mathrm{x}(t)}{V0} - \frac{\beta\,S^2\,\mathrm{x}(t)^2}{V0^2}$$

This is now simplified by using the formulas $S = \pi d^2/4$, $V_0 = \pi d^2 l_0/4$ and the constants

$$a = \sqrt{\frac{\alpha \pi}{4} \frac{d^2}{M l_0}} \qquad \text{and} \qquad b = \frac{\beta \pi}{4} \frac{d^2}{M l_0^2}.$$

```
> subs(S = Pi*d^2/4, V0 = Pi*d^2*l0/4, "):
> simplify(", {a^2*4*M*l0 = alpha*Pi*d^2,
>                b*4*M*l0^2 = beta*Pi*d^2}, [alpha, beta]);
```

$$4\frac{M\left(\frac{\partial^2}{\partial t^2} x(t)\right)}{\pi d^2} = -4\frac{x(t) M \left(a^2 + b x(t)\right)}{\pi d^2}$$

```
> eq := "*Pi*d^2/4/M;
```

$$eq := \frac{\partial^2}{\partial t^2} x(t) = -x(t)\left(a^2 + b x(t)\right)$$

```
> alpha*dV/V0 + beta*(dV/V0)^2:
> subs(dV = S*x(t), S = Pi*d^2/4, V0 = Pi*d^2*l0/4, ");
```

$$\frac{\alpha x(t)}{l0} + \frac{\beta x(t)^2}{l0^2}$$

We can now rewrite the initial value problem (13.2)-(13.3) as

$$\ddot{x}(t) + a^2 x(t) + b x(t)^2 = 0, \tag{13.4}$$
$$x(0) = 0,$$
$$\dot{x}(0) = v_0. \tag{13.5}$$

Using the solution $x(t)$ of the above initial value problem we can evaluate the volume change $\Delta V = S x(t)$. We obtain from (13.1) the value of the pressure $p = p(t)$ at time t

$$p(t) = \frac{x(t)}{l_0}\left(\alpha + \beta\frac{x(t)}{l_0}\right). \tag{13.6}$$

When measuring the hydraulic pressure pulse we record the electrical voltage $U = U(t)$ of the transducer with a digital oscilloscope. We obtain an experimental record for the transducer for some time interval $T = [0, t_{max}]$ – see Figure 13.1. For the same interval we have to find the solution $x(t)$ of the above initial value problem. Then we can compare the experimentally measured hydraulic pressure with the calculated pressure $p(t)$ from (13.6). For practical purposes it is enough to compare the measured and calculated pressures in the subinterval $T_m \subset T$, where the pressure increases up to its maximum.

The calibration of the transducer is very simple when $p(t) = kU(t)$, where k is a constant. If k is not constant we must use the method of Fourier Transform. The detailed analysis of this method is beyond the scope of this chapter.

13.3   Approximation by Splitting the Solution

Let us suppose that the solution $x(t)$ of the initial value problem (13.4 – 13.5) has the form

$$x(t) = x_a(t) + x_b(t)$$

where $x_a(t)$ is the solution of the initial value problem of the free undamped motion [4]

$$\ddot{x}_a(t) + a^2 x_a(t) = 0,$$
$$x_a(0) = 0,$$
$$\dot{x}_a(0) = v_0.$$

The above initial value problem has the analytic solution for $a \neq 0$ of

$$x_a(t) = \frac{v_0 \sin(at)}{a}. \tag{13.7}$$

Note that MAPLE could be used to derive this solution:

```
> deq := {diff(xa(t), t, t) + a^2*xa(t) = 0,
          xa(0) = 0, D(xa)(0) = v0}:
> dsolve(deq, xa(t));
```

$$\mathrm{xa}(t) = \frac{v0 \sin(a\,t)}{a}$$

The solution $x(t)$ of the initial value problem (13.4 – 13.5) has the form

$$x(t) = \frac{v_0 \sin(at)}{a} + x_b(t) \tag{13.8}$$

where it is assumed that $x_b(t)$ is a small perturbation of $x_a(t)$. Inserting $x(t)$ in (13.4) we obtain for $x_b(t)$ the initial value problem

$$\ddot{x}_b(t) + a^2 x_b(t) + b x_b(t)^2 + \frac{2bv_0 \sin(at)}{a} x_b(t) + b\left(\frac{v_0 \sin(at)}{a}\right)^2 = 0,$$
$$x_b(0) = 0,$$
$$\dot{x}(0) = 0.$$

We will solve the differential equation for $x_b(t)$ in MAPLE, using its powerful tool for the solution of differential equations – the truncated power series. In our case the truncated power series is expanded at $t = 0$. It is applied to polynomials of degrees less or equal than the order N of the truncated power series. The MAPLE statements follow (we will use Order = 12)

```
>                 # initial value problem B
> eqb := diff(xb(t), t, t) + a^2*xb(t) + b*xb(t)^2
>          + 2*b*v0*sin(a*t)*xb(t)/a + b*(v0*sin(a*t)/a)^2 = 0;
```

$$eqb :=$$
$$(\frac{\partial^2}{\partial t^2} \mathrm{xb}(t)) + a^2 \, \mathrm{xb}(t) + b \, \mathrm{xb}(t)^2 + 2 \, \frac{b \, v0 \sin(a\,t)\,\mathrm{xb}(t)}{a} + \frac{b \, v0^2 \sin(a\,t)^2}{a^2}$$
$$= 0$$

```
> incondb := xb(0) = 0,D(xb)(0) = 0:
>          # determination of Order truncated power series
> Order := 12:
> solb := dsolve({eqb, incondb}, xb(t), series);
```

$$solb := \mathrm{xb}(t) = -\frac{1}{12} b\, v0^2\, t^4 + \frac{1}{72} b\, v0^2\, a^2\, t^6 + \frac{1}{252} b^2\, v0^3\, t^7 -$$

$$\frac{1}{960} b\, v0^2\, a^4\, t^8 - \frac{5}{6048} b^2\, v0^3\, a^2\, t^9 + \frac{1}{362880} b\, v0^2\, (17\, a^6 - 60\, b^2\, v0^2)\, t^{10}$$

$$+\frac{1}{12320} a^4\, b^2\, v0^3\, t^{11} + O(t^{12})$$

```
> polb := convert(rhs(solb), polynom):
```

The polynomial approximation of the truncated power series $s_N(t)$ given by MAPLE is equivalent to

$$p_N(t) = -\frac{bv_0^2\, t^4}{12} \left(1 - \frac{a^2 t^2}{6} - \frac{v_0\, bt^3}{21} + \frac{a^4 t^4}{80} + \right.$$
$$\left. \frac{5v_0\, a^2 bt^5}{504} + \frac{v_0^2\, t^6 b^2}{504} - \frac{17t^6 a^6}{30240} - \frac{3ba^4 v_0\, t^7}{3080} \cdots \right) \quad (13.9)$$

The approximation of the solution $\tilde{x}_N(t) = x_a(t) + p_N(t)$ will enable us to determine the approximation $\tilde{p}_N(t)$ of the pressure $p(t)$ using $\tilde{x}_N(t)$ in Equation (13.6).

Prior to choosing the degree of the polynomial for obtaining a sufficiently accurate approximation of $x(t)$ we have to estimate the region of convergence of the truncated power series. For this we will use some simple heuristic arguments. From the definition of the solution $x_a(t)$ and the pressure $p(t)$ it follows that the pressure pulse is positive for

$$t < t_a = \frac{\pi}{a}. \quad (13.10)$$

This is the first condition to determine the interval T where we will find the solution $x(t)$ and the pressure $p(t)$.

Let us suppose that each term in the parentheses in (13.9) has an absolute value less than 1. We can see that fourth and further terms in the parentheses in (13.9) are the combinations of the second and third term multiplied by a constant, which has an absolute value less than 1. Consequently only the second and the third term in the parentheses in (13.9) play a role. This assumption leads to the second condition

$$t < t_{coef} = \min \left(\left| \frac{a}{\sqrt{6}} \right|^{-1}, \left| \frac{v_0 b}{21} \right|^{-1/3} \right). \quad (13.11)$$

It is clear that for these values of t, the absolute value of the term in the parenthesis in (13.9) will be less than 1. If condition (13.11) is valid condition (13.10) is also true. From (13.7) follows that $|x_a(t)| \le v_0/a$, and the calibration makes sense for the pressure $p(t) > 0$. To obtain the positive approximation $\tilde{x}_N(t)$ the solution $x(t)$ from (13.7),(13.8) and (13.11), we use the third condition:

$$t < t_b = \left(\frac{abv_0}{12}\right)^{-1/4}. \tag{13.12}$$

The conditions (13.11) and (13.12) determine the maximum interval $[0, t_m)$, where $t_m = \min(t_{coef}, t_b)$. Where the truncated power series will convergence to the solution $x_b(t)$ and the approximation of the solution $\tilde{x}_N(t)$ will converge to the solution $x(t)$ of our problem.

How do we choose the degree of the truncated power series to obtain the given accuracy? We will apply the following criterion

$$|a_N| + |a_{N-1}| \leq \varepsilon |\sum_{i=0}^{N} a_i|, \tag{13.13}$$

where a_i are the terms of the power series and ε denotes the maximum of the tolerable local relative approximation error [2]. This criterion uses the sum of the absolute values of the two last terms of the expression of the power series of order N as a measure of the truncation error, and relates it to the absolute value of the truncated power series. In our case, from the criterion (13.13) it follows that $\varepsilon = O(10^{-3})$ for the order of the power series equal to 12.

Example

To verify this approach we will solve the above initial value problem with the experimental values from the collection of J. Buchar $v_0 = 5\,[ms^{-1}]$, $\alpha = 3.8E9\,[Nm^{-2}]$, $\beta = 3E10\,[Nm^{-2}]$, $l_0 = 2E{-}2\,[m]$, $d = 1.5E{-}2\,[m]$, $M = 5\,[kg]$. The following MAPLE program will plot graphs with the numerical solution and solutions using polynomial approximation of truncated power series of different degrees of the above initial value problem.

```
>                # experimental constants
> alpha := 3.8*10^9:   beta := 30*10^9:  d  := 0.015:
> 10    := 0.02:       M    := 5:        v0 := 5:
>                # determination of parameter a, b
> a := sqrt(alpha*Pi*d^2/(4*M*10)):  b := beta*Pi*d^2/(4*M*10^2):
>                # initial value problem B
> eqb      := diff(xb(t), t, t) + a^2*xb(t) + b*xb(t)^2
>                + 2*b*v0*sin(a*t)*xb(t)/a + b*(v0*sin(a*t)/a)^2=0:
> incondb := xb(0) = 0, D(xb)(0) = 0:

>                # numerical solution
> numsol := dsolve({eqb, incondb}, xb(t), numeric):
> sol    := x -> subs(numsol(x), xb(t) + v0*sin(a*t)/a):
> solp   := t-> (sol(t)/10)*(alpha + beta*(sol(t)/10))/10^9:
>                # estimation of radius of convergence
```

FIGURE 13.2.
Numeric and Acceptable Polynomial Approximation of Solution.

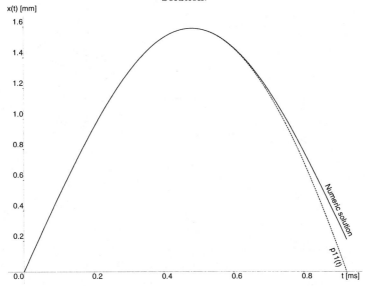

FIGURE 13.3.
Numeric and Acceptable Polynomial Approximation of Pressure.

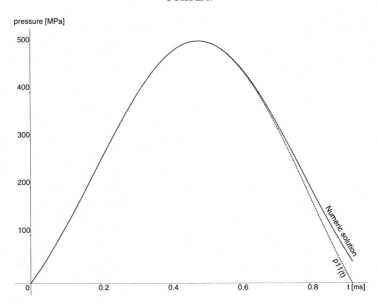

FIGURE 13.4.
Comparison of Numeric and Polynomial Approximations of Solutions.

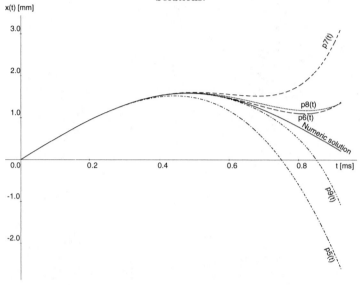

FIGURE 13.5.
Comparison of Numeric and Polynomial Approximations of Pressure.

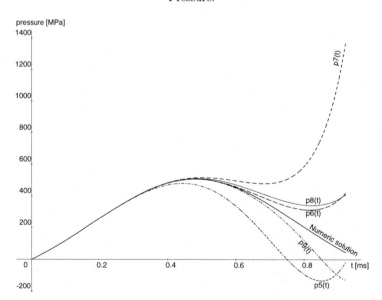

```
> tm       := min(evalf(sqrt(6)/a),evalf(21/(v0*b))^(1/3),
>                  evalf(21/(a*b*v0))^(1/4)):
>                 # saving graphs of numerical solution
> X[num] := plot(sol, 0..tm, thickness=3):
> P[num] := plot(solp, 0..tm, thickness=3):
>                 # analytic solution of first equation
> xa := v0*sin(a*t)/a:
>                 # Order of truncated power series approximation
>                 # and preparation of graphs
> ORD1 := [seq(i, i = 6..10), 12]:
> ORD2 := [seq(i, i = 6..10)]:
>                 # polynomial approximation of 5th - 9th
>                 # and 11th degree of  solution
> for ord in ORD1 do:
>     Order := ord:
>     sersol := dsolve({eqb, incondb}, xb(t), series):
>     pol := convert(rhs(sersol), polynom):
>     sol := pol + xa:
>     solp := (sol/10)*(alpha + beta*sol/10)/10^9:
>                 # saving graphs
>     X[ord] := plot(sol, t = 0..tm):
>     P[ord] := plot(solp, t = 0..tm):
> od:

>                 # display graphs of solutions and pressures
>                 # numeric and acceptable power series solution
> with(plots):
> display({X[num], X[12]});
> display({P[num], P[12]});
>                 # comparison of numeric and polynomial solution
> display({X[num], seq(X[i], i = ORD2)});
> display({P[num], seq(P[i], i = ORD2)});
```

13.4   Conclusions

From the graphs we can see that polynomials of degrees less than 8 are not acceptable because of the poor accuracy at the end of the interval. But for the purpose of calibration in a smaller interval (from zero to the maximum value of the pressure) polynomials of degrees higher than 5 are acceptable. It follows that for the accurate approximation of the solution $x(t)$ of our problem with the given experimental data a polynomial of 11th degree yields the best approximation.

References

[1] M. W. CHANG and W. M. ZHU, *Dynamic calibration of chamber pressure measuring equipment*, Proceedings 13th international symposium on ballistics, 1, 1992, pp. 443 – 450.

[2] H. J. HALIN, *The applicability of Taylor series methods in simulation*, Proceedings of 1983 Summer Computer Simulation Conference, 2, 1983, pp. 1032 – 1076.

[3] G. RESCH, *Dynamic conformity and dynamic peak pressure accuracy – to new features of pressure transducers*, Proceeding 14th Transducer Workshop, 1987, pp. 1 – 10.

[4] D. G. ZILL and M. R. CULLEN, *Advanced Engineering Mathematics*, PWS-KENT, Boston, 1992.

Chapter 14. Heat Flow Problems

S. Bartoň and J. Hřebíček

14.1 Introduction

The heat flow problems are a very important part of thermodynamics. The solution of these problems influences many other technical problems. The most important equation describing heat flow rules, is the heat equation (Fourier equation)

$$a^2 \left(\frac{\partial^2 T}{\partial x^2} + \frac{\partial^2 T}{\partial y^2} + \frac{\partial^2 T}{\partial z^2} \right) = \frac{\partial T}{\partial t}. \tag{14.1}$$

The difficulty of the solution of the Equation (14.1) depends on the difficulty of the boundary and initial conditions. We can distinguish two main groups of heat flow problems.

1. The steady state problems – in this case temperature T is not a function of the time t and Equation (14.1) becomes the Laplace equation. We shall solve this problem in the first part of this chapter.

2. Time-dependent problems – these problems are usually solved numerically. We will show, that for the very simple one-dimensional problem $T = T(x,t)$ with a constant boundary condition, $T(0,t) = $ const, and our initial condition of $T(x,0) = $ const, an analytical solution can be found.

14.2 Heat Flow through a Spherical Wall

Consider a hollow sphere of inner and outer radii r_1 and r_2 respectively. Let T_1 and T_2 be the constant temperatures over the inner and outer spherical surfaces respectively, and let k be the finite thermal conductivity of the wall material. We will find a steady state solution to the heat flow rate and temperature distribution as a function of the radius r.

Let x,y,z be the coordinates of the Euclidean space. If the center of the spherical body is at the origin $O(0,0,0)$, then the temperature $T = T(x, y, z)$ is a function of the position within the body. From the symmetry and the boundary conditions, it is clear that $T(x, y, z) = $ const for all the points (x, y, z) of the sphere $x^2 + y^2 + z^2 = r^2$, $r_1 \leq r \leq r_2$. Therefore, $T = T(r)$ is a function of r only, and our problem is one–dimensional. We shall solve it in MAPLE by two methods which differ by the physical model used.

FIGURE 14.1. *Cross Section of the Spherical Wall.*

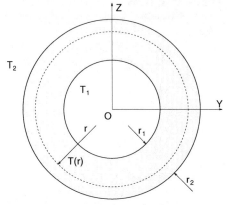

14.2.1   A Steady State Heat Flow Model

Let $\dot{q}$ be the heat flow density. According to the Fourier law

$$\dot{q} = -k \operatorname{grad} T \quad [Wm^{-2}; Wm^{-1}K^{-1}, K]. \tag{14.2}$$

Consider now the spherical surface with radius r, $r_1 \leq r \leq r_2$, with the center at the origin and area $A = 4\pi r^2$. Then the heat $\dot{Q}$ flowing through it is given by

$$\dot{Q} = \dot{q}A \quad [W]. \tag{14.3}$$

In steady state this quantity is independent of the radius r, because there are no heat sources inside the body. In the one-dimensional problem $\operatorname{grad} T$ can be replaced by dT/dr. Then from (14.2) and (14.3) we obtain

$$\dot{Q} = -k4\pi r^2 \frac{dT}{dr}. \tag{14.4}$$

Denoting

$$a = \frac{\dot{Q}}{4k\,\pi} \tag{14.5}$$

we get

$$\frac{dT}{dr} = -a\frac{1}{r^2}, \tag{14.6}$$

where the parameter a is unknown. The general solution of this equation is a function of the parameter a and that can be derived by hand.

$$T = T(r; a). \tag{14.7}$$

We now show how simple it is to solve Equation (14.6) analytically:

```
> sol := dsolve(diff(T(r), r) = -a/(r^2), T(r));
```

$$sol := \mathrm{T}(r) = \frac{a + \_C1\, r}{r}$$

```
> T := unapply(rhs(sol), r);
```

$$T := r \rightarrow \frac{a + \_C1\, r}{r}$$

We shall use the initial conditions to determine a and _C1.

$$T(r_1) = T_1, \qquad T(r_2) = T_2 \tag{14.8}$$

```
> sol := solve({T(r1) = T1, T(r2) = T2}, {a, _C1});
```

$$sol := \{\_C1 = \frac{-T2\, r2 + T1\, r1}{-r2 + r1}, \; a = \frac{r2\, r1\, (-T1 + T2)}{-r2 + r1}\}$$

```
> assign(sol);
> normal(T(r));
```

$$\frac{-r1\, r2\, T1 + r1\, r2\, T2 - r\, T2\, r2 + r\, T1\, r1}{(-r2 + r1)\, r}$$

Now the temperature function is completely known. For given temperatures on the outside and on the inside surfaces the temperature between the spherical walls is a function of r only.

Finally, we can compute the desired heat flow $\dot{Q}$ based on the definition of the parameter a using (14.5)

```
> Q := 4*k*Pi*a;
```

$$Q := 4\,\frac{k\, \pi\, r2\, r1\, (-T1 + T2)}{-r2 + r1}$$

14.2.2   Fourier Model for Steady State

In two-dimensional steady state problems we have

$$\frac{\partial^2 T}{\partial x^2} + \frac{\partial^2 T}{\partial y^2} = 0, \tag{14.9}$$

or

$$\frac{\partial^2 T}{\partial r^2} + \frac{2}{r}\frac{\partial T}{\partial r} + \frac{\cos\varphi}{r^2 \sin\varphi}\frac{\partial T}{\partial\varphi} + \frac{1}{r^2}\frac{\partial^2 T}{\partial\varphi^2} = 0. \tag{14.10}$$

in polar coordinates r, φ.

Since in our case $T = T(r)$ is only a function of r, Equation (14.10) simplifies to

$$\frac{\partial^2 T}{\partial r^2} + \frac{2}{r}\frac{\partial T}{\partial r} = 0 \qquad (14.11)$$

with initial conditions (14.8). Solving (14.11) we obtain the temperature $T(r)$ as a function of the radius r:

```
> restart;
> inicon := T(r1) = T1, T(r2) = T2:
> deq := diff(T(r), r, r) + diff(T(r) ,r)*2/r = 0;
```

$$deq := (\frac{\partial^2}{\partial r^2}\,\mathrm{T}(r)) + 2\,\frac{\frac{\partial}{\partial r}\,\mathrm{T}(r)}{r} = 0$$

```
> sol := simplify(dsolve({deq,inicon}, T(r)));
```

$$sol := \mathrm{T}(r) = \frac{r\ T2\ r2 - r\ T1\ r1 + r1\ r2\ T1 - r1\ r2\ T2}{(-r1 + r2)\,r}$$

By defining the distribution of the temperature $T = T(r)$ inside the spherical body for the given parameters r_1, T_1, r_2, T_2, we obtain the analytical solution of our problem by the Fourier model approach:

```
> T := unapply(rhs(sol), r);
```

$$T := r \to \frac{r\ T2\ r2 - r\ T1\ r1 + r1\ r2\ T1 - r1\ r2\ T2}{(-r1 + r2)\,r}$$

which is the same as in the first approach. To get the desired heal flow $\dot{Q}$ we use (14.4)

```
> Q := simplify(-4*Pi*r^2*k*diff(T(r, r1, T1, r2, T2),r));
```

$$Q := -4\,\frac{\pi\,k\,r2\,r1\,(-T1 + T2)}{-r1 + r2}$$

and again obtain the same result as with the first method.

14.2.3 MAPLE Plots

Let us plot now the distribution of the temperature $T = T(r; r_1, T_1, r_2, T_2)$, and print the heat flow $\dot{Q}$ for given values of the finite thermal conductivity of the wall material k, for inner and outer radii r_1 and r_2, and temperature T_1 and three different temperatures T_2. We will use:
$k = 12\ [Wm^{-1}K^{-1}]$, $T_1 = 400\ [K]$, $T_2 = 300, 200, 100\ [K]$, $r_1 = 0.1\ [m]$, $r_2 = 0.4\ [m]$.

We will compute the general analytical solution using the statements from the Section 14.2.1. We will then assign the values of k, T_1, r_1, r_2 and in a loop we will compute the three plots. Finally, we will plot all three graphs into one picture. Here is the complete MAPLE code for this job:

```
> r1 := 0.1: T1 := 400: r2 := 0.4: k := 12:
> TTs := [100, 200, 300]:
> for T2 in TTs do;
>       Tpl[T2] := T(r);
>       Qpl[T2] := evalf(Q);
> od:
> plot({seq(Tpl[T2], T2 = TTs)}, r = r1..r2);
```

FIGURE 14.2. *Temperature Distribution.*

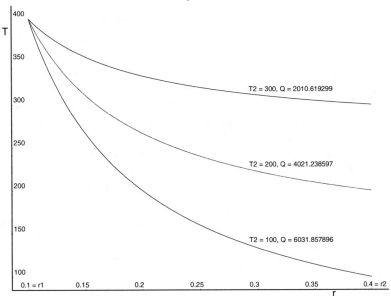

14.3   Non Stationary Heat Flow through an Agriculture Field

Consider an agriculture field that is being heated by the sun. Let us assume that at the beginning the temperature distribution in the field is given by T_f and the temperature T_s on the surface of the field is always constant (cf. Figure 14.3). Let $k > 0$ be the finite thermal conductivity of the field. The non stationary temperature distribution as function of the field depth x and the time t is to be determined.

FIGURE 14.3.
Boundary and Initial Condition for the Agricultural Field.

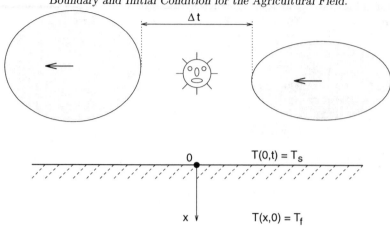

Let x, y, z be the coordinates of the Euclidean space. Let the origin be on the surface of the field and the axis x be positive oriented into the field. The temperature $T = T(x, y, z, t)$ is then a function of the point (x, y, z) inside the field and the time t. From symmetry and the boundary conditions, $T(x, y, z, t) = \mathrm{const}(t)$ for all the points (x, y, z) on the plane $x = c$, $0 \leq c \leq d_{max}$ where d_{max} is the maximum depth of the field. Therefore, we can consider $T = T(x, t)$, as a function of only two variables x and t. It is well known that the temperature distribution $T = T(x, t)$ is the solution of the heat equation [1]

$$a^2 \frac{\partial^2 T}{\partial x^2} = \frac{\partial T}{\partial t} \qquad (14.12)$$

where we denote $a^2 = k$.

Let us try to find the solution of (14.12) in the form of the function $V(u)$ of one variable u

$$V(u) = T(x, t) \qquad (14.13)$$

where

$$u = u(x, t) = \frac{x}{2a\sqrt{t}}. \qquad (14.14)$$

The Equation (14.14) substitutes two variables x and t by only one variable u. We can see, that this substitution converts the partial differential Equation (14.12) into a second order differential equation of one variable. This idea can be realized in MAPLE by:

```
> restart;
> Subs := u = x/(2*a*sqrt(t)):
> T(x,t):=V(rhs(Subs));
```

$$T(x,\ t) := \mathrm{V}(\frac{1}{2}\frac{x}{a\sqrt{t}})$$

```
> eq := a^2*diff(T(x,t), x, x) = diff(T(x, t), t);
```

$$eq := \frac{1}{4} \frac{(D^{(2)})(V)(\frac{1}{2}\frac{x}{a\sqrt{t}})}{t} = -\frac{1}{4} \frac{D(V)(\frac{1}{2}\frac{x}{a\sqrt{t}})x}{a\,t^{3/2}}$$

```
> eq := subs(x = solve(Subs, x), eq)*t*4;
```

$$eq := (D^{(2)})(V)(u) = -2\,D(V)(u)\,u$$

Thus we obtain a ordinary differential equation of the second order

$$\frac{d^2V(u)}{du^2} = -2u\frac{dV(u)}{du}, \qquad (14.15)$$

which can be solved by hand using $\frac{dV(u)}{du} = W$.

$$\frac{d^2V(u)}{du^2} = -2u\frac{dV(u)}{du} \implies \frac{dW}{du} = -2uW$$

$$\implies \frac{dW}{W} = -2u\,du \implies \text{can be integrated}$$

$$\implies \ln|W| = -u^2 + C_1' \implies \frac{dV(u)}{du} = C_1e^{-u^2}$$

$$\implies dV(u) = C_1e^{-u^2}du \implies V(u) = C_1\int_0^u e^{-p^2}dp + C_2.$$

The solution of (14.15) with MAPLE is

```
> Sol := dsolve(eq, V(u));
```

$$Sol := V(u) = \_C1 + \_C2\,\mathrm{erf}(u)$$

As we see, MAPLE is able to solve this equation. Both solutions are the same, because erf(u) (*The Gauss or error function*) is defined as

$$\mathrm{erf}(u) = \frac{2}{\sqrt{\pi}}\int_0^u e^{-p^2}dp. \qquad (14.16)$$

Now the variables x and t can be substituted back in the solution:

```
> T := unapply(rhs(subs(u=rhs(Subs),Sol)), x, t);
```

$$T := (x, t) \to \_C1 + \_C2\,\mathrm{erf}(\frac{1}{2}\frac{x}{a\sqrt{t}})$$

To determine the constants _C1 and _C2 the initial and boundary conditions (cf. Figure 14.3) are used. Let us prescribe the boundary condition $T(0,t) = \lim_{x\to0_+} T(x,t) = T_s$ and the initial condition $T(x,0) = \lim_{t\to0_+} T(x,t) = T_f$ i.e. for $t = 0$ the temperature of the whole field is constant and equal to T_f. For MAPLE we assume $a \geq 0$ and $x \geq 0$.

```
> assume(a >= 0): assume(x >= 0):
> BounCon := Ts = limit(T(x,t), x = 0, right);
```

$$BounCon := Ts = \_C1$$

FIGURE 14.4.
Field Temperature for Depths 0.0, 0.01, 0.05, 0.1, 0.2, and 0.3[m].

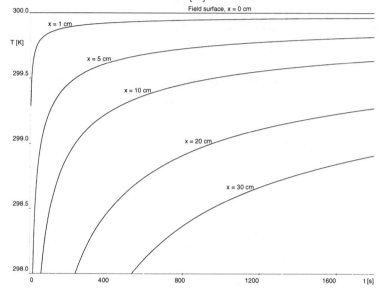

FIGURE 14.5.
Field Temperature as the Function of the Depth and Time.

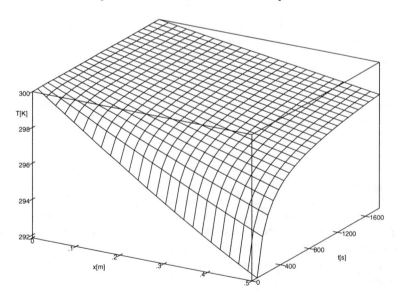

```
> IniCon := Tf = limit(T(x, t), t = 0, right);
```

$$IniCon := Tf = \_C1 + \_C2$$

```
> Sol := solve({IniCon, BounCon}, {_C1, _C2});
```

$$Sol := \{\_C1 = Ts, \_C2 = Tf - Ts\}$$

```
> assign(Sol);
> T(x,t);
```

$$Ts + (Tf - Ts)\,\mathrm{erf}(\frac{1}{2}\,\frac{x}{a}\,\frac{}{\sqrt{t}})$$

```
> evalb(diff(T(x,t),x,x)  - diff(T(x,t), t)/a^2=0);
```

$$true$$

By rewriting, the final form is obtained,

$$T(x,t) = \frac{2(T_f - T_s)}{\sqrt{\pi}} \int_0^{\frac{x}{2a\sqrt{t}}} e^{-p^2}\,dp + T_s.$$

14.3.1 MAPLE Plots

Let us compute and plot the temperature distribution inside the agriculture field which has the maximum depth $d_{max} = 0.5[m]$ and the thermal conductivity $k = a^2 = 0.003[m^2 s^{-1}]$. The sun is shining for half an hour $\Delta t = 1800[s]$ during which the temperature on the surface of the field becomes $T_s = 300[K]$. The initial temperature of the field is $T_f = 285[K]$.

```
> Ts := 300: Tf := 285: a := sqrt(0.003):
> DVEC := [0, 0.01, 0.05, 0.1, 0.2, 0.3]:
> for d in DVEC do  Td[d] := T(d, t) od:
> plot({seq(Td[d], d = DVEC)}, t = 0..1800, 298..300);
> plot3d(T(x, t), x = 0..0.5, t = 0..1800,
>        axes = boxed, orientation = [-50, 50]);
```

References

[1] D. G. ZILL and M. R. CULLEN, *Advanced Engineering Mathematics*, PWS-KENT, Boston, 1992.

Chapter 15. Modeling Penetration Phenomena

J. Buchar and J. Hřebíček

15.1 Introduction

Penetration phenomena are of interest in numerous areas (cf. [3]). They are often associated with the problem of nuclear waste containment and with the protection of spacecraft or satellites from debris and/or meteorite impact. Formally the penetration is defined as the entrance of a projectile into the target without completing its passage through the body. The penetration phenomenon can be characterized according to the impact angle, the geometry and material characteristics of the target and the projectile and the striking velocity. In this chapter we limit our considerations to the normal incidence impact of a long rod on a semi-infinite target. This model corresponds for example to the situation in which a very thick armor is penetrated by a high kinetic energy projectile. The most efficient method for the solution of this problem is the numerical modeling by the finite element method. Many finite element computer programs are capable of handling very complex material constitutive relations. These programs are expensive and often require a substantial amount of execution time. This is the main reason why simple one-dimensional theories still have considerable value. Such theories also provide insight into the interactions between the physical parameters and their relationship to the outcome of the event. These interactions are usually difficult to ascertain from the computer analyses mentioned above. As a result, simple theories often provide the basis for the design of experiments, refining the areas in which numerical modeling by finite element methods is applied. In this spirit, we will investigate some penetration models which are treated using MAPLE.

15.2 Short description of the penetration theory

The list of publications on the study of penetration is long, although the first models of this process were formulated during the sixties [8]. The model of the penetration of a semi-infinite target consists of the following four regimes:

1. Transient shock regime: On initial contact, shock waves are formed both in the target and the projectile as schematically shown in Figure 15.1. The material response is mainly governed by its compressibility and density. The shock pressures are usually high enough to cause extensive plastic flow, melting and vaporization.

FIGURE 15.1.
A scheme of wave pattern during the transient shock regime.

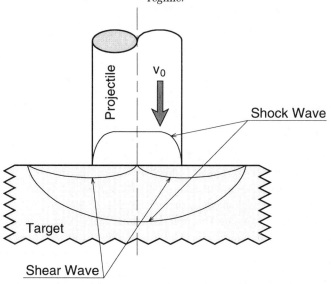

2. Steady state regime: During this period of penetration the projectile erodes, while at the same time causing the formation of a crater. This stage is usually described by a modification of Bernoulli's equation.

3. Cavitation regime: After the complete consumption of the projectile, the crater continues to expand as a result of its own inertia and the energy trapped when the target pressures reduce to the order of the material strength. This reduction is a consequence of the multiple wave reflections [9].

4. Recovery regime: The dimension of the crater reduces slightly as a result of the elastic rebound. This rebound may produce tensile stresses high enough to cause spallation, as can be often observed on the recovered targets [5].

The main features of this model have been verified computationally [11]. The numerical simulation yields the most complex analysis of the penetration process. Simulation is extremely expensive because it requires many hours of CPU time on state-of-the-art machines. This procedure is thus neither convenient for the evaluation of series of ballistic experiments nor for an extensive study of the role of the penetration characteristics which were mentioned in the introduction. These facts strongly encourage the development of new analytical models of the penetration process. There are many analytical models of the penetration, for a summary cf. [2]. The most widely used is a one-dimensional model which is known as the Tate model [10] even though it was independently proposed by Alekseevskii [1]. The main features of this model which has become a standard reference for long-rod penetration of thick targets are reviewed in the following sections.

15.3   The Tate-Alekseevskii model

This model assumes that the projectile is rigid, except for a thin region near the target-projectile interface where erosion can occur. This region has no spatial extent but contains the interface between target and projectile. The behavior of this interface is described by a modification of the Bernoulli equation

$$\frac{1}{2}\rho_p(v-u)^2 + Y_p = \frac{1}{2}\rho_t u^2 + R_t \tag{15.1}$$

where ρ_p and ρ_t are the projectile and target densities, respectively, v is the velocity of the rear part of the projectile, u is the penetration velocity, Y_p is the strength of the projectile and R_t is defined as the target resistance in the one-dimensional formulation.

The penetrated projectile is decelerated by a force transmitted from the interface along the length of the projectile. The deceleration is given by

$$\rho_p l \frac{dv}{dt} = -Y_p \tag{15.2}$$

where $l = l(t)$ is the current length of the projectile. The change of the rod length in consequence of the erosion process is given by Newton's equation of motion for the undeformed section of the rod

$$\frac{dl}{dt} = -(v-u). \tag{15.3}$$

Equations (15.1)–(15.3) represent a set of equations for the three unknown functions $l = l(t), v = v(t)$ and $u = u(t)$ with the initial conditions

$$v(0) = v_0, \qquad l(0) = L \tag{15.4}$$

where v_0 is an impact velocity and L is the initial length of the projectile. The penetration velocity u can be obtained from Equation (15.1):

```
>      # Bernoulli equation
> eq := (1/2)*rho[p]*(v-u)^2+Yp=(1/2)*rho[t]*u^2+Rt:
> pen := solve(eq, u);
```

$$pen := \frac{1}{2}\frac{2\,\rho_p\,v + 2\,\sqrt{-2\,\rho_t\,Rt + 2\,\rho_t\,Yp + \rho_t\,\rho_p\,v^2 + 2\,\rho_p\,Rt - 2\,\rho_p\,Yp}}{-\rho_t + \rho_p},$$

$$\frac{1}{2}\frac{2\,\rho_p\,v - 2\,\sqrt{-2\,\rho_t\,Rt + 2\,\rho_t\,Yp + \rho_t\,\rho_p\,v^2 + 2\,\rho_p\,Rt - 2\,\rho_p\,Yp}}{-\rho_t + \rho_p}$$

```
>           # We take the second solution
>           # because of u < v and simplify it
> u_v[1] := simplify(pen[2]):
```

The value of u for the same density of the projectile and target, i.e. $\rho_p = \rho_t$, may be computed as follows

```
> simplify(series(u_v[1],rho[p]=rho[t]),symbolic):
> u_v[2] := subs({rho[t]=rho,rho[p]=rho},op(1,")));
```

$$u\_v_2 := \frac{1}{2} \frac{\rho v^2 - 2Rt + 2Yp}{\rho v}$$

Thus, for $\rho_p \neq \rho_t$ the penetration velocity u is given by

$$u = \frac{\rho_p v - \sqrt{2(R_t - Y_p)(\rho_p - \rho_t) + \rho_p \rho_t v^2}}{\rho_p - \rho_t} \tag{15.5}$$

and for $\rho_p = \rho_t$ by

$$u = \frac{\rho v^2 + 2(Y_p - R_t)}{2\rho v} . \tag{15.6}$$

Inserting Equation (15.2) into (15.3) we obtain

$$\frac{dl}{dv} = \frac{l(v)\rho_p(v - u)}{Y_p}. \tag{15.7}$$

Integration of differential Equation (15.7) with the initial condition $l(v_0) = L$ by MAPLE

```
>           # Definition of initial value problem
> deq := diff(l(v), v)=l(v)*rho[p]*(v-u(v))/Yp:
> incond := l(v0) = L:
> len[1] := simplify (rhs (dsolve({deq, incond}, l(v))));
```

$$len_1 := e^{\left(-\frac{\rho_p \int_{v0}^{v} -u+u(u)\,du}{Yp}\right)} L$$

leads to the current length l of the projectile, which may be simplified for equal penetrator and target densities, i.e., $\rho_p = \rho_t$:

```
> subs(u=unapply(u_v[2],v),{rho[p]=rho,rho[t]=rho},len[1]):
> len[2] := simplify (expand ("));
```

$$len_2 := e^{\left(-1/4 \frac{\rho(v0-v)(v0+v)}{Yp}\right)} v^{\left(-\frac{-Rt+Yp}{Yp}\right)} v0^{\left(\frac{-Rt+Yp}{Yp}\right)} L$$

The penetration depth pe of the rod at time T is

$$pe = \int_0^T u(t)\,dt. \tag{15.8}$$

Using (15.2) yields

$$pe = \frac{\rho_p}{Y_p} \int_{v(T)}^{v_0} u\, l\, dv \, . \tag{15.9}$$

Inserting the formulas for $u(v)$ and $l(v)$ into Equation (15.9), we obtain the general penetration equation

$$\frac{pe}{L} = F(v_0, R_t, Y_p, \rho_t, \rho_p). \tag{15.10}$$

The above integral cannot be solved analytically in the general case, but we describe in the following subsections the exact solution of (15.10) for the choice of special parameter values.

15.3.1   Special case $R_t = Y_p$

For equal target resistance R_t and projectile strength Y_p, we compute $\lim_{T\to\infty} p(T)$. Obviously $T \to \infty$ corresponds to $v \to 0$.

```
> u_v[3] := simplify (subs(Rt=Yp, u_v[1]), symbolic):
> subs(u=unapply(u_v[3],v),len[1]):
> l := simplify(", symbolic):
> pe := simplify(int(l*u_v[3], v=0..v0)):
```

The normalized penetration depth pe/L is then given by

$$\frac{pe}{L} = \frac{Y_p}{\rho_p} \frac{\left(\rho_p - \sqrt{\rho_p \rho_t}\right)}{\left(\rho_t - \sqrt{\rho_p \rho_t}\right)} \left(-1 + e^{\frac{\rho_p\left(\rho_t - \sqrt{\rho_p \rho_t}\right)v_0^2}{2(\rho_p - \rho_t)Y_p}}\right). \tag{15.11}$$

For $\rho_p = \rho_t = \rho$, Equation (15.11) may be further simplified by

```
> simplify(limit(pe,rho[t]=rho[p]),symbolic):
```

Thus we obtain

$$\frac{pe}{L} = \frac{Y_p}{\rho}\left(1 - e^{-\frac{\rho v_0^2}{4 Y_p}}\right) \, . \tag{15.12}$$

15.3.2   Special case $\rho_p = \rho_t = \rho$

For equal penetrator and target densities, we distinguish the following cases

Case 1: $R_t \geq Y_p$. According to (15.1), penetration occurs only if

$$\frac{1}{2}\rho_p v_0^2 + Y_p \geq R_t. \tag{15.13}$$

For $R_t = mY_p$, the velocity v ranges therefore from v_0 to v_m, where

$$v_m = \sqrt{\frac{2(m-1)Y_p}{\rho_p}}. \tag{15.14}$$

The normalized penetration depth can be determined as follows:

```
> vm := sqrt(2*(m-1)*Yp/rho):
> simplify (subs(Rt=m*Yp,len[2]*u_v[2])):
> pe := (rho/Yp)*int(", v=vm..v0):
```

which may be neatly written as

$$\frac{pe}{L} = \frac{v_0 L}{2Y_p} \int_{vm}^{v0} e^{-\frac{\rho\,(v_0-v)(v_0+v)}{4\,Y_p}} \left(\rho\,v^2 + 2Y_p - 2\,mY_p\right) \left(\frac{v}{v_0}\right)^m v^{-2} dv.$$

For $m \in \mathbf{N}$, the above expression may be further simplified, see for example p_1 and p_3:

```
> pe1 := simplify(eval(subs(m=1, pe)));
```

$$pe1 := -L\left(-1 + e^{(-1/4\,\frac{\rho\,v0^2}{Y_p})}\right)$$

```
> pe3 := simplify(eval(subs(m=3, pe)));
```

$$pe3 := \frac{L\left(v0^2\,\rho - 8\,Yp + 4\,Yp\,e^{(-1/4\,\frac{-4\,Yp+v0^2\,\rho}{Yp})}\right)}{\rho\,v0^2}$$

Case 2: $R_t < Y_p$. There are two velocity regimes with different penetration processes described by the inequalities

$$Y_p \geq \frac{1}{2}\rho_t u^2 + R_t, \qquad \text{for } u = v_0, \tag{15.15}$$

$$Y_p < \frac{1}{2}\rho_t u^2 + R_t, \qquad \text{for } u < v_0. \tag{15.16}$$

If (15.15) holds, then the rod penetrates like a rigid body with the impact penetration velocity v_0. Equality in (15.15) holds for the limit velocity

$$v_l = \sqrt{\frac{2(Y_p - R_t)}{\rho_t}}. \tag{15.17}$$

For $v_0 \leq v_l$ the decelerating pressure $\tilde{P}$ at the tip of the rod is given by $\tilde{P} = \frac{1}{2}\rho_t v_0^2 + R_t$. Inserting pressure $\tilde{P}$ into Equation (15.2) we obtain

$$\rho_p L \frac{dv}{dt} = -\left(\frac{1}{2}\rho_t v_0^2 + R_t\right). \tag{15.18}$$

Integration within boundaries v_0 and zero velocity leads to

$$p_r = L\left(\frac{\rho_p}{\rho_t}\right)^2 \ln\left(1 + \frac{\rho_t v_0^2}{2R_t}\right). \tag{15.19}$$

For $v_0 > v_l$ the penetration depth p_f is calculated by integrating (15.9) from v_0 to v_l. The total penetration pe is then given by

$$pe = p_r + p_f. \tag{15.20}$$

The basic Equations (15.1)–(15.3) of this penetration theory are highly non-linear. Generally, their solution can only be computed numerically.

MAPLE enables us to obtain an approximation of the solution in the form of a series. A detailed analysis performed in [4] reveals that an approximation of the penetration velocity u by a 5th order polynomial u_{appr} is sufficient.

```
> restart:
>    # Bernoulli equation for same densities rho=rho[t]=rho[p]
> eqe := (1/2)*rho*(v(t)-u)^2+Yp=(1/2)*rho*u^2+Rt:
> u := solve(eqe,u):
> deqe1 := rho*l(t)*diff(v(t), t)=-Yp:
> deqe2 := diff(l(t), t)=-(v(t)-u):
> incond := l(0)=L, v(0)=v0:
> sol := dsolve({deqe1, deqe2, incond}, {l(t), v(t)}, series):
>    # Polynomial approximation of v(t) and l(t)
> v := op(2,op(select(x->op(1,x)='v(t)',sol))):
> l := op(2,op(select(x->op(1,x)='l(t)',sol))):
>    # Polynomial approximation of u(t)
> uappr := convert(series(v/2+(Yp-Rt)/(rho*v),t=0,6),polynom):
>    # First few coefficients of uappr
> series(uappr,t=0,2);
```

$$\left(\frac{1}{2} v0 + \frac{Yp - Rt}{\rho \, v0}\right) + \left(-\frac{1}{2} \frac{Yp}{\rho \, L} - \frac{(-Yp + Rt) \, Yp}{\rho^2 \, v0^2 \, L}\right) t + \mathrm{O}(t^2)$$

We shall use this model in the following example

```
> rho := 7810: # Target and penetrator density
> L   := 0.25: # Length of rod
>    # Target resistance Rt = 900*10^6
>    # Strength of projectile Yp = 900*10^6
> u := subs(Rt=900*10^6,Yp=900*10^6,uappr):
> plot3d(u,t=0..0.0004,v0=1000..2500,
>          axes=boxed,orientation=[-30,60]);
```

where the experimental data was taken from the shot No. B19, see [4]: $\rho = \rho_t = \rho_p = 7810[kgm^{-3}]$, $L = 0.25[m]$, $R_t = Y_p = 9.10^8[Nm^{-2}]$. The results of this computation (displayed in Figure 15.2) were compared with results of two-dimensional numerical simulations by the finite element program AUTODYN 2D. It is shown in [4] that the results are essentially the same. The following penetration model takes into account especially the erosion processes at the rod nose.

15.4   The eroding rod penetration model

Assume that a cylindrical rod with initial cross-sectional area A_p, diameter d_p and length l_p, normally impacts the semi-infinite target with an impact velocity v_0 (cf. Figure 15.3). The target is made with a material of strength Y_t and density ρ_t, and the rod is made with a material of strength Y_p and density ρ_p.

FIGURE 15.2.
Penetration velocity versus time and impact velocity v_0.

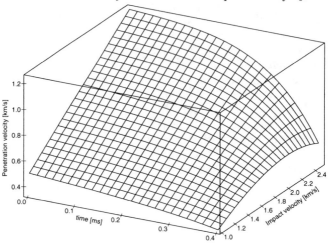

FIGURE 15.3. *Penetration Process.*

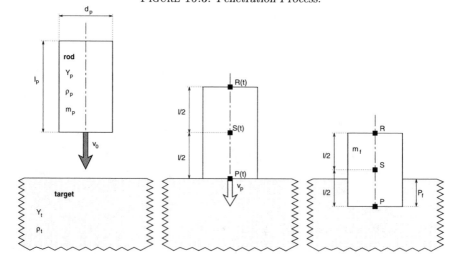

It was experimentally found that the rod erodes during the penetration. This erosion process occurs at the rod head, and leads to the decrease of an initial mass of the rod, $m_p = \rho_p A_p l_p$, and to a reduction of the rod length l_p. The mass of the rod lost per unit time can be expressed as

$$\dot{m} = \rho_p A_p \dot{l},$$

where $m(t)$ and $l(t)$ are the instantaneous mass and length of the rod, and the dots represent differentiation with respect to time t. Since $\dot{m} < 0$, the use of the symbol $\dot{m}_p = |\dot{m}|$ is convenient, i.e.

$$\dot{l} = -\frac{\dot{m}_p}{\rho_p A_p}. \tag{15.21}$$

Let $S(t)$ be the position of the center of the rod which is, at any time t, the mid point between the head position $P(t)$, and the rear position $R(t)$. Let us neglect the stress propagation phenomena in the impacting rod. The head and the rear velocity are given by

$$\dot{P} = \dot{S} + \frac{\dot{l}}{2} = \dot{S} - \frac{\dot{m}_p}{2\rho_p A_p}, \tag{15.22}$$

$$\dot{R} = \dot{S} - \frac{\dot{l}}{2} = \dot{S} + \frac{\dot{m}_p}{2\rho_p A_p}. \tag{15.23}$$

Consider the situation at the head of the rod. The mass Δm_p is lost while the velocity of the center of the rod, (the center of mass), becomes $\dot{S} + \Delta \dot{S}$. The momentum balance is

$$[(m - \Delta m_p)(\dot{S} + \Delta \dot{S}) + \dot{P}\Delta m_p] - m\dot{S} = -F\Delta t, \tag{15.24}$$

where F is the force exerted by the target opposing the motion of the rod. In the limit, for $\Delta t \to 0$, the Equation (15.24) has the form

$$m\frac{d\dot{S}}{dt} = (\dot{S} - \dot{P})\dot{m}_p - F. \tag{15.25}$$

Inserting Equation (15.22) into (15.25) we obtain

$$m\frac{d\dot{S}}{dt} = a_p \dot{m}_p^2 - F, \quad \text{where we denote} \quad a_p = \frac{1}{2\rho_p A_p}.$$

In [7] it is shown that the force F can be expressed as

$$F = a + b\dot{P}^2. \tag{15.26}$$

The constant, a, depends on the target strength, $a = 3Y_t A_p$. It express the target resistance against to the projectile immediately after impact, when the projectile head stops for a certain time period [6]. The second term, b, is the kinetic energy per unit volume or force per unit area, i.e., $b = \rho_t A_p/2$.

The erosion of the rod can be described, (see [4]), as

$$\dot{m} = -\mu_0 \dot{P} \tag{15.27}$$

or

$$\dot{m}_p = \mu_p \dot{P}, \tag{15.28}$$

respectively, where μ_p is an unknown parameter. Since the penetration depth $P(t)$ at $t = 0$ is zero, i.e. $P(0) = 0$, the solution of the differential Equation (15.27) is

$$m(t) = m_p - \mu_p P(t). \tag{15.29}$$

Let P_f be the final penetration depth of the rod head and let m_f be its residual mass at time t_f when the penetration is completed. Then the parameter μ_p has the value

$$\mu_p = \frac{m_p - m_f}{P_f}, \tag{15.30}$$

and it follows that the maximum value of μ_p is $\mu_{max} = m_p/P_f$. Putting Equation (15.28) into (15.22) we obtain

$$\dot{S} = (1 + \mu_p a_p)\dot{P} = k\dot{P}, \quad \text{where} \quad k = k(\mu_p) = 1 + \mu_p a_p. \tag{15.31}$$

Using Equations (15.26), (15.28), (15.29), and (15.31) the Equation (15.25) can be modified to the form

$$(m_p - \mu_p P)k\frac{d\dot{P}}{dt} = -(a + c\dot{P}^2), \quad \text{where} \quad c = c(\mu_p) = b - a_p\mu_p^2. \tag{15.32}$$

The solution of the differential Equation (15.32) with respect to the initial conditions

$$P(0) = 0, \tag{15.33}$$
$$\dot{P}(0) = v_p, \tag{15.34}$$

gives the time dependency of the penetration depth, i.e. $P = P(t)$. The initial penetration velocity v_p in (15.34) is determined according to Tate theory [6]

$$v_p = \frac{v_0}{2} + \frac{Y_t - Y_p}{\rho_t v_0}. \tag{15.35}$$

Because we cannot measure the function $P(t)$ experimentally it is necessary to look for some simplifications of Equation (15.32). If we take into account that

$$\frac{d\dot{P}}{dt} = \frac{d\dot{P}}{dP}\frac{dP}{dt}$$

the Equation (15.32) can be transformed to

$$-\frac{k}{a + c\dot{P}^2}\dot{P}\frac{d\dot{P}}{dP} = \frac{1}{m_p - \mu_p P}.$$

with the initial condition (15.34).

The above initial value problem can be easily solved by MAPLE.

```
> restart;
> de := -k*DP(P)*diff(DP(P), P)/(a + c*DP(P)^2) =1/(mp - mu*P);
```

$$de := -\frac{k\,\mathrm{DP}(P)\,(\frac{\partial}{\partial P}\mathrm{DP}(P))}{a + c\,\mathrm{DP}(P)^2} = \frac{1}{mp - \mu\,P}$$

```
> ic := DP(0) = vp:
> pen := dsolve({de,ic},DP(P));
```

$$pen := \mathrm{DP}(P) = -\frac{\sqrt{c\left(-a + \frac{e^{(\frac{c\ln(mp^2 - 2\,mp\,\mu\,P + \mu^2\,P^2)}{\mu\,k})}(a + vp^2\,c)}{(mp^{(\frac{c}{\mu\,k})})^2}\right)}}{c},$$

$$\mathrm{DP}(P) = \frac{\sqrt{c\left(-a + \frac{e^{(\frac{c\ln(mp^2 - 2\,mp\,\mu\,P + \mu^2\,P^2)}{\mu\,k})}(a + vp^2\,c)}{(mp^{(\frac{c}{\mu\,k})})^2}\right)}}{c}$$

```
> sol := op(2,pen[2]);
```

$$sol := \frac{\sqrt{c\left(-a + \frac{e^{(\frac{c\ln(mp^2 - 2\,mp\,\mu\,P + \mu^2\,P^2)}{\mu\,k})}(a + vp^2\,c)}{(mp^{(\frac{c}{\mu\,k})})^2}\right)}}{c}$$

```
> sol1 := subs(2*c/(mu*k)=1/e,sol);
```

$$sol1 := \frac{\sqrt{c\left(-a + \frac{e^{(\frac{c\ln(mp^2 - 2\,mp\,\mu\,P + \mu^2\,P^2)}{\mu\,k})}(a + vp^2\,c)}{(mp^{(\frac{c}{\mu\,k})})^2}\right)}}{c}$$

After some simple manipulations with the solution we obtain

$$\dot{P} = \sqrt{\frac{a}{c}\left((1 + \frac{c}{a}v_p^2)(1 - \frac{\mu_p}{m_p}P)^{1/\varepsilon} - 1\right)}, \qquad (15.36)$$

where

$$\varepsilon = \varepsilon(\mu_p) = \frac{\mu_p k}{2c}$$

is the dimension less erosion rate.

When the embedment occurs in a semi-infinite target, $\dot{P} = 0$ at time t_f, and from (15.36) it follows that

$$P_f = \frac{m_p}{\mu_p}\left(1 - (1 + \frac{c(\mu_p)}{a}v_p^2)^{-\varepsilon(\mu_p)}\right). \qquad (15.37)$$

Equation (15.37) is more convenient for the evaluation of the parameter μ_p than Equation (15.30) because of the difficulties to determine the residual rod mass m_f. The data $[P_f, v_0]$ are then easily available during each terminal-ballistic experiment.

The simulation of the ballistic experiments involves the following steps:

1. The determination of the parameter μ_p. If the theory of penetration outlined in the this chapter is valid, then μ_p should be a constant for each rod-target material combination and independent of the striking velocity v_0. We will solve the Equation (15.37) in MAPLE for μ_p numerically and also evaluate the limit of (15.37) for $\mu_p \to 0$. This limit corresponds to the penetration of a rigid projectile. We will define the quantities k, c, and ε as functions of the variable μ. Also P_f is defined as a function of μ. We will denote μ_p the solution of the equation $P_f(\mu) = P_f$.

2. The evaluation of the velocity $\dot{P}$ as a function of P along the penetration path $P \in [0, P_f]$.

3. The solution of the initial value problem (15.32), (15.33), and (15.34) in order to obtain the function $P(t)$ and its graph.

4. The investigation of the influence of the target strength Y_t on the parameters and functions determined at points 1–3. The value of Y_t may be affected by the strain rate which is much higher than that occurring at a conventional test of materials.

15.5   Numerical Example

We will use MAPLE for the solution of a concrete problem. Let us consider the following experimental data from the shot No. C06 from the collection of J. Buchar [4]: $m_p = 1.491$ $[kg]$, $v_0 = 1457.9$ $[ms^{-1}]$, $\rho_t = \rho_p = 7810$ $[kgm^{-3}]$, $d_p = 0.0242$ $[m]$, $Y_t = 9.7 \cdot 10^8$ $[Nm^{-2}]$, $Y_p = 15.81 \cdot 10^8$ $[Nm^{-2}]$, $P_f = 0.245$ $[m]$.

We now present the MAPLE program for the solution of the above example, and the four steps of the ballistics experiment analysis. In the first step we will find μ_p, and then we will plot the Figure 15.4, the graph of $P_f(\mu)$ on the interval $[0, \mu_{max}]$. In the second step we will plot the Figure 15.5, the graph of the penetration velocity $\dot{P}$ as a function of P and the determined parameter μ_p, see (15.37). In the third step we will solve the above initial value problem numerically and plot the Figure 15.6, the graph of the solution $P(t)$ in the chosen interval $(0, t_f)$, $t_f = 0.0006$. In the fourth step we will plot the Figure 15.7, the graph of the penetration depth as a function of Y_t and v_0.

FIGURE 15.4.
The Influence of μ on the Penetration Depth.

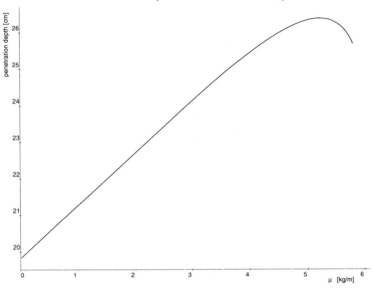

FIGURE 15.5.
Dependence of Penetration Velocity on Depth.

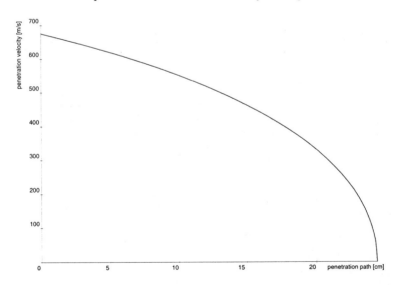

FIGURE 15.6. *Dependence of Depth on Time.*

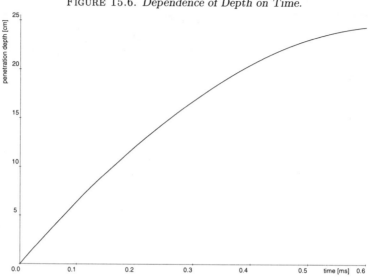

FIGURE 15.7.
The Influence of Target Strength and Impact Velocity.

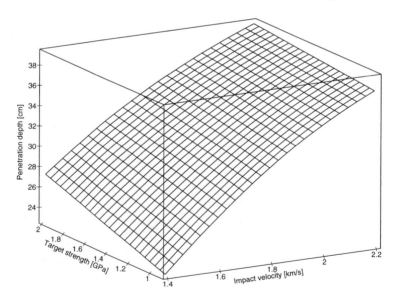

```
>       # Projectile and target data and the final penetration depth
>  rho[t] := 7810:        mp   := 1.491:  Yt   := 9.70*10^8:
>  rho[p] := rho[t]:      dp   := 0.0242: Yp   := 15.81*10^8:
>  Ap     := Pi*dp^2/4: v0   := 1457.9:
>  pf     := 0.245:
>
>       # 1st step - Determination of mup and graph penetration depth
>       # vs erosion rate mu, the definition of constants ap, a, b
>       # and initial penetration velocity
>  ap  := 1/(2*rho[p]*Ap): a   := 3*Yt*Ap:
>  b   := rho[t]*Ap/2:       vp := v0/2+(Yt-Yp)/(rho[t]*v0):
>       # The definition of functions k, c, eps and Pf of mu
>  k   := mu->1+ap*mu:      c := mu->b-ap*mu^2:
>  eps := mu->mu*k(mu)/(2*c(mu)):
>  Pf  := mu->mp*(1-(1+c(mu)*vp^2/a)^(-eps(mu)))/mu:
>       # - Graph of influence of mu on the penetration depth
>  plot(Pf(mu),mu=0..6);
>       # Determination of limit
>  Pfzero := limit(Pf(mu),mu=0):
>       # Solution of the equation Pf(mu)=pf
>  mup := fsolve(Pf(mu)=pf,mu):
>
>       # 2nd step - Graph of the penetration velocity
>       # along the penetration path
>  dP := P -> sqrt(a*((1 + c(mup)*vp^2/a)*
>                 (1 - mup*P/mp)^(1/eps(mup)) - 1)/c(mup)):
>  plot(dP(P),P=0..pf);
>
>       # 3rd step - Graph of  the time dependence of the penetration
>  tf := 0.0006:
>       # The evaluation of the time dependence
>       # of the penetration and its velocity
>  deq := (mp - mup*p(t))*k(mup)*diff(p(t), t, t) =
>         -(a + c(mup)*diff(p(t),t )^2):
>  incond := p(0) = 0, D(p)(0) = vp:
>  F := dsolve({deq, incond}, p(t), numeric):
>  plots[odeplot](F, [t, p(t)], 0..tf);
>
>       # 4th step - The evaluation of the influence
>       # of target strength and impact velocity
>  vp1 := (vi,y) -> vi/2 + (y - Yp)/(rho[t]*vi):
>  depth := (vi,y) ->
>               mp*(1 - (1 + c(mup)*vp1(vi,y)^2/a)^(-eps(mup)))/mup:
>  opt := orientation=[-145,60], axes=boxed:
>  plot3d(depth(vi,y),vi=1400..2200,y=900*10^6..2000*10^6,opt);
```

15.6   Conclusions

It has been demonstrated that methods of symbolic computation which are represented by MAPLE are very effective for the study of the penetration phenomena. The given examples show the usefulness of MAPLE for the solution of penetration problems. Namely, the described procedure enables us to evaluate the influence of both target and projectile strength on the penetration depth. Previously, this problem was solved numerically. The analysis of the projectile's erosion is also very easy using MAPLE, other advantages of this approach are described in [4].

Acknowledgment

The authors wish to thank the Czech Found Agency for supporting this work under contract No 106/94/0567.

References

[1] V. P. ALEKSEEVSKII, *Penetration of a rod into a target at high velocity*, Combustion, explosion and shock waves, 2, 1966, pp. 63 – 66.

[2] C. E. ANDERSON JR., S. R, BODNER, *Ballistic impact: the status of analytical and numerical modeling*, Int. J. Impact Engng., 7, 1988, pp. 9 – 35.

[3] Z. BÍLEK and J. BUCHAR, *The behavior of metals under high rates of strain (in Czech)*, Academia, Praha, 1984.

[4] J. BUCHAR, M. LAZAR, S. ROLC, *On the penetration of steel targets by long rods*, Acta Techn. CSAV 39, 1994, pp. 193 – 220.

[5] D. R. CHRISTMAN, J. W. GEHRING, *Analysis of high-velocity projectile penetration mechanics*, J. Appl. Phys., 27, 1966, pp. 63 – 68.

[6] J. D. CINNAMON ET AL., *A one - dimensional analysis of rod penetration*, Int. J. Impact Engng., 12, pp. 145 – 166, 1992.

[7] J.DEHN, *A unified theory of penetration*, Int. J. Impact Engng., 5, 1987, pp. 239 – 248.

[8] W. HERRMANN, J. S. WILBECK, *Review of hypervelocity penetration theories*, Int. J. Impact Engng., 5, 1987, pp. 307 – 322,

[9] V. K. LUK, M. J. FORRESTAL, D.E. AMOS, *Dynamical spherical cavity-expansion of strain-hardening materials*, J. Appl. Phys., 58, 1991, pp. 1 – 6.

[10] A. TATE, *A theory of the deceleration of long rods after impact*, J. Mech. Phys. Solids 15, 1967, pp. 287 – 399.

[11] J. A. ZUKAS, *High velocity impact dynamics*, Wiley Inter science, New York, 1990.

Chapter 16. Heat Capacity of System of Bose Particles

F. Klvaňa

16.1 Introduction

In this chapter we will study a system of Bose particles with nonzero mass, (for example He^4), in low temperature near absolute zero, when an interesting effect of superfluidity, (or also superconductivity for electrons) appears.

Let us consider a quantum system of N noninteracting boson particles, (particles with integer spin), with nonzero mass m. Let the system be in a volume V and in equilibrium with its surroundings, with absolute temperature T. Then the mean value of number of particles with energies in interval $[\epsilon, \epsilon + d\epsilon]$ is given by the eqn. [1, 2]

$$< n(\epsilon) > \cdot d\epsilon = \frac{g(\epsilon) \cdot d\epsilon}{e^{(\epsilon - \mu)/\theta} - 1}, \qquad \epsilon \in [0, \infty), \tag{16.1}$$

where $< n(\epsilon) >$ has the meaning of an energy density of number of particles, $g(\epsilon)$ is a density of states, which in our case is given by eqn.

$$g(\epsilon) = g_0 V \frac{4\pi}{h^3} 2m^{3/2} \cdot \epsilon^{1/2} = \lambda \cdot \sqrt{\epsilon}, \tag{16.2}$$

where $\theta = kT$, (k is the Boltzmann constant), is a statistical temperature.

The quantity $\mu = \mu(T, V, N)$ is the chemical potential, and its dependence on the thermodynamics parameters (T, V, N) is given by the normalization condition

$$N = \int_0^\infty < n(\epsilon) > d\epsilon = \lambda \int_0^\infty \frac{\epsilon^{1/2} d\epsilon}{e^{(\epsilon - \mu)/\theta} - 1} = \lambda \cdot \theta^{3/2} \int_0^\infty \frac{x^{1/2} \cdot dx}{e^{x - \mu/\theta} - 1} \tag{16.3}$$

For a system of bosons it can be proved that $\mu \leq 0$ and $\mu = 0$ for $T = 0$. But it follows from Equation (16.3) that $\mu = 0$ for $\theta = \theta_c > 0$ (θ_c is called critical temperature). So for $\theta = \theta_c$

$$N = \lambda \cdot \theta_c^{3/2} \int_0^\infty \frac{x^{1/2} \cdot dx}{e^x - 1}. \tag{16.4}$$

Using a definition of the Riemann zeta function

$$\zeta(\nu) = \frac{1}{\Gamma(\nu)} \int_0^\infty \frac{x^{\nu-1}dx}{e^x - 1}$$

(16.5)

we can write (16.4) in the form

$$N = \lambda \cdot \Gamma\left(\frac{3}{2}\right) \cdot \theta_c^{3/2} \cdot \zeta\left(\frac{3}{2}\right)$$

(16.6)

For $\theta < \theta_c$ we have to use a different model for our system. In this case μ has to be zero and the normalization condition has the form

$$N - N' = \lambda \int_0^\infty \frac{\epsilon^{1/2}d\epsilon}{e^{\epsilon/\theta} - 1}$$

(16.7)

where N' is a number of particles, which are "condensed" in an one-particle ground state with $\epsilon = 0$. These condensed particles have a special behavior called *superfluidity*. So μ as a function of θ has two analytically different regions: $\mu = 0$ for $\theta \leq \theta_c$, and for $\theta \geq \theta_c$, μ is given as a solution of Equation (16.3).

Our task will be to find temperature dependency of the constant volume heat capacity per particle C for such a system. By the definition

$$C = \frac{\partial E_p}{\partial T} = k \cdot \frac{\partial E_p}{\partial \theta}$$

(16.8)

where E_p is an internal energy per particle given by equation

$$E_p = \frac{\lambda}{N} \int_0^\infty \epsilon < n(\epsilon) > d\epsilon = \frac{\lambda}{N} \cdot \theta^{5/2} \int_0^\infty \frac{x^{3/2}dx}{e^{x-\mu/\theta} - 1}.$$

(16.9)

Let us define the function (usually called Bose-Einstein's integral – see [2])

$$F_{be}(z, \nu) = \frac{1}{\Gamma(\nu)} \int_0^\infty \frac{x^{\nu-1}dx}{e^{x+z} - 1}, \qquad z \geq 0, \quad \nu > 0;$$

(16.10)

as we can see, $F_{be}(0, \nu) = \zeta(\nu)$.

We can now formulate the following steps of solution for our problem:

1. Calculation of temperature dependence of μ:

   for $\theta \leq \theta_c$: set $\mu = 0$,

   for $\theta \geq \theta_c$: μ is the solution of the Equation (16.3 and 16.6)

$$N = \lambda \cdot \theta_c^{3/2} \cdot \Gamma\left(\frac{3}{2}\right) \cdot \zeta\left(\frac{3}{2}\right) = \lambda \cdot \theta^{3/2} \cdot \Gamma\left(\frac{3}{2}\right) \cdot F_{be}\left(-\frac{\mu}{\theta}, \frac{3}{2}\right),$$

(16.11)

which leads to equation

$$F_{be}\left(-\frac{\mu}{\theta}, \frac{3}{2}\right) = \left(\frac{\theta_c}{\theta}\right)^{3/2} \cdot \zeta\left(\frac{3}{2}\right).$$

(16.12)

It is useful to transform the previous equation into canonical form using the dimensionless variables

$$y = \frac{\theta}{\theta_c}, \qquad \mu_d = \frac{\mu}{\theta_c}.$$

Then Equation (16.12) becomes

$$F_{be}\left(-\frac{\mu_d}{y}, \frac{3}{2}\right) = \frac{\varsigma\left(\frac{3}{2}\right)}{y^{3/2}} \tag{16.13}$$

2. Calculation of the energy per particle, which is by (16.9) (and using 16.6) equal to

$$Ep = \frac{\Gamma\left(\frac{5}{2}\right)}{\Gamma\left(\frac{3}{2}\right)} \cdot \frac{\theta^{5/2}}{\theta_c^{3/2}} \cdot \frac{1}{\varsigma\left(\frac{3}{2}\right)} \begin{cases} \varsigma\left(\frac{5}{2}\right) & \text{for } \theta \leq \theta_c \\ F_{be}\left(-\frac{\mu}{\theta}, \frac{5}{2}\right) & \text{for } \theta \geq \theta_c \end{cases} \tag{16.14}$$

Using a dimensionless quantity $\epsilon_d = E_p/\theta_c$ instead of E_p we can rewrite (16.14) in the form

$$\epsilon_d = \frac{\Gamma\left(\frac{5}{2}\right)}{\Gamma\left(\frac{3}{2}\right)} \cdot \frac{y^{5/2}}{\varsigma\left(\frac{3}{2}\right)} \begin{cases} \varsigma\left(\frac{5}{2}\right) & \text{for } y \leq 1 \\ F_{be}\left(z, \frac{5}{2}\right) & \text{for } y \geq 1 \end{cases} \tag{16.15}$$

where $z = -\mu_d/y$ is given as a solution of Equation (16.13)

$$y^{3/2} \cdot F_{be}\left(z, \frac{3}{2}\right) = \varsigma\left(\frac{3}{2}\right) \tag{16.16}$$

3. Calculation of heat capacity in units of k

$$c = \frac{C}{k} = \frac{d\epsilon_d}{dy} \tag{16.17}$$

Let us note, that for $y \gg 1$ a classical expression for energy per particle, $E_p = 3\theta/2$, holds, from which there follows that $c = 3/2$.

16.2 MAPLE Solution

The basic problem for solution of our task is to represent the function

$$F_{be}(z, \nu) = \frac{1}{\Gamma(\nu)} \int_0^\infty \frac{x^{\nu-1} dx}{e^{x+z} - 1}, \qquad z \geq 0, \quad \nu > 0; \tag{16.18}$$

This function is a generalization of the Riemann zeta function and is not implemented in MAPLE. Because we are interested in the behavior of our system for low temperatures when $z \to 0$, we have to use the power expansion of F_{be}

ALGORITHM 16.1. *Function* $F_{be}(z, \nu)$.

```
'series/Fbe' := proc (z, nu::numeric, t) local n;
        # calculation of Bose-Einstein's  integral
        # for nu noninteger
        # 1/GAMMA(nu)*int(x^(nu - 1)/(e^(z + x) - 1), x = 0..inf)
    if type(nu, integer) or not (nu > 1) then
        ERROR('2-nd arg. expected to be noninteger and > 1');
    else
        # power expansion in z to order Order
        series((z)^(nu - 1)*GAMMA(1 - nu)
          + sum((-z)^n*Zeta(nu-n)/n!,n=0..Order -1)
          + O(t^Order),t);
    fi
end;
```

in z. MAPLE cannot derive this expansion directly from Equation (16.18). We have therefore to use the following expansion ([2, 3]) for noninteger $\nu > 1$:

$$F_{be}(z, \nu) = z^{\nu-1}\, \Gamma(1-\nu) + \sum_{n=0}^{\infty} (-z)^n\, \frac{\zeta(\nu-n)}{n!} \qquad (16.19)$$

which is convergent for all $z \geq 0$. Let us define a function series/Fbe(z,nu,t) (see Algorithm 16.1), which implements the evaluation of this series expansion of $F_{be}(z, n)$ in t using the standard MAPLE function *series* (see Help in MAPLE).

The second problem is to compute functions like $\mu(\theta)$ or $E_p(\theta)$, which have two analytically different regions. Physicists like to use one object to represent one physical quantity. So for the definition of the above quantities the best way is to define a functional analog of a statement $if - then - else$, for example of the form

$$If(< relation >, < vthen >, < velse >),$$

which returns the expression $< vthen >$ or $< velse >$ depending on a boolean value of the relation $< relation >$ (support of the definition of piecewise functions will be in the next version of MAPLE). Because we will need to calculate derivatives of this function, it will again be a good idea to use the standard functions *diff* by defining functions diff/If (see Help in MAPLE). These functions are defined in Algorithm 16.2 .

The first step in the solution of our problem is to solve Equation (16.16) for z. In this equation $F_{be}(z, 3/2)$ is expressed in the form of series in z, which includes also the term $z^{1/2}$. But MAPLE can solve equations with pure power series only. For this reason we make the transformation $z \to w^2$ and solve the transformed equation

$$y^{3/2} \cdot F_{be}\left(w^2, \frac{3}{2}\right) = \zeta\left(\frac{3}{2}\right) \qquad (16.20)$$

for w. MAPLE can solve such equations and gives z as a series of $x = 1 - y^{-3/2}$. The chemical potential μ_d we can then express using the function If in the form

$$If\left(y \leq 1, 0, -y * z\right).$$

ALGORITHM 16.2. *Functions for Evaluated If.*

```
# syntax of calling:  If(bool,then,else)

'diff/If' := proc(bool, vthen, velse, x) ;
   If(bool, diff(vthen, x), diff(velse, x))
end;

If:= proc(bool, vthen, velse)
   if type(bool, 'relation'(numeric)) then
      if bool then vthen else velse fi
   else 'If(args)'
   fi
end;
```

A graph of this function is in Figure 16.1. The needed MAPLE statements follows:

```
>           # Basic equation for the chem. potential mud
>           # for y >= 1; z = -mud/y
> Eq := series(Fbe(z, 3/2), z, 4) = y^(-3/2)*Zeta(3/2);
```

$$Eq := (-2\sqrt{z}\,\sqrt{\pi} + \zeta(\frac{3}{2}) - z\,\zeta(\frac{1}{2}) + \frac{1}{2}\,z^2\,\zeta(\frac{-1}{2}) - \frac{1}{6}\,z^3\,\zeta(\frac{-3}{2}) + O(z^4)) = \frac{\zeta(\frac{3}{2})}{y^{3/2}}$$

```
>           # we choose the substitution
>           # y^(-3/2) -> 1 - x  and z-> w^2
>           #  and invert the series
>   solw := solve(series(Fbe(w^2, 3/2), w, 7)
>        = (1 - x)*Zeta(3/2), w):
>           # then
> z := series(solw^2, x, 4); # because  z = w^2 (z = -mu/theta);
```

$$z := \frac{1}{4}\frac{\zeta(\frac{3}{2})^2}{\pi}\,x^2 - \frac{1}{8}\frac{\zeta(\frac{3}{2})^3\,\zeta(\frac{1}{2})}{\pi^2}\,x^3 + O(x^4)$$

```
> z:= convert(z, polynom): # for next evaluation
>           # graf of mud=-z*y on y
> plot(If(y <= 1, 0, -y*subs(x = 1 - y^(-3/2), z)), y = 0..3);
```

The dimensionless internal energy is given by Equation (16.15), which we can express in the form

$$\epsilon_d = \frac{3}{2}\frac{y^{5/2}}{\zeta\left(\frac{3}{2}\right)} \cdot If\left(y \le 1, \zeta\left(\frac{5}{2}\right), F_{be}\left(z, \frac{5}{2}\right)\right) \tag{16.21}$$

where $F_{be}(z, 5/2)$ is given using z expressed in the previous step as a series in $x = 1 - y^{-3/2}$. Then the heat capacity c is given by Equation (16.17) and the

FIGURE 16.1.
Chemical Potential μ as a Function of $y = \theta/\theta_c$.

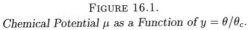
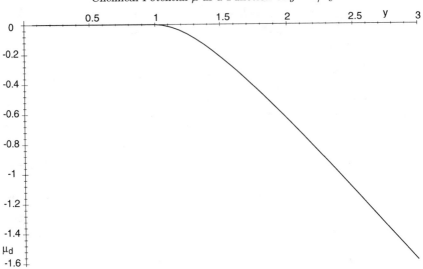

FIGURE 16.2. Heat Capacity c as a Function of $y = \theta/\theta_c$.

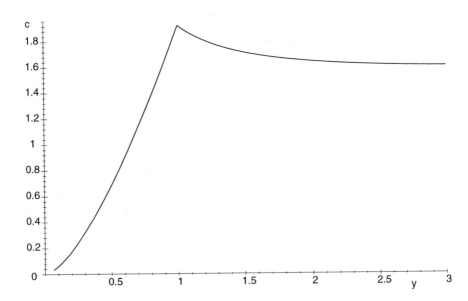

graph of this function $c(y)$ is shown in Figure 16.2. As we should expect, the derivative of $c(y)$ has a discontinuity at the point $y = 1$; this indicates, that this point $(\theta = \theta_c)$ is a point of the phase transition of the second order. The rest of MAPLE's program follows:

```
>                 # calculation of dimensionless internal energy
>                 # per particle Ed, expansion of Fbe(z,5/2) in y
> Fy := subs(x = 1 - y^(-3/2),
>                 convert(series(Fbe(z, 5/2), x, 4), polynom));
```

$$Fy := \zeta(\tfrac{5}{2}) - \frac{1}{4} \frac{\zeta(\tfrac{3}{2})^3 (1 - \frac{1}{y^{3/2}})^2}{\pi} + \left(\frac{1}{6} \frac{\zeta(\tfrac{3}{2})^3}{\pi} + \frac{1}{8} \frac{\zeta(\tfrac{3}{2})^4 \zeta(\tfrac{1}{2})}{\pi^2} \right) (1 - \frac{1}{y^{3/2}})^3$$

```
>                 # using the function If we can express energy for all y
> Ed := 3/2/Zeta(3/2)*y^(5/2)*If(y <= 1, Zeta(5/2), Fy):
>                 #heat capacity is (in units of k)
> C := diff(Ed, y):
>                 #and than we make graph of heat capacity
> plot(C, y = 0..3);
```

References

[1] C. KITTEL, *Elementary Statistical Physics*, John Wiley & Sons, Inc., 1958.

[2] F. LONDON, *Superfluids II.*, John Wiley & Sons, Inc., 1954.

[3] J.E. ROBINSON, *Note on the Bose-Einstein integral functions*, Physical Review, 83, 1951, pp. 678–679.

Chapter 17. Free Metal Compression

S. Bartoň

17.1 Introduction[1]

Compression is a widely used basic process in metal forming. If compression is performed using two plane platens (a special case of compression), the lateral face is distorted (see Figure 17.1). This successive forming operation is called die forming, and it is necessary to predict the distortion in advance in order to provide enough space to fit the distorted body into a specific die. In the following we will restrict our attention to metal rods with constant cross section, since nonconstant cross sections are not relevant in practice.

FIGURE 17.1. *Free Compression.*

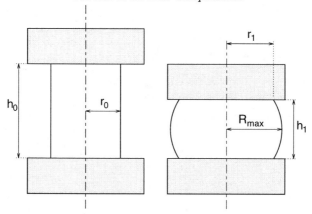

We will study the shape of the compressed body under the following assumptions:

1. The body volume V remains constant

$$V = S_0 H = \text{const.}, \tag{17.1}$$

where S_0 = initial base area, H = initial height of the rod.

---

[1]This chapter is a generalization of Chapter 17 in the previous editions of this book.

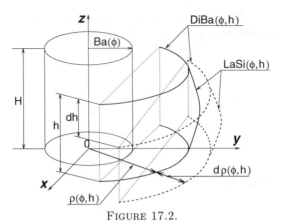

FIGURE 17.2.
The coordinate system and the base expansion

2. The initial rod is described in cylindrical coordinates. The length of the rod is parallel to the z-axis, and the origin of the coordinate system is at the center of mass of the base of the rod (see Figure 17.2).

3. The material flow has only radial and axial components, i.e., there is no tangential flow of the material.

4. If the initial is body is symmetric about the z-axis, the symmetry will be maintained for the distorted body.

5. The radial distance $\rho(\phi, h)$, which describes the perimeter of the base, is assumed to fulfill the differential equation

$$\rho_h(\phi, h) \equiv \frac{\partial \rho(\phi, h)}{\partial h} = C(h)\, k\, \frac{1}{\rho(\phi, h)}\ , \tag{17.2}$$

where h is the current height of the stamped body, $C(h)$ is a proportionality factor yet to be determined, and k is a measure of the friction between the platens and the rod:

(a) $k = 1$: No friction. The lateral side is parallel to the z-axis. The bases expand maximally. In this case we have:

$$S = S_0 \sqrt{H/h}\ ,$$

where S denotes the current base area and H is the initial height.

(b) $k = 0$: Unlimited friction. The bases keep their initial shape:

$$S = S_0 = \text{const.}\ .$$

The distortion of the lateral side is maximal.

(c) $0 < k < 1$. The two base areas expand and the lateral side is distorted.

$$S_0 < S < S_0\sqrt{H/h} \ .$$

Typical values for k are $0.3 \leq k \leq 0.7$.

6. Any axial section of the distorted body yields a symmetric curve; we will approximate it by a parabola (cf. [4]).

17.2   The Base expansion

Before we solve equation (17.2) with MAPLE, we will use equation (17.1) to determine the unknown function $C(h)$.

```
> de := diff(rho(h), h) = C(h)*k/rho(h):
> dsolve(de, rho(h));
```

$$\rho(h)^2 = 2\,k \int C(h)\,dh + \_C1$$

```
> Sol_de := solve(", rho(h));
```

$$Sol\_de := \sqrt{2\,k \int C(h)\,dh + \_C1}, \ -\sqrt{2\,k \int C(h)\,dh + \_C1}$$

```
> DiBa := Sol_de[1]:
```

To simplify further computations, we introduce a new variable ICh for $\int C(h)dh$. The constant of integration $\_C1$ is determined by looking at the uncompressed rod:

$$\mathrm{DiBa}(\phi, H) = \mathrm{Ba}(\phi) \ ,$$

where $\mathrm{Ba}(\phi)$ describes the perimeter of the base in polar coordinates (see Figure 17.2). Let's substitute

$$\mathrm{NH} := \int_0^H C(h)\,dh$$

```
> DiBa := Sol_de[1]:
> DiBa := subs(int(C(h), h) = ICh, DiBa):
> E1 := subs(ICh = NH, DiBa) = Ba(phi):
> _C1:=solve(E1,_C1);
```

$$\_C1 := -2\,k\,NH + \mathrm{Ba}(\phi)^2$$

If there is no friction between the platens and the rod, Equation (17.1) can be written in the form

$$h\int_{-\pi}^{\pi} \frac{\mathrm{DiBa}(\phi, h)^2}{2}\,d\phi = H\int_{-\pi}^{\pi} \frac{\mathrm{Ba}(\phi)^2}{2}\,d\phi \ ,$$

because the lateral side is not distorted and the cross section area is constant. ICh can now be determined with this equation:

```
> Eq := h*Int(subs(k = 1, DiBa)^2/2, phi = -Pi..Pi)
        = H*Int(Ba(phi)^2/2, phi = -Pi..Pi):
> EI1 := ICh = value(solve(Eq, ICh));
```

$$EI1 := ICh = \frac{1}{4} \frac{4\,h\,NH\,\pi - h\int_{-\pi}^{\pi} \mathrm{Ba}(\phi)^2\,d\phi + H\int_{-\pi}^{\pi} \mathrm{Ba}(\phi)^2\,d\phi}{h\,\pi}$$

Now we use the well known formula for the area of a closed curve given in polar coordinates

$$S_0 = \int_{-\pi}^{\pi} \frac{\mathrm{Ba}(\phi)^2}{2}\,d\phi$$

to simplify our problem, since S_0 is the initial cross section area of the rod. In a second step, we compute the derivative of the last result and obtain an expression for $C(h)$:

```
> EI2 := subs(int(Ba(phi)^2, phi = -Pi..Pi) = 2*So, EI1):
> EI3 := subs(ICh = int(C(h), h), EI2);
```

$$EI3 := \int C(h)\,dh = \frac{1}{4} \frac{4\,h\,NH\,\pi - 2\,h\,So + 2\,H\,So}{h\,\pi}$$

```
> EI4 := normal(diff(EI3, h));
```

$$EI4 := C(h) = -\frac{1}{2} \frac{H\,So}{h^2\,\pi}$$

Now it is simple to compute $NH = C(H)$, and thereby $\mathrm{DiBa}(\phi, h)$ is determined.

```
> NH := subs(h = H, rhs(EI4));
```

$$NH := -\frac{1}{2} \frac{So}{H\,\pi}$$

Our solution is verified by substituting it into the differential equation:

```
> DiBa:=simplify(subs(ICh = rhs(EI3), DiBa));
```

$$DiBa := \sqrt{\frac{-k\,So\,h + k\,So\,H + \mathrm{Ba}(\phi)^2\,h\,\pi}{h\,\pi}}$$

```
> de_test := simplify(subs(rho(h) = DiBa, C(h) = rhs(EI4), de),
>               symbolic):
> evalb(");
```

$$true$$

Let's apply this solution to the special case of a metal cylinder without friction: In this case, Equation (17.1) leads to

$$\mathrm{DiBa}(h) = \sqrt{n}\,r\ ,$$

where $r =$ the initial cylinder radius and $n = H/h =$ the compression ratio.

```
> DiBa:=subsop(1 = collect(expand(op(1, DiBa)), [So, Pi, k]),
>                 DiBa);
```

$$DiBa := \sqrt{\dfrac{(-1 + \dfrac{H}{h})\, k\, So}{\pi} + \text{Ba}(\phi)^2}$$

```
> Cylinder_test := evalb(r*sqrt(n) = simplify(subs(So = Pi*r^2,
>                 Ba(phi) = r, h = H/n, k = 1, DiBa),symbolic)));
```

$$Cylinder\_test := true$$

```
> save(DiBa, 'DiBa.sav');
```

17.3   Base Described by One and Several Functions

In this section we shall consider functions $\text{Ba}(\phi)$ other than $\text{Ba}(\phi) = \text{const.}$ as before. In this generalization, the perimeter of the base may be ellipse-shaped or triangle-shaped; note that the solution for the triangle can be easily adapted for any segmented base.

A common technical description of a triangle is by the length of its sides a, b, c. But for our purpose, we have to describe the vertexes of the triangle in polar coordinates and the origin of the system has to be at center of mass of the triangle. This can be done as follows: Rotate the triangle so that c is

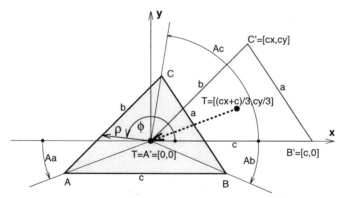

FIGURE 17.3. *Triangle at coordinate system*

parallel to the x-axis; then shift the triangle so that its center of mass is at the origin of the coordinate system. Finally compute the angles of the vertexes to determine the intervals of ϕ describing separate triangle sides. See Figure 17.3 for an illustration of these transformations.

```
> restart;
> with(linalg): with(plots):
> e1 := cx^2 + cy^2 = b^2:
> e2 := (cx - b)^2 + cy^2 = a^2:
> allvalues(solve({e1, e2}, {cx, cy}));
```

$$\{cy = \frac{1}{2}\frac{\sqrt{4b^2 - a^2}\,a}{b},\ cx = \frac{1}{2}\frac{2b^2 - a^2}{b}\},$$
$$\{cy = -\frac{1}{2}\frac{\sqrt{4b^2 - a^2}\,a}{b},\ cx = \frac{1}{2}\frac{2b^2 - a^2}{b}\}$$

```
> assign("[1]);
> A := [0, 0]:  B := [c, 0]:  C := [cx, cy]:
> T := (A + B+ C)/3:
> A := normal(A - T):  B := normal(B - T):  C := normal(C - T):
> Aa := abs(arctan(A[2]/A[1])):
> Ab := abs(arctan(B[2]/B[1])):
> Ac := Pi/2 - arctan(C[1]/C[2]):
```

To derive the equations for the sides of the triangle, we use the general equation of a line given by two vertexes $P \equiv [P_x, P_y]$ and $Q \equiv [Q_x, Q_y]$ in polar coordinates. The line segments describing the sides are further specified by the angles Aa, Ab and Ac.

```
> rho := (Q[y]*P[x] - Q[x]*P[y])/
>          ((P[x] - Q[x])*sin(phi) -(P[y] - Q[y])*cos(phi));
```

$$\rho := \frac{P_y\,Q_x - P_x\,Q_y}{\sin(\phi)\,Q_x - \sin(\phi)\,P_x - \cos(\phi)\,Q_y + \cos(\phi)\,P_y}$$

```
> r1 := normal(subs(P[x] = A[1], Q[x] = B[1], P[y] = A[2],
>                    Q[y] = B[2], rho));
> r2 := normal(subs(P[x] =B[1], Q[x] = C[1], P[y] = B[2],
>                    Q[y] = C[2], rho));
> r3 := normal(subs(P[x] = C[1], Q[x] = A[1], P[y] = C[2],
>                    Q[y] = A[2], rho));
```

$$r1 := -\frac{1}{6}\frac{\sqrt{4b^2 - a^2}\,a}{b\sin(\phi)}$$

$$r2 := \frac{1}{3}\frac{\sqrt{4b^2 - a^2}\,a\,c}{-2\sin(\phi)\,b^2 + \sin(\phi)\,a^2 + 2\sin(\phi)\,cb + \cos(\phi)\,\sqrt{4b^2 - a^2}\,a}$$

$$r3 := -\frac{1}{3}\frac{\sqrt{4b^2 - a^2}\,a\,c}{-2\sin(\phi)\,b^2 + \sin(\phi)\,a^2 + \cos(\phi)\,\sqrt{4b^2 - a^2}\,a}$$

The function $Ba(\phi)$ is composed of the three lines r_1, r_2 and r_3. To compute S_0 we make use of Heron's rule.

```
> o := (a + b + c)/2:
> Sb := sqrt(o*(o - a)*(o - b)*(o - c)):
> save(r1, r2, r3, Sb, Aa, Ab, Ac, 'Tri.sav');
```

The important variables are saved in the file `Tri.sav` for later use.

To continue, we have to substitute numerical values for a, b and c. Let us observe the expansion of the triangle rod described by $a = 8$, $b = 7$, $c = 6$. We shall plot the expanded triangle base for the following values of the inverse compression ratio $h/H := n = 1, 0.9, \ldots, 0.1$ for three friction coefficients $k = 1/3$, $2/3$ and 1. To compare this distortion we also plot the base expansion of the rod with an elliptic cross section with major semi-axis $a = 3.5$ and minor semi-axis $b = 1.85$, so that the cross section of both rods are equal. Heights, friction and compression coefficients for both rods are equal.

To describe an ellipse in the polar coordinate system, we use

$$\frac{x^2}{a^2} + \frac{y^2}{b^2} = 1 \quad \Longrightarrow \quad \rho = \frac{a\,b}{\sqrt{a^2\,\sin(\phi)^2 + b^2\,\cos(\phi)^2}}$$

```
> Sus := a = 8, b = 7, c = 6:   Suf := h = N*H, So = Sb:
> Aa := evalf(subs(Sus, Aa)):
> Ab := evalf(subs(Sus, Ab)):
> Ac := evalf(subs(Sus, Ac)):
> R1 := evalf(subs(Suf, Ba(phi) = r1, Sus, DiBa)):
> R2 := evalf(subs(Suf, Ba(phi) = r2, Sus, DiBa)):
> R3 := evalf(subs(Suf, Ba(phi) = r3, Sus, DiBa)):
> RE := subs(h = N*H, Ba(phi) = a*b/sqrt(a^2*sin(phi)^2
>           + b^2*cos(phi)^2), So = Pi*a*b, a = 3.5, b = 1.85, DiBa):
> SqN := [seq(1 - i*0.1, i = 0..8)]:
> with(plots):
```

The following MAPLE program generates the plots shown in Figure 17.4.

```
> for k from 1/3 by 1/3 to 1 do:
>    P1 := polarplot({seq(R1, N = SqN)}, phi= - Pi + Aa..-Ab):
>    P2 := polarplot({seq(R2, N = SqN)}, phi = -Ab..Ac):
>    P3 := polarplot({seq(R3, N = SqN)}, phi = Ac..Pi + Aa):
>    print(display({P1, P2 ,P3}, scaling = constrained,
>            view = [-6.5..6.5, -6..7]));
>    print(polarplot({seq(RE, N = SqN)}, phi = -Pi..Pi,
>              scaling = constrained, view = [-6.5..6.5, -6..7]));
> od:
> k := 'k':
```

FIGURE 17.4. *Distortion of triangle and elliptic bases*

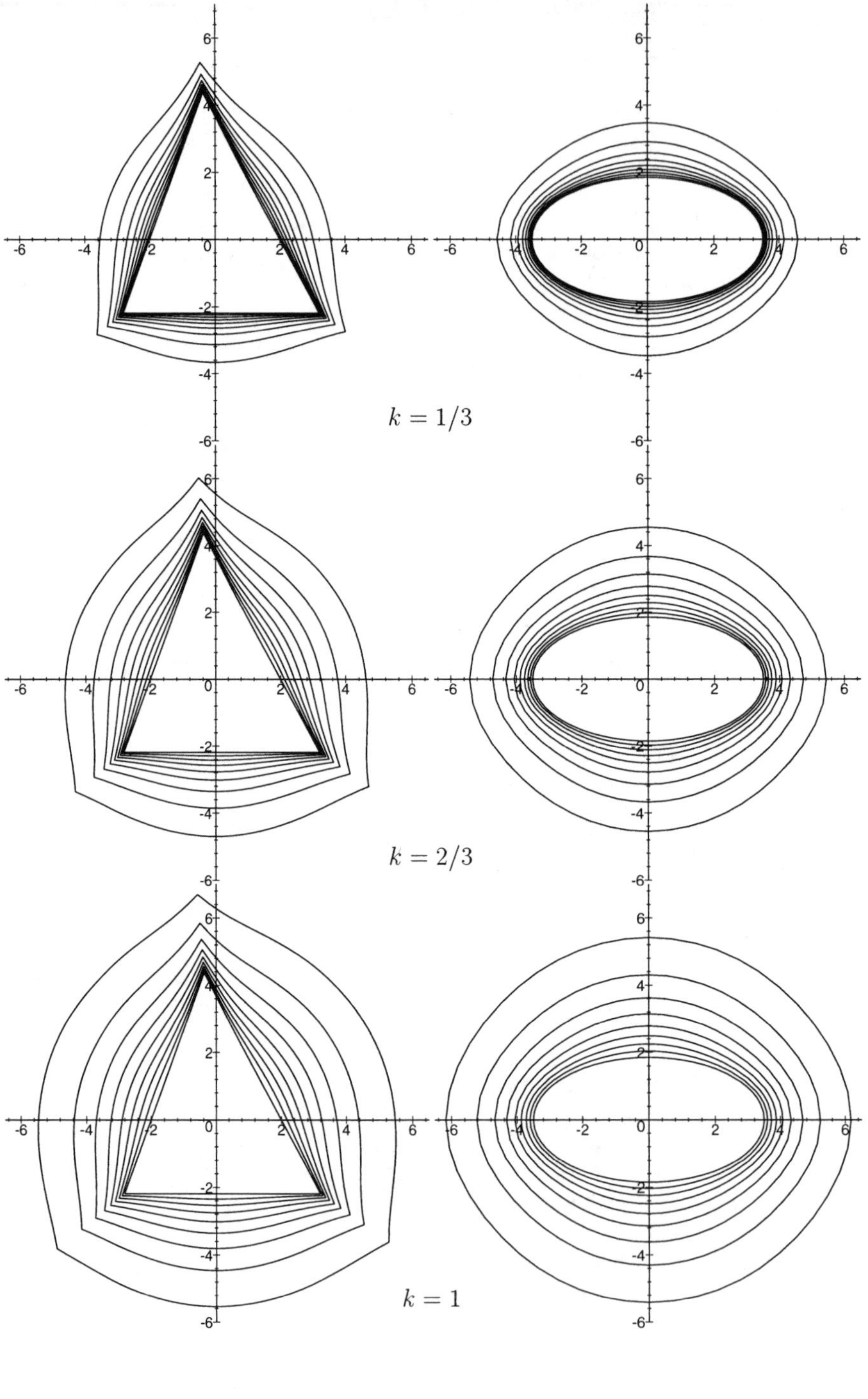

$k = 1/3$

$k = 2/3$

$k = 1$

17.4 The Lateral Side Distortion

Since the material flows radially, as we have assumed, we can simplify the computation of the shape of the distorted lateral side. Only the border line of the radial cross section has to be determined. The functions we will use are described in Figure 17.5. In cylindrical coordinates the shape of the lateral

FIGURE 17.5. *Shape of the lateral side*

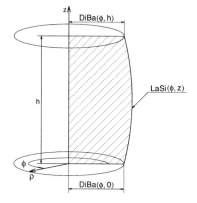

surface is assumed to be a parabola

$$\mathrm{LaSi}(\phi,\, z) = Q_2(\phi)z^2 + Q_1(\phi)z + Q_0(\phi)\,, \tag{17.3}$$

where we need to determine the functions $Q_2(\phi)$, $Q_1(\phi)$ and $Q_0(\phi)$. From Figure 17.5 we see that

$$\begin{aligned}
\mathrm{LaSi}(\phi,\, 0) &= Q_0(\phi) \\
\mathrm{LaSi}(\phi,\, h) &= Q_0(\phi)\,.
\end{aligned}$$

Furthermore, the constant in Equation (17.1) states

$$\iiint dV = S_0 H\,.$$

Since the function $\mathrm{Ba}(\phi)$ is still not specified, we interchange the usual order of integration by applying Fubini's law:

$$\iiint dV = \int_{-\pi}^{\pi} \left(\int_0^h \frac{\mathrm{LaSi}(\phi,\, z)^2}{2}\, dz \right) d\phi$$

For better readability, we will substitute $DiBa$ in the derivation below by

$$DiBa \equiv \sqrt{\frac{\left(-1 + \dfrac{H}{h}\right) k\, So}{\pi} + \mathrm{Ba}(\phi)^2}$$

in the final expression only.

```
> restart;
> LaSi := Q2*z^2 + Q1*z + Q0:
> E1 := subs(z = 0, LaSi) = DiBa;
```

$$E1 := Q0 = DiBa$$

```
> assign(");
> E2 := subs(z = h, LaSi) = DiBa:
> Q1 := solve(E2, Q1);
```

$$Q1 := -Q2\,h$$

```
> E3 := Int(Int(LaSi^2/2, z = 0..h), phi = -Pi..Pi) = H*So;
```

$$E3 := \int_{-\pi}^{\pi}\int_{0}^{h}\frac{1}{2}\,(Q2\,z^2 - Q2\,h\,z + DiBa)^2\,dz\,d\phi = H\,So$$

To evaluate the double integral we proceed in several steps. First we integrate for z:

```
> I1 := int(LaSi^2/2, z = 0..h);
```

$$I1 := \frac{1}{60}\,Q2^2\,h^5 - \frac{1}{6}\,DiBa\,Q2\,h^3 + \frac{1}{2}\,DiBa^2\,h$$

Then we split the sum into three integrals:

```
> I2 := subsop(seq(i = Int(op(i, I1), phi = -Pi..Pi),
>                i = 1..nops(I1)), I1);
```

$$I2 := \int_{-\pi}^{\pi}\frac{1}{60}\,Q2^2\,h^5\,d\phi + \int_{-\pi}^{\pi}-\frac{1}{6}\,DiBa\,Q2\,h^3\,d\phi + \int_{-\pi}^{\pi}\frac{1}{2}\,DiBa^2\,h\,d\phi$$

```
> I21 := subsop(2 = h/2*Int(DiBa^2, phi = -Pi..Pi),
>          3 = -Q2*h^3/6*Int(DiBa, phi = -Pi..Pi), I2);
```

$$I21 := \int_{-\pi}^{\pi}\frac{1}{60}\,Q2^2\,h^5\,d\phi + \frac{1}{2}\,h\int_{-\pi}^{\pi}DiBa^2\,d\phi - \frac{1}{6}\,Q2\,h^3\int_{-\pi}^{\pi}DiBa\,d\phi$$

We substitute the distorted base function in the integrals and simplify:

```
> read('DiBa.sav'):
> I22 := subsop(1 = value(op(1, I21)),
>                2 = -So*(k*h - k*H - h), I21);
```

$$I22 := \frac{1}{30}\,\pi\,Q2^2\,h^5 - So\,(k\,h - k\,H - h)$$

$$-\frac{1}{6}\,Q2\,h^3\int_{-\pi}^{\pi}\sqrt{\frac{(-1+\dfrac{H}{h})\,k\,So}{\pi} + \mathrm{Ba}(\phi)^2}\,d\phi$$

```
> E3 := subs(op(4, op(3, I22)) = IDiBa, I22) = H*So;
```

$$E3 := \frac{1}{30} \pi \, Q2^2 \, h^5 - So \, (k \, h - k \, H - h) - \frac{1}{6} \, Q2 \, h^3 \, IDiBa = H \, So$$

We thus obtain a quadratic equation in $Q2$, which has two solutions. In order to determine the physically correct one, we will substitute both solutions $Q2_1$ and $Q2_2$ into $LaSi$, and decide by inspection of Figure 17.6.

```
> Sol := solve(E3, Q2):
> T1 := Sol[1]: T2 := Sol[2]:
> LaSi_1:=[subs(Q2 = T1, LaSi), subs(Q2 = T2, LaSi)]:
> Cylinder_Subs := IDiBa = int(DiBa, phi = -Pi..Pi),
>       So = Pi*r^2, Ba(phi) = r, r=1 , H = 4, k = kappa, h = 1:
> LaSi_2 := simplify(subs(Cylinder_Subs, LaSi_1)):
> Kappa := [seq(1 - i/2, i = 0..2)]:
> with(plots):
> P1 := plot({seq([LaSi_2[1], z, z = 0..1], kappa = Kappa)},
>               thickness = 3):
> kappa := 'kappa':
> P2 := plot({seq([LaSi_2[2], z, z = 0..1], kappa = Kappa)},
>               thickness = 1):
> kappa := 'kappa':
> display({P1, P2});
```

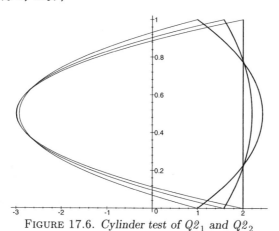

FIGURE 17.6. Cylinder test of $Q2_1$ and $Q2_2$

We substitute the correct root $Q2_1$ into $LaSi$ and we obtain a rather complicated expression. After letting MAPLE simplify some subexpressions, we obtain the final result, which can be written in the following form

```
> LaSi := -z*(-z + h)/(2*h^2*Pi)*(5*Int(DiBa, phi = A..B)
>             -sqrt(5)*sqrt(5*Int(DiBa, phi = A..B)^2
>             - 24*Pi*So*(k - 1)*(-1 + H/h)))  + DiBa:
> subs (DiBa='DiBa', LaSi);
```

$$LaSi := -\frac{1}{2}z\,(-z+h)$$

$$(5\int_A^B DiBa\,d\phi - \sqrt{5}\,\sqrt{5\,(\int_A^B DiBa\,d\phi)^2 - 24\,\pi\,So\,(k-1)\,(-1+\frac{H}{h}))}/(h^2\,\pi)$$

$$+\,DiBa\,,$$

where A, B are the angles describing the segment.

```
> save(LaSi, 'LaSi.sav'):
```

17.5 Non-centered bases

Let us define *Non-centered Bases* to be bases where the center of mass is not at the origin of the coordinate system. For such bases we need to perform a transformation of the coordinate system in order to move the center of mass to the origin of the coordinate system. This is trivial in Cartesian coordinates, but not so simple in polar coordinates (see Figure 17.7). The perimeter of the

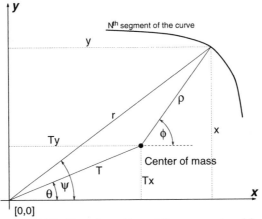

FIGURE 17.7. *Transformation of the non centered base*

non-centered base is described by $r(\psi)$, but we need the function $\rho(\phi)$. We have to express the integral

$$\int_A^B \sqrt{\frac{(-1+\frac{H}{h})\,k\,So}{\pi} + Ba(\phi)^2}\,d\phi\,. \qquad (17.4)$$

in the new variables. So we have to find the functions r_x, ψ_x and $d\psi_x$ satisfying the equations $\rho(\phi) = r_x(\psi)$, $\phi = \psi_x(\psi)$ and $d\phi = d\psi_x(\psi)$.

```
> restart;
> Tx := T*cos(theta):  Ty := T*sin(theta):
> r1x := rho(phi)*cos(phi) + Tx = r(psi)*cos(psi):
> r1y := rho(phi)*sin(phi) + Ty = r(psi)*sin(psi):
> r2x := map(t -> t - Tx, r1x): r2y := map(t -> t - Ty, r1y):
> R1 := combine(map(t -> t^2, r2x) + map(t -> t^2, r2y)):
> sol := rho(phi) = combine(solve(R1, rho(phi))[1]);
```

$$sol := \rho(\phi) = \sqrt{\mathrm{r}(\psi)^2 - 2\,\mathrm{r}(\psi)\,T\cos(-\psi + \theta) + T^2}$$

We have obtained the well known expression for the third side of a triangle given
an angle and the two adjacent sides (*cosine rule*). Now we will express ϕ and $d\phi$
as functions of ψ. It is convenient to define $\cos(\phi) = \mathrm{Cf}(\psi)$ and $\sin(\phi) = \mathrm{Sf}(\psi)$
because we need these expressions for the plots.

```
> R2 := phi = solve(lhs(r2y)/lhs(r2x) = rhs(r2y)/rhs(r2x), phi):
> dR2 := D(phi) = normal(diff(rhs(R2),psi)):
> df := factor(combine(rhs(dR2)));
```

$$df := \frac{-T\left(\frac{\partial}{\partial\psi}\,\mathrm{r}(\psi)\right)\sin(-\psi + \theta) - \mathrm{r}(\psi)^2 + \mathrm{r}(\psi)\,T\cos(-\psi + \theta)}{-\mathrm{r}(\psi)^2 + 2\,\mathrm{r}(\psi)\,T\cos(-\psi + \theta) - T^2}$$

```
> Cf := solve(r2x, cos(phi)):  Sf := solve(r2y, sin(phi)):
> Cf := subs(sol, Cf);  Sf := subs(sol, Sf);
```

$$Cf := \frac{\mathrm{r}(\psi)\cos(\psi) - T\cos(\theta)}{\sqrt{\mathrm{r}(\psi)^2 - 2\,\mathrm{r}(\psi)\,T\cos(-\psi + \theta) + T^2}}$$

$$Sf := \frac{\mathrm{r}(\psi)\sin(\psi) - T\sin(\theta)}{\sqrt{\mathrm{r}(\psi)^2 - 2\,\mathrm{r}(\psi)\,T\cos(-\psi + \theta) + T^2}}$$

Now we can insert the expressions for Ba and $d\phi$ into the integral (17.4). If the
perimeter of the base is composed of several segments with different function
descriptions, we have to split and compute the integral (17.4) for each segment.

To optimize the computations for a segment, we use the following MAPLE
statements (F0 optimizes the code outside of the integral, FI does the same
with the integrand):

```
> read('LaSi.sav'):  read('DiBa.sav'):
> FRzS := subs(Int(DiBa, phi = A .. B) =
>         Int(DiBa*df, psi = A ..B), Ba(phi) = Ba(psi), LaSi):
> readlib(optimize):
> F0 := optimize([W1 = FRzS,  W2 = Cf,  W3 = Sf]):
> FI := optimize(op(1,rhs(op(1, select(has, [F0], Int)))));
```

After eliminating some code common to all segments, we obtain the final Algorithm 17.1. It is important to assign the variable $EIco$ before calling procedure 17.1. This variable defines the numerical precision and the numerical method used for the computation of the integrals. Usually $EIco$ has the following values:

```
> EIco := NDigits:
> EIco := NDigits, _NumMethod;
```

where NDigits determines the desired numerical precision and _NumMethod defines the quadrature method with the following alternatives:

$$\_NumMethod = \begin{array}{ll} \_CCquad & \text{(Clenshaw-Curtis quadrature)} \\ \_Dexp & \text{(double exponential routine)} \\ \_NCrule & \text{(Newton-Cotes rule)} \end{array}$$

The optimal choice is application dependent depend. See

```
> ?evalf[int]
```

for detailed informations. Algorithm 17.1 requires as input

1. $R::list$ is a list of functions describing the segments of the perimeter of the base. The sequence of the segments must be ordered counterclockwise.

$$R := [u \to f_1(u),\ u \to f_2(u),\ \cdots,\ u \to f_n(u)]$$

2. $Bo::list$ is a list of the position angles of the initial point of each segment plus the position angle of the last point of the last segment.

$$Bo := [\alpha_1, \alpha_2, \cdots, \alpha_n, \alpha_{n+1}] \qquad \alpha_{n+1} \equiv \alpha_1 + 2\pi$$

3. Hi, hf, kf are the initial *uncompressed height*, the final *compressed height* and the *measure of the friction*.

If R and Bo and So are defined and if the base curve is centered, we could call Algorithm 17.1:

```
> T := 0:   theta := 0:
> EIco := 12, _Dexp:
> n := nops(R):
> Body(R, Bo, H, h, k):
```

The procedure returns a list of functions describing the distortion of each lateral side in cylindrical coordinates

$$[(\phi, z) \to f_i(\phi, z)] \quad i = 1, \cdots, n$$

if the base is centered. If not, we obtain a list of parameterized functions

$$[(\psi, z) \to f_i(\phi(\psi), z)] \quad i = 1, \cdots, n.$$

ALGORITHM 17.1.

The optimized algorithm for a non-centered n segmented curve

```
Body := proc (R::list, Bo::list, Hi, hf, kf)
  local i, q1, q2, q3, q4, q5, q6, w1, w2, w3, w4, w5,
       w6, w7, w11, w8, w9, w10, w12, w13, w14;
  global Rho, Cf, Sf, X, Y;
  q3 := -u + theta;  q5 := T^2;  w2 := 0;
  for i to n do
    q1[i] := R(u)[i];  q2[i] := q1[i]^2;
    q4[i] := q1[i]*T*cos(q3);
    q6[i] := ((1/hf*Hi - 1)/Pi*kf*So + q2[i] - 2*q4[i]
            + q5)^(1/2);
    w2 := w2 + evalf(Int(q6[i]*(T*diff(R(u)[i], u)*sin(q3)
            + q2[i] - q4[i])/(q2[i] - 2*q4[i] + q5),
            u = Bo[i].. Bo[i+1], EIco));
  od;
  w1 := hf^2;  w3 := 5^(1/2);  w4 := w2^2;
  w5 := -1 + 1/hf*Hi;
  w6 := (5*w4 - 24*Pi*So*w5*(-1 + kf))^(1/2);
  w7 := 1/Pi;  w11 := T^2;
  for i to n do
    w8[i] := R(u)[i];  w9[i] := w8[i]^2;
    w10[i] := w8[i]*T*cos(q3);
    w12[i] := (w5*w7*kf*So + w9[i] - 2*w10[i] + w11)^(1/2);
    w13[i] := (w9[i] - 2*w10[i] + w11)^(1/2);
    w14[i] := 1/w13[i];
  od;
  Rho := [seq(unapply(1/2*z*(-z + hf)/w1*(-5*w2 + w3*w6)*w7
            + w12[i], u, z), i = 1 .. n)];
  Cf := [seq(unapply((w8[i]*cos(u) - T*cos(theta))*w14[i], u),
            i = 1 .. n)];
  Sf := [seq(unapply((w8[i]*sin(u) - T*sin(theta))*w14[i], u),
            i = 1 .. n)];
  X := [seq(unapply(Rho(u, z)[i]*Cf(u)[i], u, z), i = 1 .. n)];
  Y := [seq(unapply(Rho(u, z)[i]*Sf(u)[i], u, z), i = 1 .. n)];
  print('To display the compressed body use the Maple command:');
  if T = 0 then
    print('cylinderplot(r(f,z), f=a..b, z=0..h)');
  else
    print('plot3d([x(f,z), y(f,z), z], f=a..b, z=0..h)');
  fi;
end;
```

In the second case we can also use the list of the functions $[X_i]$, $[Y_i]$, $i = 1, \cdots, n$, which are parameterized functions of ψ describing the distorted body in rectangular coordinates.

If the base is non-centered, we first have to compute the polar coordinates (T, θ) of the center of mass. This can be done with Procedure 17.2, the input parameters $\rho{::}list$ and $bords{::}list$ being the same as for the preceding Algorithm 17.1. If the third parameter *Plot* is 1 the procedure will plot the shape of the base and position of the mass center. The variable EIco has to be given in advance. The procedure returns T, θ and So – the size of the initial base area. We will show in the next section how to use these procedures.

17.6   Three Dimensional Graphical Representation of the Distorted Body

17.6.1   Centered base

An example with a triangle-shaped base will illustrate the usage of Procedure Body without calling Procedure MsCe.

```
> restart;
> read('Tri.sav'):
> a := 8:  b := 7:   c := 6:
> Tr := [unapply(r1, phi), unapply(r2, phi), unapply(r3, phi)];
```

$$Tr := [\phi \rightarrow -\frac{4}{21} \frac{\sqrt{132}}{\sin(\phi)}, \phi \rightarrow 16 \frac{\sqrt{132}}{50\sin(\phi) + 8\sqrt{132}\cos(\phi)},$$
$$\phi \rightarrow -16 \frac{\sqrt{132}}{-34\sin(\phi) + 8\sqrt{132}\cos(\phi)}]$$

```
> TrBo := [-Pi + Aa, -Ab, Ac, Pi + Aa];
```

$$TrBo := [-\pi + \arctan(\frac{4}{59}\sqrt{132}), -\arctan(\frac{4}{67}\sqrt{132}),$$
$$\frac{1}{2}\pi + \arctan(\frac{1}{132}\sqrt{132}), \pi + \arctan(\frac{4}{59}\sqrt{132})]$$

```
> So := simplify(Sb);
```

$$So := \frac{21}{4}\sqrt{15}$$

```
> T := 0:  theta := 0:  n:= nops(Tr): EIco := 12:
```

For the triangle we don't need the variable EIco for numerical integration because the integrals can be computed analytically.

ALGORITHM 17.2. *Computation of the center of mass*

```
MsCe := proc (rho::list, bords::list, Plot)
  local i, Mx, My, Tx, Ty, PT, pl;
  global So, T, theta;
  So := 0;
  Mx := 0;
  My := 0;
  for i to n do;
      So := So + evalf(Int(1/2*rho(u)[i]^2,
                        u = bords[i] .. bords[i + 1], EIco));
      Mx := Mx + evalf(Int(1/3*rho(u)[i]^3*cos(u),
                        u = bords[i] .. bords[i + 1], EIco));
      My := My + evalf(Int(1/3*rho(u)[i]^3*sin(u),
                        u = bords[i] .. bords[i+1], EIco));
  od;
  Tx := Mx/So;
  Ty := My/So;
  readlib(polar);
  PT := polar(Tx + (-1)^(1/2)*Ty);
  T := op(1, PT);
  theta := op(2, PT);
  if T < 1/10000 then
      T := 0;
      theta := 0;
  fi;
  if abs(theta) < 1/10000 then theta := 0; fi;
  if Plot = 1 then
      for i to n do
          pl[i] := plots[polarplot](rho(u)[i],
                        u = bords[i] .. bords[i + 1]);
      od;
      pl[0] := plot([[Tx, Ty]], style = point, symbol = box);
      pl[-1] := plot([[Tx, Ty], [0, 0]]);
      print(plots[display]({seq(pl[i], i = -1 .. n)}));
  fi;
  print('So = ',So,'   T = ',T,'   ','theta',' = ',theta);
end;
```

```
> read('Body.map'):
> with(plots):
> for i from 0 to 3 do;
>    h := 16 - 4*i;
>    Body(Tr, TrBo, 16, h, 0.4);
>    for j from 1 to n do:
>      PL[j] := cylinderplot(Rho(u, z)[j], u = TrBo[j]..TrBo[j + 1],
>              z = 0..h, grid=[3^i + 1, 3^i + 1]);
>    od:
>    SPL[i] := display({seq(PL[j], j = 1..n)}, color = black,
>              style = hidden):
> od:
```

> *To display the compressed body you can use the Maple command :*
> $$cylinderplot(r(f, z),\ f = a..b,\ z = 0..h)$$

```
> display({seq(SPL[i], i = 0..3)}, axes=boxed,
>         scaling = constrained);
```

The results are displayed in Figure 17.9.

17.6.2   Non-centered, Segmented Base

Now we shall consider a more complicated (though technically not equally relevant) base, as described in Figure 17.8.

FIGURE 17.8. *Non-centered, segmented base*

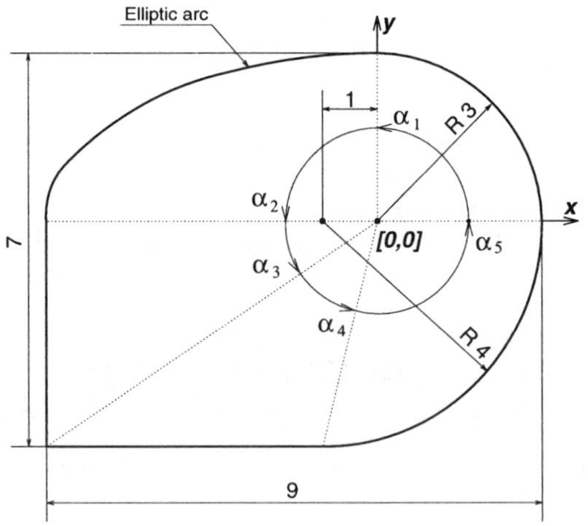

```
> restart;
> EL := 18/sqrt(36*sin(u)^2 + 9*cos(u)^2):
> p1 := -6/cos(u):
> p2 := -4/sin(u):
```

FIGURE 17.9. *The triangle rod compression*

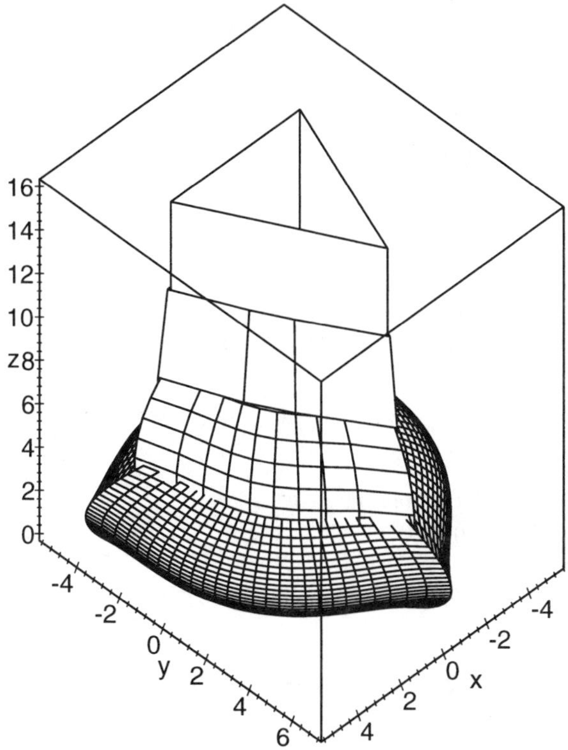

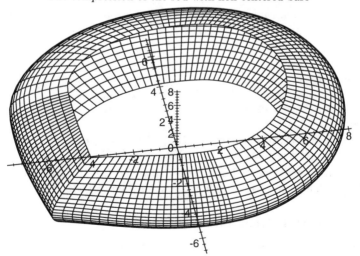

FIGURE 17.10.
The compression of the rod with non centered base

```
> k2 := solve(subs(x = r*cos(u), y = r*sin(u),
>               (x+1)^2 + y^2 = 16) ,r)[1]:
> CC := [u -> 3, unapply(EL, u), unapply(p1, u), unapply(p2, u),
>           unapply(k2, u)];
```

$$CC := [3, u \to \frac{6}{\sqrt{4\sin(u)^2 + \cos(u)^2}}, u \to -\frac{6}{\cos(u)}, u \to -\frac{4}{\sin(u)},$$
$$u \to -\cos(u) + \sqrt{\cos(u)^2 + 15}]$$

```
> alpha[3] := Pi + arctan(4/6):
> alpha[4] := 3/2*Pi - arctan(1/4):
> CB := [0, Pi/2, Pi, alpha[3], alpha[4], 2*Pi];
```

$$CB := [0, \frac{1}{2}\pi, \pi, \pi + \arctan(\frac{2}{3}), \frac{3}{2}\pi - \arctan(\frac{1}{4}), 2\pi]$$

```
> save(CC,CB, 'crazy.sav');
> EIco := 10, _CCquad:
> n := nops(CC):
> read('MsCe.map'):
> MsCe(CC, CB, 0);
```

$$So = , 53.77212102, \quad T = , 1.760718963, \quad , \theta, = , -2.770499706$$

```
> read('Body.map'):
> Body(CC, CB, 20, 8, 0.2);
```

To display the compressed body you can use the Maple command :
$$plot3d([x(f, z), y(f, z), z], f = a..b, z = 0..h)$$

```
> with(plots):
> for i from 1 to n do:
>    pl[i] := plot3d([X(u, z)[i], Y(u, z)[i], z],
>           u = CB[i]..CB[i + 1], z = 0..8);
> od:
> display({seq(pl[i], i = 1..n)}, color = black,
>         style = hidden, orientation = [-100, 15], axes = normal);
```

It is easy to use the procedures MsCe and Body and to visualize the results graphically. The compressed body is shown in Figure 17.10.

17.6.3 Convex Polygon Base

We have seen in Sections 17.3 and 17.6 how to compress triangle-shaped rods. We described the triangle by giving the length of its sides. A polygon described by the coordinates of its vertexes requires a slight modification of the algorithm. The solution is shown with the following MAPLE example for 5 vertexes. There is no limit on the number of vertexes, which are given counterclockwise as seen from the center of mass of the polygon.

```
> restart;
> PNTi := [[1,1], [4,3], [3,4], [2,4], [0,2]]:
> PNTo := [PNTi[], PNTi[1]]:
> n := nops(PNTi):
> Tx := sum(op(i, PNTi)[1], i=1..n)/n;
> Ty := sum(op(i, PNTi)[2], i= 1..n)/n;
```

$$Tx := 2$$
$$Ty := \frac{14}{5}$$

The center of mass of the vertexes is used to compute the center of mass of the polygon. We shall describe the polygon in the new coordinate system with the origin at $[T_x, T_y]$.

```
> readlib(polar): read('MsCe.map'):  EIco := 10, _CCquad:
> T := op(1, polar(Tx + I*Ty)): theta := op(2, polar(Tx + I*Ty)):
> while evalf(T) > 0 do;
>    Tx := T*cos(theta):
>    Ty := T*sin(theta):
>    PNTo := [seq([op(i, PNTo)[1] - Tx, op(i, PNTo)[2] - Ty],
>            i = 1..n+1)];
>    i := 'i';  j := i + 1;
>    r := [seq(unapply(normal((PNTo[j, 1]*PNTo[i, 2]
>          -PNTo[i, 1]*PNTo[j, 2])/
>          (sin(u)*(PNTo[j, 1] - PNTo[i, 1]) +
>          cos(u)*(PNTo[i, 2]  - PNTo[j, 2]))), u), i = 1..n)];
>    bo := [seq(op(2, polar(PNTo[i, 1] + I*PNTo[i,2 ])),
>            i = 1..n + 1)]:
>    for i from 1 to n do;
>        if evalf(bo[j]) < evalf(bo[i]) then
>            bo[j] := bo[j] + 2*Pi;
>        fi; od:
>    MsCe(r, bo, 0);
> od:
```
$So \ = , 6.000000000, \quad T \ = , .1704206850, \quad , \theta, \ = , -1.902855794$
$So \ = , 6.000000001, \quad T \ = , 0, \quad , \theta, \ = , 0$

```
> read('Body.map'):
> Body(r, bo, 4, 2, 0.3):
```

To display the compressed body you can use the Maple command :
$$cylinderplot(r(f,z), f = a..b, z = 0..h)$$

```
> with(plots):
> for i from 1 to n do;
>    pl[i] := cylinderplot(Rho(u, z)[i], u=bo[i]..bo[j], z = 0..2);
> od:
> display({seq(pl[i], i = 1..n)});
```

The compressed body is not displayed here, but it can be generated as before.

17.7   Three Dimensional Animation

If our computer is equipped with enough memory, we can display the whole process of the compression as a movie. Let us consider the compression of the rod whose base is described in Figure 17.8. We will reuse the computed variables which were saved before in the file crazy.sav. The MAPLE commands will be the same but we call the procedure Body for various time-steps, and obtain a sequence of frames of the compressed body.

```
> restart;
> Grid := [5, 5, 3, 3, 5]:
> Frames := 20: Hi := 20: hf := 6: k := 1/3:
> read('crazy.sav'): read('MsCe.map'):
> n := nops(CC): EIco := 10,_Dexp:
> MsCe(CC,CB,1):
```

$$So = , 53.77212102, \quad T = , 1.760718963, \quad , \theta, = , -2.770499706$$

```
> read('Body.map'):
> for j from 0 to Frames do;
>    h := Hi - (Hi - hf)*j/Frames;
>    Body(CC, CB, Hi, h, k);
>    An[j] := [seq(op(plot3d([X(u, z)[i], Y(u, z)[i], z],
>               u = CB[i]..CB[i+1], z = 0..h, grid = [Grid[i], 15])),
>               i = 1..n)];
>    print(j);
> od:
```

$$\vdots \quad \vdots \quad \vdots$$

To display the compressed body you can use the Maple command :
$$plot3d([x(f,z), y(f,z), z], f = a..b, z = 0..h)$$
$$20$$

To save space, we did not print all the 20 similar outputs for each frame. These commands create 21 partial plots (see Figure 17.11), which are used to generate data for the animation. The collected frames in An can finally be animated with

```
> PLOT3D(ANIMATE(seq(An[i], i = 0..Frames)),TITLE('Animation'),
>        AXESLABELS(x,y,z));
```

FIGURE 17.11. *Partial plots collection*

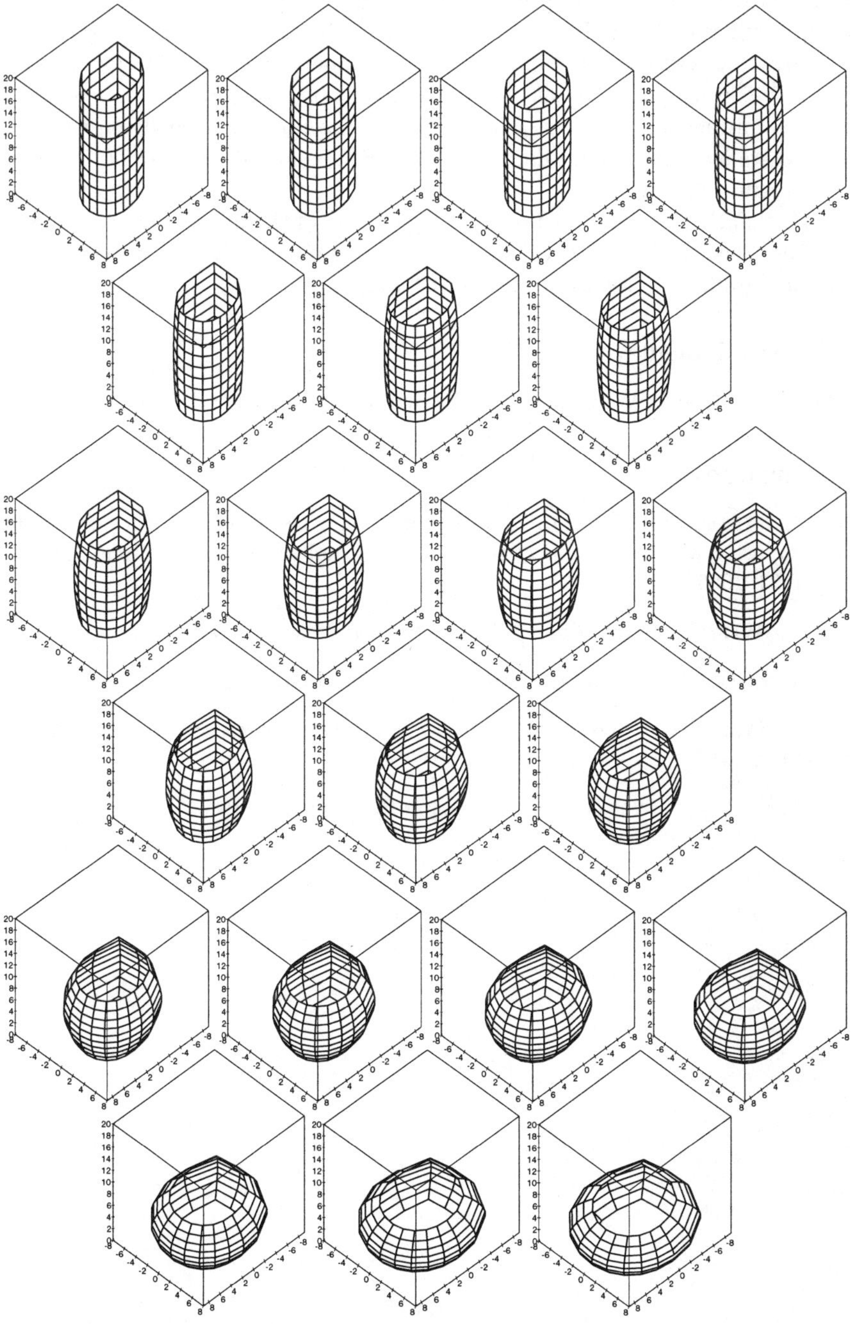

17.8 Limitations and Conclusions

Our model is only a description of the geometry of the lateral surface of the compressed rod. Thus it is unable to model the flow of the material, and is limited to convex cross sections. Our simple model is appropriate for homogeneous and isotropic materials with constant (*especially pressure independent*) physical properties, however. The accuracy of the model depends on the ratio b/a, where $b =$ the *shortest diameter* and $a =$ the *longest diameter* of the initial cross section of the rod.

This mathematical model of the deformation of the metal body can be successfully used for technical calculations. The number of technical experiments can be significantly reduced, thus saving time and money.

Acknowledgments

The author thanks W. Gander and R. Strebel for their helpful suggestions.

References

[1] S. BARTOŇ, *Free Metal Compression*, in Solving Problems in Scientific Computing using MAPLE and MATLAB, W. Gander and J. Hřebíček, editors, Springer, 1993, 1995

[2] B. AVITZUR, *Metal Forming*, Marcel Dekker, New York, 1980.

[3] T. Z. BLAZYNSKI, *Metal Forming*, MacMillan Press LTD, London, 1976.

[4] A. FARLÍK and E. ONDRÁČEK, *Theory of the dynamic forming (in czech)*, SNTL, Praha, 1968.

[5] S. R. REID, *Metal Forming*, Pergamon Press, Oxford, 1985.

Chapter 18. Gauss Quadrature

U. von Matt

18.1 Introduction

In this chapter, we study how to compute Gauss quadrature rules with the help of MAPLE. We consider the integral

$$\int_a^b f(x)\omega(x)\,dx, \tag{18.1}$$

where $\omega(x)$ denotes a nonnegative weight function. We assume that the integrals

$$\int_a^b |x|^k \omega(x)\,dx \tag{18.2}$$

exist for all $k \geq 0$. Additionally, we assume that $\omega(x)$ has only a finite number of zeroes in the interval $[a, b]$ (cf. [20, p. 18]).

The purpose of Gauss quadrature is to approximate the integral (18.1) by the finite sum

$$\int_a^b f(x)\omega(x)\,dx \approx \sum_{i=1}^{m+n} w_i f(x_i), \tag{18.3}$$

where the abscissas x_i and the weights w_i are determined such that all polynomials to as high a degree as possible are integrated exactly.

We also consider the cases of Gauss-Radau and Gauss-Lobatto quadrature where m abscissas on the boundary of the interval $[a, b]$ are prescribed. We denote these prescribed abscissas by $x_1, \ldots, x_m$, where m can take the values $m = 0$, $m = 1$, and $m = 2$.

The theory of Gauss quadrature is a large and varied field. In this chapter we will concentrate on those parts that are necessary to derive the quadrature rules. The reader is also referred to the relevant literature [4, 20, 31].

Orthogonal polynomials play a key role in Gauss quadrature. As soon as the three-term recurrence relationship of the orthogonal polynomials is known the abscissas and the weights of the Gauss quadrature rule can be computed by means of an eigenvalue decomposition. In Section 18.2 we will therefore derive two alternative ways of computing the orthogonal polynomials.

In the first case the interval $[a, b]$ and the weight function $\omega(x)$ have to be known explicitly. Then we can use the Lanczos algorithm to compute the three-term recurrence relationship of the orthogonal polynomials by symbolic integration.

In the other case only the moment corresponding to the interval $[a, b]$ and the weight function $\omega(x)$ need to be available. We present an algorithm to compute the three-term recurrence relationship of the orthogonal polynomials based on the Cholesky decomposition of the Gram matrix.

In Section 18.3 we prove a theorem which is fundamental to the calculation of Gauss quadrature rules. On the basis of this theorem we can give algorithms to compute Gauss, Gauss-Radau, and Gauss-Lobatto quadrature rules in Sections 18.4, 18.5, and 18.6. The calculation of the weights w_i is addressed in Section 18.7. Finally, we derive an expression for the quadrature error in Section 18.8.

The calculation of Gauss quadrature rules is a rewarding application of MAPLE. We can use its symbolic capabilities to compute the coefficients of the three-term recurrence relationship exactly. Since MAPLE also supports floating-point arithmetic with an arbitrary precision we can compute the abscissas and weights of the quadrature rules by an eigenvalue decomposition to any desired accuracy.

18.2 Orthogonal Polynomials

In this section we discuss several approaches to computing the orthogonal polynomials corresponding to the interval $[a, b]$ and the weight function $\omega(x)$.

DEFINITION 18.0.1. *We denote by*

$$(f, g) := \int_a^b f(x) g(x) \omega(x) \, dx \tag{18.4}$$

the inner product with respect to the weight function $\omega(x)$ and the interval $[a, b]$.
This inner product has also the important property that

$$(xf, g) = (f, xg), \tag{18.5}$$

which will be used frequently throughout this chapter.

DEFINITION 18.0.2. *The expression*

$$\mu_k := (x^k, 1) = \int_a^b x^k \omega(x) \, dx \tag{18.6}$$

is called the kth moment with respect to the weight function $\omega(x)$.

Many books are concerned with calculating Gauss quadrature rules from these ordinary moments [1, 6, 28, 33]. This approach, however, is cursed with numerical instabilities, as Gautschi has shown in [9]. As a remedy, so-called modified moments are introduced (cf. [10, p. 245] and [27, p. 466]).

DEFINITION 18.0.3. *The expression*

$$\nu_k := (\pi_k, 1) = \int_a^b \pi_k(x) \omega(x) \, dx, \tag{18.7}$$

where π_k denotes a polynomial with exact degree k, is called the kth modified moment with respect to the weight function $\omega(x)$.

We assume that the polynomials $\pi_0(x), \pi_1(x), \ldots$ satisfy the three-term recurrence relationship

$$x\pi_{k-1} = \hat{\gamma}_{k-1}\pi_{k-2} + \hat{\alpha}_k\pi_{k-1} + \hat{\beta}_k\pi_k, \qquad k = 1, 2, \ldots \qquad (18.8)$$

where $\hat{\gamma}_0 := 0$ and $\pi_{-1}(x) := 0$. The coefficients $\hat{\beta}_k$ are supposed to be non-zero.

The ordinary moments μ_k represent a special case of the modified moments ν_k. They can be obtained by setting $\pi_k(x) = x^k$. In this case the three-term recurrence relationship (18.8) is given by $\hat{\alpha}_k = \hat{\gamma}_k = 0$ and $\hat{\beta}_k = 1$.

DEFINITION 18.0.4. *We define the matrix*

$$M := \begin{bmatrix} m_{00} & \cdots & \cdots & m_{0,n-1} \\ \vdots & & & \vdots \\ \vdots & & & \vdots \\ m_{n-1,0} & \cdots & \cdots & m_{n-1,n-1} \end{bmatrix} \qquad (18.9)$$

to be the Gram matrix of order n, where each entry

$$m_{ij} := (\pi_i, \pi_j) \qquad (18.10)$$

is an inner product.

Not only is the matrix M symmetric, but because of

$$\mathbf{v}^T M \mathbf{v} = \sum_{i=0}^{n-1} \sum_{j=0}^{n-1} v_i m_{ij} v_j = \left(\sum_{i=0}^{n-1} v_i \pi_i, \sum_{j=0}^{n-1} v_j \pi_j \right) > 0$$

for $\mathbf{v} \neq \mathbf{0}$ it is also positive definite.

Now we show how the matrix M can be computed from the modified moments ν_k and the three-term recurrence relationship (18.8). A similar derivation can also be found in [10, pp. 255–256].

Since $\pi_0(x)$ is a constant, we immediately have

$$m_{i0} = (\pi_i, \pi_0) = \pi_0 \nu_i, \qquad m_{0j} = (\pi_0, \pi_j) = \pi_0 \nu_j. \qquad (18.11)$$

Consequently, we will assume $i > 0$ and $j > 0$ in the following. If we substitute in (18.10) the polynomial $\pi_j(x)$ by the recurrence relationship (18.8), we get

$$m_{ij} = (\pi_i, \pi_j) = \frac{1}{\hat{\beta}_j}(\pi_i, x\pi_{j-1} - \hat{\gamma}_{j-1}\pi_{j-2} - \hat{\alpha}_j\pi_{j-1})$$

$$= \frac{1}{\hat{\beta}_j}((x\pi_i, \pi_{j-1}) - \hat{\gamma}_{j-1}m_{i,j-2} - \hat{\alpha}_j m_{i,j-1}).$$

Applying the recurrence relationship (18.8) once more for substituting $x\pi_i(x)$ one has

$$m_{ij} = \frac{1}{\hat{\beta}_j}((\hat{\gamma}_i\pi_{i-1} + \hat{\alpha}_{i+1}\pi_i + \hat{\beta}_{i+1}\pi_{i+1}, \pi_{j-1}) - \hat{\gamma}_{j-1}m_{i,j-2} - \hat{\alpha}_j m_{i,j-1})$$

or

$$m_{ij} = \frac{1}{\hat{\beta}_j}(\hat{\gamma}_i m_{i-1,j-1} + (\hat{\alpha}_{i+1} - \hat{\alpha}_j)m_{i,j-1} + \hat{\beta}_{i+1}m_{i+1,j-1} - \hat{\gamma}_{j-1}m_{i,j-2}), \quad (18.12)$$

where $m_{i,-1} = 0$. Consequently, we have in (18.12) a recursive scheme to progressively build up the matrix M from the initial values (18.11).

The dependencies in (18.12) can also be described by the following stencil:

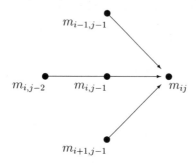

This picture indicates which entries of the matrix M have to be known before the value of m_{ij} can be computed. The calculation of the Gram matrix M is also shown in MAPLE as Algorithm 18.1.

DEFINITION 18.0.5. *The polynomials $p_0(x), p_1(x), \ldots$ are called orthogonal polynomials with respect to the inner product (18.4), if the following conditions are met:*

1. *$p_k(x)$ is of degree k.*

2. *The polynomial $p_k(x)$ satisfies the orthogonality condition*

$$(p_k, p) = 0 \qquad (18.13)$$

   for all polynomials $p(x)$ of degree less than k.

3. *The polynomial $p_k(x)$ satisfies the normalization condition*

$$(p_k, p_k) = 1. \qquad (18.14)$$

Because of condition (18.14) the polynomial p_0 must be a constant different from zero. Therefore a simple corollary of (18.13) consists in

$$(p_k, 1) = 0 \qquad (18.15)$$

for $k \geq 1$.

The following theorem ensures the existence of orthogonal polynomials with respect to the inner product (18.4), and it shows a first way of computing them.

ALGORITHM 18.1.
Calculation of the Gram matrix M from Modified
Moments.

```
Gram := proc (pi0, alphahat, betahat, gammahat, nu, n)
  local i, j, M;

  M := array (symmetric, 0..2*n-1, 0..2*n-1);
  for i from 0 to 2*n-1 do
    M[i,0] := simplify (pi0 * nu[i]);
  od;
  for i from 1 to 2*n-2 do
    M[i,1] := simplify (
              (gammahat[i] * M[i-1,0] +
               (alphahat[i+1] - alphahat[1]) * M[i,0] +
               betahat[i+1] * M[i+1,0]) / betahat[1]);
  od;
  for j from 2 to n-1 do
    for i from j to 2*n-1-j do
      M[i,j] := simplify (
              (gammahat[i] * M[i-1,j-1] +
               (alphahat[i+1] - alphahat[j]) * M[i,j-1] +
               betahat[i+1] * M[i+1,j-1] -
               gammahat[j-1] * M[i,j-2]) / betahat[j]);
    od;
  od;

  RETURN (M);
end:
```

THEOREM 18.1. *For any given admissible inner product there exists a sequence $\{p_k(x)\}_{k=0}^{\infty}$ of orthogonal polynomials.*

PROOF. According to a paper of Mysovskih [22], we set up $p_k(x)$ as

$$p_k(x) = \sum_{i=0}^{k} s_{ki}\pi_i(x). \tag{18.16}$$

We define the lower triangular matrix

$$S := \begin{bmatrix} s_{00} & & & \\ \vdots & \ddots & & \\ \vdots & & \ddots & \\ s_{n-1,0} & \cdots & \cdots & s_{n-1,n-1} \end{bmatrix}. \tag{18.17}$$

The orthogonality condition (18.13) implies

$$\left(\sum_{i=0}^{k} s_{ki}\pi_i, \sum_{j=0}^{l} s_{lj}\pi_j\right) = \sum_{i=0}^{k}\sum_{j=0}^{l} s_{ki}s_{lj}m_{ij} = 0 \tag{18.18}$$

for $k \neq l$. The normalization condition (18.14) means that

$$\Big(\sum_{i=0}^{k} s_{ki}\pi_i, \sum_{i=0}^{k} s_{ki}\pi_i\Big) = \sum_{i=0}^{k}\sum_{j=0}^{k} s_{ki}s_{kj}m_{ij} = 1. \tag{18.19}$$

In matrix terms conditions (18.18) and (18.19) can be represented by

$$SMS^{\mathrm{T}} = I,$$

· where M denotes the Gram matrix from (18.9). Because M is a symmetric and positive definite matrix we can compute the Cholesky decomposition

$$M = LL^{\mathrm{T}} \tag{18.20}$$

with the lower triangular matrix L. Since the inverse of L is a lower triangular matrix as well, we have found the desired matrix S in $S := L^{-1}$. $\square$

In contrast to (18.16) the following representation of the orthogonal polynomials is usually more useful.

THEOREM 18.2. *The orthogonal polynomials $p_0(x), p_1(x), \ldots$ satisfy the three-term recurrence relationship*

$$xp_{k-1} = \beta_{k-1}p_{k-2} + \alpha_k p_{k-1} + \beta_k p_k, \qquad k = 1, 2, \ldots \tag{18.21}$$

with

$$\alpha_k = (xp_{k-1}, p_{k-1}), \tag{18.22}$$
$$\beta_k = (xp_{k-1}, p_k), \tag{18.23}$$

where $\beta_0 := 0$, $p_{-1}(x) := 0$, and $p_0(x) := \pm\sqrt{1/\mu_0} = \pm\sqrt{\pi_0/\nu_0}$.

PROOF. Obviously, xp_{k-1} can be written as

$$xp_{k-1} = \sum_{i=0}^{k} c_i p_i$$

with $c_i = (p_i, xp_{k-1}) = (xp_i, p_{k-1})$. Because of (18.13) we have $c_0 = \cdots = c_{k-3} = 0$. If we define α_k and β_k according to (18.22) and (18.23), we can immediately derive (18.21). $\square$

This theorem allows us to derive a first recursive algorithm for the calculation of the orthogonal polynomials. If the coefficients $\alpha_1, \ldots, \alpha_k$ and $\beta_1, \ldots, \beta_{k-1}$ as well as the polynomials $p_0, \ldots, p_{k-1}$ are known, then p_k is already determined up to a multiple due to Equation (18.21). We have

$$q_k := \beta_k p_k = (x - \alpha_k)p_{k-1} - \beta_{k-1}p_{k-2}. \tag{18.24}$$

The normalization condition (18.14) implies

$$\beta_k = \pm\sqrt{(q_k, q_k)}, \tag{18.25}$$

and thus $p_k = q_k/\beta_k$. Finally, the value of α_{k+1} is given by (18.22).

ALGORITHM 18.2. *Lanczos Algorithm.*

```
Lanczos := proc (iprod, alpha, beta, n)
  local k, p, q, x;

  alpha := array (1..n);
  if n > 1 then beta := array (1..n-1) else beta := 'beta' fi;

  p[0] := collect (1 / sqrt (iprod (1, 1, x)), x, simplify);
  alpha[1] := normal (iprod (x*p[0], p[0], x));
  q := collect ((x - alpha[1]) * p[0], x, simplify);
  for k from 2 to n do
    beta[k-1] := normal (sqrt (iprod (q, q, x)));
    p[k-1] := collect (q / beta[k-1], x, simplify);
    alpha[k] := normal (iprod (x*p[k-1], p[k-1], x));
    q := collect ((x - alpha[k]) * p[k-1] - beta[k-1]*p[k-2],
                  x, simplify);
  od;

  RETURN (NULL);
end:
```

This recursive scheme is also known as the Lanczos algorithm. We present its implementation in MAPLE as Algorithm 18.2. It should be noted that the inner product (18.4) is passed as the parameter iprod. As an example, we can use the statements

```
> a := 0:  b := infinity:
> omega := t -> exp (-t):
> iprod := (f, g, x) -> int (f * g * omega (x), x=a..b):
> n := 10:
> Lanczos (iprod, 'alpha', 'beta', n):
```

to compute the recurrence coefficients $\alpha_k = 2k - 1$ and $\beta_k = k$ of the Laguerre polynomials.

The Lanczos algorithm requires the capability of evaluating the inner product (18.4) with arbitrary polynomials. It is also possible to derive an alternative algorithm based only on the knowledge of the modified moments (18.7).

In the proof of Theorem 18.1 we have presented a way of computing the orthogonal polynomials which is based on the Cholesky decomposition (18.20) of the Gram matrix M. The matrix S, which is used to represent the p_k's in (18.16), has been obtained as the inverse of the Cholesky factor L. The direct evaluation of the Cholesky decomposition (18.20) requires $O(n^3)$ operations [17, Section 4.2]. However, we will now show that the entries of the matrix L satisfy a similar recurrence relationship as those of the Gram matrix M. This observation will allow us to devise a modified algorithm which only requires $O(n^2)$ operations.

Because of $S = L^{-1}$ we have

$$L = MS^{\mathrm{T}}$$

from (18.20). The entry l_{ij} of L can thus be written as

$$l_{ij} = \sum_{k=0}^{j} m_{ik}s_{jk} = \left(\pi_i, \sum_{k=0}^{j} s_{jk}\pi_k\right),$$

and because of (18.16) we get

$$l_{ij} = (\pi_i, p_j). \qquad (18.26)$$

Due to (18.20) the value of l_{00} is given by

$$l_{00} = \sqrt{m_{00}}. \qquad (18.27)$$

We will assume $i > 0$ in the following. If we substitute in (18.26) the polynomial π_i by means of the recurrence relationship (18.8), we get

$$l_{ij} = (\pi_i, p_j) = \frac{1}{\hat{\beta}_i}(x\pi_{i-1} - \hat{\gamma}_{i-1}\pi_{i-2} - \hat{\alpha}_i\pi_{i-1}, p_j)$$

$$= \frac{1}{\hat{\beta}_i}((x\pi_{i-1}, p_j) - \hat{\gamma}_{i-1}l_{i-2,j} - \hat{\alpha}_i l_{i-1,j}).$$

Now, we can use the recurrence relationship (18.21) to substitute the polynomial $xp_j(x)$ in the expression $(x\pi_{i-1}, p_j) = (\pi_{i-1}, xp_j)$. This leads to

$$l_{ij} = \frac{1}{\hat{\beta}_i}((\pi_{i-1}, \beta_j p_{j-1} + \alpha_{j+1}p_j + \beta_{j+1}p_{j+1}) - \hat{\gamma}_{i-1}l_{i-2,j} - \hat{\alpha}_i l_{i-1,j})$$

or

$$l_{ij} = \frac{1}{\hat{\beta}_i}(\beta_j l_{i-1,j-1} + (\alpha_{j+1} - \hat{\alpha}_i)l_{i-1,j} + \beta_{j+1}l_{i-1,j+1} - \hat{\gamma}_{i-1}l_{i-2,j}), \qquad (18.28)$$

where $l_{-1,j} = 0$. We have derived a recurrence relationship for the elements of the matrix L. The dependencies in L can also be described by the following stencil:

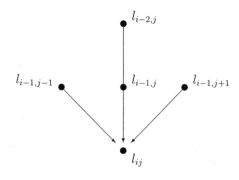

One should note, however, that the unknown coefficients α_k and β_k appear as well in Equation (18.28). Therefore, we cannot compute all the entries in L according to (18.28).

If $\alpha_1, \ldots, \alpha_{k-1}$ and $\beta_1, \ldots, \beta_{k-1}$ as well as all the l_{ij}'s with $i < k$ are known, we can use (18.28) to compute l_{kj} for $j = 0, \ldots, k-2$. The values of

$$l_{k,k-1} = \frac{1}{l_{k-1,k-1}}(m_{k,k-1} - \sum_{j=0}^{k-2} l_{kj}l_{k-1,j}) \tag{18.29}$$

and

$$l_{kk} = \sqrt{m_{kk} - \sum_{j=0}^{k-1} l_{kj}^2} \tag{18.30}$$

are available from the Cholesky decomposition (18.20) of M. On the other hand the recurrence relationship (18.28) gives the expression

$$l_{k,k-1} = \frac{1}{\hat{\beta}_k}(\beta_{k-1}l_{k-1,k-2} + (\alpha_k - \hat{\alpha}_k)l_{k-1,k-1}) \tag{18.31}$$

for $l_{k,k-1}$. If we solve (18.31) for α_k we get

$$\alpha_k = \hat{\alpha}_k + \frac{\hat{\beta}_k l_{k,k-1} - \beta_{k-1}l_{k-1,k-2}}{l_{k-1,k-1}}. \tag{18.32}$$

In the same way we get

$$l_{kk} = \frac{1}{\hat{\beta}_k}\beta_k l_{k-1,k-1} \tag{18.33}$$

from (18.28), such that

$$\beta_k = \hat{\beta}_k \frac{l_{kk}}{l_{k-1,k-1}}. \tag{18.34}$$

In [27] Sack and Donovan introduced a similar approach to computing the three-term recurrence relationship (18.21) from modified moments. They called it the "long quotient—modified difference (LQMD) algorithm". Gautschi presented in [10] a related algorithm based on the Cholesky decomposition (18.20) of the Gram matrix M. Unfortunately, he did not notice the recurrence relationship (18.28) of the entries in the Cholesky factor L. Although his presentation is much clearer than that by Sack and Donovan, the amount of work required by his algorithm is of the order $O(n^3)$.

Let us now present the implementation of our approach in MAPLE as Algorithm 18.3. For a given n, the constant polynomial π_0, the coefficients $\hat{\alpha}_k$, $k = 1, \ldots, 2n-1$, $\hat{\beta}_k$, $k = 1, \ldots, 2n-1$, and $\hat{\gamma}_k$, $k = 1, \ldots, 2n-2$, as well as the modified moments ν_k, $k = 0, \ldots, 2n-1$, must be supplied as input parameters. Then, Algorithm 18.3 builds up the Gram matrix M based on the recursive formulas (18.11) and (18.12). Afterwards, it uses Equations (18.27), (18.28), (18.29), (18.30), (18.32), and (18.34) to construct the Cholesky factor L as well

ALGORITHM 18.3.
Calculation of the Three-Term Recurrence
Relationship (18.21)
from Modified Moments.

```
SackDonovan := proc (pi0, alphahat, betahat, gammahat, nu, n,
                     alpha, beta)
  local j, k, L, M;

  M := Gram (pi0, alphahat, betahat, gammahat, nu, n);
  L := array (0..n, 0..n);
  alpha := array (1..n);
  if n > 1 then beta := array (1..n-1) else beta := 'beta' fi;
  L[0,0] := simplify (sqrt (M[0,0]));
  L[1,0] := simplify (M[1,0] / L[0,0]);
  alpha[1] := simplify (
                alphahat[1] + betahat[1] * L[1,0] / L[0,0]);
  k := 1;
  while k < n do
    L[k,k] := simplify (
                sqrt (M[k,k] - sum ('L[k,j]^2', 'j'=0..k-1)));
    beta[k] := simplify (betahat[k] * L[k,k] / L[k-1,k-1]);
    k := k+1;
    L[k,0] := M[k,0] / L[0,0];
    for j from 1 to k-2 do
      L[k,j] := simplify (
                  (beta[j] * L[k-1,j-1] +
                  (alpha[j+1] - alphahat[k]) * L[k-1,j] +
                  beta[j+1] * L[k-1,j+1] -
                  gammahat[k-1] * L[k-2,j]) / betahat[k]);
    od;
    L[k,k-1] := simplify (
                  (M[k,k-1] -
                  sum ('L[k,j] * L[k-1,j]', 'j'=0..k-2)) /
                  L[k-1,k-1]);
    alpha[k] := simplify (
                  alphahat[k] +
                  (betahat[k] * L[k,k-1] -
                  beta[k-1] * L[k-1,k-2]) / L[k-1,k-1]);
  od;

  RETURN (NULL);
end:
```

as the coefficients α_k, $k = 1, \ldots, n$, and β_k, $k = 1, \ldots, n - 1$ of the three-term recurrence relationship (18.21). The amount of work required by this algorithm is of the order $O(n^2)$.

The amount of storage needed by Algorithm 18.3 is also of the order $O(n^2)$. A more elaborate implementation, like that given by Sack and Donovan, could even get by with storage requirements proportional to n. For the sake of clarity we do not pursue this any further.

The reader should notice that the savings of Algorithm 18.3 over the algorithm given by Gautschi can only be attained in floating-point arithmetic. In a symbolic calculation, as it is performed in MAPLE, not only the number of operations but also the size of the operands affects the total execution time. Although Algorithm 18.3 needs less arithmetic operations compared to the algorithm by Gautschi the operands tend to be more complex. Consequently, we do not observe significant speedups by our algorithm over that given by Gautschi.

As a sample application of Algorithm 18.3 we consider the calculation of the orthogonal polynomials that correspond to the interval $[0, \infty]$ and the two weight functions

$$\omega_{\cos}(x) := e^{-x}(1 + \cos \varphi x), \tag{18.35}$$

$$\omega_{\sin}(x) := e^{-x}(1 + \sin \varphi x). \tag{18.36}$$

It is well-known that the Laguerre polynomials $L_n(x)$ are the orthogonal polynomials with respect to the same interval and the weight function $\omega(x) = e^{-x}$. Therefore, it is sensible to consider the modified moments

$$\nu_n^{\cos} := \int_0^\infty L_n(x)\omega_{\cos}(x)\,dx, \tag{18.37}$$

$$\nu_n^{\sin} := \int_0^\infty L_n(x)\omega_{\sin}(x)\,dx. \tag{18.38}$$

We now show how the modified moments $\nu_n^{\cos}$ and $\nu_n^{\sin}$ can be computed analytically.

THEOREM 18.3. *The polynomial*

$$L_n(x) := \sum_{k=0}^n (-1)^{n-k}\binom{n}{k}\frac{x^k}{k!} \tag{18.39}$$

is the nth Laguerre polynomial.

PROOF. The Laguerre polynomials are known to satisfy the three-term recurrence relationship (18.21) with $p_0 = 1$, $\alpha_k = 2k - 1$, and $\beta_k = k$. Since the polynomials defined by (18.39) satisfy the same recurrence relationship they must be identical to the Laguerre polynomials. $\square$

For any complex constant c with $\mathbb{R}(c) < 0$ we have

$$\int_0^\infty x^n e^{cx}\,dx = n!\left(\frac{-1}{c}\right)^{n+1}. \tag{18.40}$$

This identity can be proved by partial integration and by induction on n. Because of Equations (18.39) and (18.40) we can also show that

$$\int_0^\infty \tilde{L}_n(x)e^{cx}\,dx = (-1)^{n+1}\frac{(c+1)^n}{c^{n+1}}. \tag{18.41}$$

Consequently, we have

$$\int_0^\infty \tilde{L}_n(x)e^{-x}\cos\varphi x\,dx = \mathrm{I\!R}\Big(\int_0^\infty \tilde{L}_n(x)e^{(-1+i\varphi)x}\,dx\Big)$$

$$= (-1)^{n+1}\mathrm{I\!R}\Big(\frac{(i\varphi)^n}{(-1+i\varphi)^{n+1}}\Big)$$

$$= \frac{(-1)^n\varphi^n}{(1+\varphi^2)^{n+1}}\sum_{k=0}^{\lfloor\frac{n}{2}\rfloor}\binom{n+1}{2k+1}(-1)^k\varphi^{n-2k},$$

and

$$\int_0^\infty \tilde{L}_n(x)e^{-x}\sin\varphi x\,dx = \Im\Big(\int_0^\infty \tilde{L}_n(x)e^{(-1+i\varphi)x}\,dx\Big)$$

$$= (-1)^{n+1}\Im\Big(\frac{(i\varphi)^n}{(-1+i\varphi)^{n+1}}\Big)$$

$$= \frac{(-1)^n\varphi^n}{(1+\varphi^2)^{n+1}}\sum_{k=0}^{\lfloor\frac{n+1}{2}\rfloor}\binom{n+1}{2k}(-1)^k\varphi^{n+1-2k}.$$

We can now express the modified moments (18.37) and (18.38) by

$$\nu_0^{\cos} = 1 + \frac{1}{1+\varphi^2}, \tag{18.42}$$

$$\nu_n^{\cos} = \frac{(-1)^n\varphi^n}{(1+\varphi^2)^{n+1}}\sum_{k=0}^{\lfloor\frac{n}{2}\rfloor}\binom{n+1}{2k+1}(-1)^k\varphi^{n-2k}, \qquad n>0, \tag{18.43}$$

$$\nu_0^{\sin} = 1 + \frac{\varphi}{1+\varphi^2}, \tag{18.44}$$

$$\nu_n^{\sin} = \frac{(-1)^n\varphi^n}{(1+\varphi^2)^{n+1}}\sum_{k=0}^{\lfloor\frac{n+1}{2}\rfloor}\binom{n+1}{2k}(-1)^k\varphi^{n+1-2k}, \qquad n>0. \tag{18.45}$$

By the MAPLE statements

```
> N := 10:
> phi := 1:

> alphahat := array (1..2*N-1, [seq (2*k-1, k=1..2*N-1)]):
> betahat := array (1..2*N-1, [seq (k, k=1..2*N-1)]):
> gammahat := array (1..2*N-2, [seq (k, k=1..2*N-2)]):
> L := (n, x) -> sum ('(-1)^(n-k) * binomial (n,k) * x^k / k!',
>                       'k'=0..n):
```

TABLE 18.1. *Values of* $\alpha_n^{\cos}$ *and* $\beta_n^{\cos}$ *for* $\varphi = 1$.

n	$\alpha_n^{\cos}$	$\beta_n^{\cos}$
1	$\frac{2}{3} \approx 0.66667$	$\frac{\sqrt{5}}{3} \approx 0.74536$
2	$\frac{53}{15} \approx 3.5333$	$\frac{12}{5} \approx 2.4000$
3	$\frac{329}{80} \approx 4.1125$	$\frac{\sqrt{1655}}{16} \approx 2.5426$
4	$\frac{44489}{5296} \approx 8.4005$	$\frac{8\sqrt{29499}}{331} \approx 4.1511$
5	$\frac{26238754}{3254723} \approx 8.0617$	$\frac{5\sqrt{91073657}}{9833} \approx 4.8527$
6	$\frac{31146028765}{2705520451} \approx 11.512$	$\frac{18\sqrt{1040328450 1}}{275147} \approx 6.6726$
7	$\frac{41962386991493}{3493256406708} \approx 12.012$	$\frac{7\sqrt{123806905335947}}{12695964} \approx 6.1349$
8	$\frac{99084727782033173}{5712757228305564} \approx 17.344$	$\frac{8\sqrt{213374749185568311}}{449966401} \approx 8.2126$
9	$\frac{465045058223400793942}{30249445557756103921} \approx 15.374$	$\frac{15\sqrt{158546041858245604 1029}}{67226009521} \approx 8.8845$

```
> nucos[0] := 1 + 1 / (1+phi^2):
> for n from 1 to 2*N-1 do
>    nucos[n] := (-1)^n * phi^n / (1+phi^2)^(n+1) *
>                sum ('binomial (n+1,2*k+1) *
>                    (-1)^k * phi^(n-2*k)', 'k'=0..floor (n/2));
> od:
> nusin[0] := 1 + phi / (1+phi^2):
> for n from 1 to 2*N-1 do
>    nusin[n] := (-1)^n * phi^n / (1+phi^2)^(n+1) *
>                sum ('binomial (n+1,2*k) * (-1)^k *
>                    phi^(n+1-2*k)', 'k'=0..floor ((n+1)/2));
> od:

> SackDonovan (L (0, x), alphahat, betahat, gammahat, nucos, N,
>                'alphacos', 'betacos');
> SackDonovan (L (0, x), alphahat, betahat, gammahat, nusin, N,
>                'alphasin', 'betasin');
```

we can compute the three-term recurrence relationship (18.21) of the orthogonal polynomials corresponding to the interval $[0, \infty]$ and the two weight functions $\omega_{\cos}$ and $\omega_{\sin}$. We get the results shown as Tables 18.1 and 18.2.

THEOREM 18.4. *The orthogonal polynomial* $p_k(x)$ *has exactly* k *distinct zeroes* $a < x_1 < \cdots < x_k < b$.

PROOF. The proof is by contradiction (cf. [20, p. 21]). Let us assume that the polynomial $p_k(x)$ has only $k' < k$ zeroes ζ_i with odd multiplicity in the interval $]a, b[$. Consequently, the polynomial

$$q(x) := \prod_{i=1}^{k'} (x - \zeta_i)$$

TABLE 18.2. *Values of $\alpha_n^{\sin}$ and $\beta_n^{\sin}$ for $\varphi = 1$.*

n	$\alpha_n^{\sin}$	$\beta_n^{\sin}$
1	$1 \approx 1.0000$	$\frac{\sqrt{6}}{3} \approx 0.81650$
2	$\frac{5}{2} \approx 2.5000$	$\frac{\sqrt{165}}{6} \approx 2.1409$
3	$\frac{127}{22} \approx 5.7727$	$\frac{18\sqrt{65}}{55} \approx 2.6386$
4	$\frac{2578}{429} \approx 6.0093$	$\frac{\sqrt{784190}}{195} \approx 4.5413$
5	$\frac{2747911}{278031} \approx 9.8835$	$\frac{830\sqrt{1365}}{7129} \approx 4.3015$
6	$\frac{14684038321}{1375127068} \approx 10.678$	$\frac{3\sqrt{176313765003}}{192892} \approx 6.5306$
7	$\frac{21638014309259}{1590195668348} \approx 13.607$	$\frac{16268\sqrt{11567141}}{8243969} \approx 6.7114$
8	$\frac{9020385319743068}{667514062758403} \approx 13.513$	$\frac{54\sqrt{165092626746123}}{80969987} \approx 8.5691$
9	$\frac{2469307519278826001}{131341029442952049} \approx 18.801$	$\frac{\sqrt{161559474255180636194}}{1622095227} \approx 7.8359$

has the same sign changes as $p_k(x)$ in this interval. This means that the product $p_k(x)q(x)$ has no sign change in the interval $]a, b[$, that

$$(p_k, q) \neq 0,$$

which contradicts the orthogonality property (18.13) of $p_k(x)$. □

The following theorem establishes a link between a sequence $\{p_k\}_{k=0}^n$ of polynomials, which are defined by the three-term recurrence relationship (18.21), on the one hand and the eigenvalue decomposition of the tridiagonal matrix T with the coefficients α_k and β_k on the other hand.

THEOREM 18.5. *A polynomial $p_n(x)$ of degree n, which can be represented by a three-term recurrence relationship (18.21) with $\beta_k \neq 0$, has as zeroes the eigenvalues $x_1 < \cdots < x_n$ of the tridiagonal matrix*

$$T_n := \begin{bmatrix} \alpha_1 & \beta_1 & & \\ \beta_1 & \ddots & \ddots & \\ & \ddots & \ddots & \beta_{n-1} \\ & & \beta_{n-1} & \alpha_n \end{bmatrix}. \tag{18.46}$$

Furthermore, let

$$T_n = QXQ^{\mathrm{T}} \tag{18.47}$$

be the eigenvalue decomposition of T_n, where X denotes the eigenvalue matrix with $x_1, \ldots, x_n$ as diagonal entries and Q denotes the orthogonal eigenvector matrix. Assume the definitions

$$P := \begin{bmatrix} p_0(x_1) & \cdots & p_0(x_n) \\ \vdots & & \vdots \\ p_{n-1}(x_1) & \cdots & p_{n-1}(x_n) \end{bmatrix} \tag{18.48}$$

and

$$W := \frac{1}{p_0} \begin{bmatrix} q_{11} & & 0 \\ & \ddots & \\ 0 & & q_{1n} \end{bmatrix}. \tag{18.49}$$

Then, we have

$$Q = PW. \tag{18.50}$$

PROOF. The three-term recurrence relationship (18.21) can be written in matrix terms as

$$x \begin{bmatrix} p_0 \\ \vdots \\ \vdots \\ p_{n-1} \end{bmatrix} = \begin{bmatrix} \alpha_1 & \beta_1 & & \\ \beta_1 & \ddots & \ddots & \\ & \ddots & \ddots & \beta_{n-1} \\ & & \beta_{n-1} & \alpha_n \end{bmatrix} \begin{bmatrix} p_0 \\ \vdots \\ \vdots \\ p_{n-1} \end{bmatrix} + \begin{bmatrix} 0 \\ \vdots \\ 0 \\ \beta_n p_n \end{bmatrix}. \tag{18.51}$$

It is well-known [23, p. 124], that T_n possesses n real, distinct eigenvalues $x_1 < \cdots < x_n$. Let us now consider an individual eigenvector $\mathbf{q}$ corresponding to an eigenvalue λ:

$$T_n \mathbf{q} = \lambda \mathbf{q}. \tag{18.52}$$

This is equivalent to the n equations

$$\begin{aligned} \alpha_1 q_1 + \beta_1 q_2 &= \lambda q_1, \\ \beta_{k-1} q_{k-1} + \alpha_k q_k + \beta_k q_{k+1} &= \lambda q_k, \qquad k = 2, \ldots, n-1, \\ \beta_{n-1} q_{n-1} + \alpha_n q_n &= \lambda q_n. \end{aligned} \tag{18.53}$$

Therefore, the elements of the vector $\mathbf{q}$ satisfy the three-term recurrence relationship

$$\begin{aligned} q_2 &= \frac{\lambda - \alpha_1}{\beta_1} q_1, \\ q_{k+1} &= \frac{\lambda - \alpha_k}{\beta_k} q_k - \frac{\beta_{k-1}}{\beta_k} q_{k-1}, \qquad k = 2, \ldots, n-1, \end{aligned} \tag{18.54}$$

where q_1 is chosen so that $\|\mathbf{q}\|_2 = 1$. Since $\beta_k \neq 0$, it is clear that the first element of each eigenvector must be different from zero. Otherwise, the whole vector would consist of zeroes—an obvious contradiction to the definition of an eigenvector.

On the other hand, we define

$$\mathbf{p}(x) := \begin{bmatrix} p_0(x) \\ \vdots \\ p_{n-1}(x) \end{bmatrix}. \tag{18.55}$$

From (18.21) we can see that the elements of the vector $\mathbf{p}(\lambda)$ obey the three-term recurrence relationship

$$p_1(\lambda) = \frac{\lambda - \alpha_1}{\beta_1} p_0(\lambda),$$

$$p_k(\lambda) = \frac{\lambda - \alpha_k}{\beta_k} p_{k-1}(\lambda) - \frac{\beta_{k-1}}{\beta_k} p_{k-2}(\lambda), \qquad k = 2, \ldots, n, \tag{18.56}$$

which is indeed the same as (18.54) except for a different initial value. Consequently, the vectors $\mathbf{q}$ and $\mathbf{p}(\lambda)$ must be equal up to a scaling factor:

$$\mathbf{q} = \frac{q_1}{p_0} \mathbf{p}(\lambda). \tag{18.57}$$

Because of $\beta_n \neq 0$, Equation (18.51) implies $p_n(\lambda) = 0$. If we evaluate (18.57) for each eigenvalue x_k, we get (18.50). $\square$

18.3   Quadrature Rule

Let us now consider the problem of constructing a quadrature rule

$$\int_a^b f(x)\omega(x)\, dx \approx \sum_{i=1}^{m+n} w_i f(x_i) \tag{18.58}$$

which integrates polynomials exactly up to the maximal degree possible. We assume that the m abscissas $x_1 < \cdots < x_m$ on the boundary of the interval $[a, b]$ are prescribed. Of course, m can only take the values $m = 0$, $m = 1$, and $m = 2$. Let

$$r(x) := \prod_{i=1}^{m}(x - x_i) \tag{18.59}$$

be a polynomial of degree m with the prescribed abscissas as zeroes. Furthermore let

$$s(x) := \prod_{i=m+1}^{m+n}(x - x_i) \tag{18.60}$$

be the unknown polynomial of degree n with the abscissas $x_{m+1}, \ldots, x_{m+n}$ as zeroes.

The following theorem represents the key for computing a Gauss quadrature rule (cf. [20, p. 161]).

THEOREM 18.6. *The Gauss quadrature rule (18.58) is exact for all polynomials p up to degree $m + 2n - 1$ if and only if the following two conditions are met:*

1. *The Gauss quadrature rule (18.58) is exact for all polynomials p up to degree $m + n - 1$.*

2. *The equation*

$$\int_a^b r(x)s(x)p(x)\omega(x)\, dx = 0 \tag{18.61}$$

   applies for all polynomials p up to degree $n - 1$.

PROOF. First let the Gauss quadrature rule (18.58) be exact for all polynomials up to degree $m + 2n - 1$. Trivially it will be exact for all polynomials up to degree $m + n - 1$. The polynomial $r(x)s(x)p(x)$ is of degree $\leq m + 2n - 1$ and thus is integrated exactly:

$$\int_a^b r(x)s(x)p(x)\omega(x)\,dx = \sum_{i=1}^{m+n} w_i r(x_i)s(x_i)p(x_i) = 0.$$

On the other hand assume that the two conditions 1 and 2 are satisfied. A polynomial $t(x)$ of degree $\leq m + 2n - 1$ can be factored according to

$$t(x) = r(x)s(x)p(x) + q(x),$$

where p and q are polynomials of degrees $\leq n - 1$ and $\leq m + n - 1$ respectively. We have

$$\int_a^b t(x)\omega(x)\,dx = \int_a^b r(x)s(x)p(x)\omega(x)\,dx + \int_a^b q(x)\omega(x)\,dx$$
$$= \int_a^b q(x)\omega(x)\,dx.$$

But for polynomials of degree $\leq m + n - 1$ the quadrature rule is exact, and we have

$$\int_a^b q(x)\omega(x)\,dx = \sum_{i=1}^{m+n} w_i q(x_i) = \sum_{i=1}^{m+n} w_i t(x_i),$$

since $t(x_i) = q(x_i)$. Consequently

$$\int_a^b t(x)\omega(x)\,dx = \sum_{i=1}^{m+n} w_i t(x_i),$$

and the quadrature rule is exact for all polynomials up to degree $m + 2n - 1$. □

Now, let us consider the construction of quadrature rules for the interesting cases $m = 0$, $m = 1$, and $m = 2$. To this end we have to determine the $m + n$ abscissas x_i in such a way that condition 2 of Theorem 18.6 is satisfied. This calls for a different procedure in each of the three cases.

After this we will have to determine the weights w_i such that condition 1 of Theorem 18.6 is satisfied. This problem is tackled in Section 18.7. The results are presented in MAPLE as Algorithms 18.4, 18.5, and 18.6.

18.4 Gauss Quadrature Rule

In the case of a Gauss quadrature rule no abscissa is prescribed, i.e. $m = 0$ and $r(x) = 1$. If we choose $s(x) = p_n(x)$ condition 2 of Theorem 18.6 is satisfied. Therefore, the zeroes of $p_n(x)$, i.e. the eigenvalues of the matrix T_n, represent the desired abscissas x_i (cf. [18]).

ALGORITHM 18.4. *Gauss Quadrature Rule.*

```
Gauss := proc (mu0, alpha, beta, n, x, w)
    local D, i, Q, T;

    T := array (1..n, 1..n, symmetric, [[0$n]$n]);
    for i from 1 to n do
      T[i,i] := alpha[i];
    od;
    for i from 1 to n-1 do
      T[i,i+1] := beta[i];
      T[i+1,i] := beta[i];
    od;

    D := evalf (Eigenvals (T, Q));
    x := array (1..n, [seq (D[i], i=1..n)]);
    w := array (1..n, [seq (evalf (mu0) * Q[1,i]^2, i=1..n)]);

    RETURN (NULL);
  end:
```

18.5   Gauss-Radau Quadrature Rule

In the case of a Gauss-Radau quadrature rule we prescribe exactly one abscissa x_1 on the boundary of the interval $[a, b]$, i.e. $m = 1$ and $r(x) = x - a$ or $r(x) = x - b$. Instead of directly computing a polynomial $s(x)$ that satisfies condition 2 of Theorem 18.6, we construct the polynomial $\tilde{p}_{n+1}(x) = r(x)s(x)$ of degree $n + 1$. We will determine this polynomial from the following two requirements:

1. $\tilde{p}_{n+1}$ contains s as a factor:

$$\tilde{p}_{n+1}(x_1) = 0. \tag{18.62}$$

2. The equation

$$\int_a^b \tilde{p}_{n+1}(x)p(x)\omega(x)\, dx = 0 \tag{18.63}$$

applies for all polynomials p up to degree $n - 1$.

We start with the following implicit representation of $\tilde{p}_{n+1}$ (cf. [16]):

$$x \begin{bmatrix} p_0 \\ \vdots \\ \vdots \\ p_n \end{bmatrix} = \begin{bmatrix} \alpha_1 & \beta_1 & & \\ \beta_1 & \ddots & \ddots & \\ & \ddots & \alpha_n & \beta_n \\ & & \beta_n & \tilde{\alpha}_{n+1} \end{bmatrix} \begin{bmatrix} p_0 \\ \vdots \\ \vdots \\ p_n \end{bmatrix} + \begin{bmatrix} 0 \\ \vdots \\ 0 \\ \tilde{p}_{n+1} \end{bmatrix}, \tag{18.64}$$

where the value of $\tilde{\alpha}_{n+1}$ has yet to be determined.

First we demonstrate that this choice of $\tilde{p}_{n+1}$ satisfies (18.63) for an arbitrary $\tilde{\alpha}_{n+1}$. Because of the last Equation (18.64)

$$x p_n = \beta_n p_{n-1} + \tilde{\alpha}_{n+1} p_n + \tilde{p}_{n+1}$$

and the recurrence relationship (18.21)

$$x p_n = \beta_n p_{n-1} + \alpha_{n+1} p_n + \beta_{n+1} p_{n+1}$$

we can represent the polynomial $\tilde{p}_{n+1}$ as

$$\tilde{p}_{n+1} = (\alpha_{n+1} - \tilde{\alpha}_{n+1}) p_n + \beta_{n+1} p_{n+1}. \tag{18.65}$$

Obviously, the polynomial $\tilde{p}_{n+1}$ is a linear combination of p_n and p_{n+1} and always satisfies (18.63).

We only have to determine the value of $\tilde{\alpha}_{n+1}$ such that $\tilde{p}_{n+1}(x_1) = 0$, or, equivalently, that x_1 is an eigenvalue of the matrix in Equation (18.64). For this, consider the eigenvalue equation

$$x_1 \begin{bmatrix} \mathbf{y} \\ \hline \eta \end{bmatrix} = \left[\begin{array}{c|c} T_n & \\ & \beta_n \\ \hline \beta_n & \tilde{\alpha}_{n+1} \end{array} \right] \begin{bmatrix} \mathbf{y} \\ \hline \eta \end{bmatrix},$$

or, equivalently,

$$x_1 \mathbf{y} = T_n \mathbf{y} + \beta_n \eta \mathbf{e}_n, \tag{18.66}$$
$$x_1 \eta = \beta_n y_n + \tilde{\alpha}_{n+1} \eta. \tag{18.67}$$

From (18.66) we see that $\mathbf{y}$ is a multiple of $(T_n - x_1 I)^{-1} \mathbf{e}_n$. Therefore, let

$$\mathbf{y} := (T_n - x_1 I)^{-1} \mathbf{e}_n. \tag{18.68}$$

Then, it follows from (18.66) that

$$\eta = -\frac{1}{\beta_n}. \tag{18.69}$$

Note that the matrix $T_n - x_1 I$ is positive or negative definite depending on the choice of x_1. In each case the value of the vector $\mathbf{y}$ is uniquely determined.

After substituting (18.69) into (18.67) we get

$$\tilde{\alpha}_{n+1} = x_1 + \beta_n^2 y_n. \tag{18.70}$$

Consequently, we have determined the polynomial $\tilde{p}_{n+1}(x)$ whose zeroes represent the abscissas of the Gauss-Radau quadrature rule.

ALGORITHM 18.5. *Gauss-Radau Quadrature Rule.*

```
Radau := proc (mu0, alpha, beta, n, x1, x, w)
  local alphan1tilde, D, e, i, Q, T, y;

  T := array (1..n, 1..n, symmetric, [[0$n]$n]);
  for i from 1 to n do
    T[i,i] := alpha[i] - x1;
  od;
  for i from 1 to n-1 do
    T[i,i+1] := beta[i];
    T[i+1,i] := beta[i];
  od;

  e := linalg[vector] (n, 0);  e[n] := 1;
  y := linalg[linsolve] (T, e);
  alphan1tilde := simplify (x1 + beta[n]^2 * y[n]);

  T := array (1..n+1, 1..n+1, symmetric, [[0$n+1]$n+1]);
  for i from 1 to n do
    T[i,i] := alpha[i];
  od;
  T[n+1,n+1] := alphan1tilde;
  for i from 1 to n do
    T[i,i+1] := beta[i];
    T[i+1,i] := beta[i];
  od;

  D := evalf (Eigenvals (T, Q));
  x := array (1..n+1, [seq (D[i], i=1..n+1)]);
  w := array (1..n+1,
              [seq (evalf (mu0) * Q[1,i]^2, i=1..n+1)]);

  RETURN (NULL);
end:
```

18.6 Gauss-Lobatto Quadrature Rule

In the case of a Gauss-Lobatto quadrature rule we prescribe the two abscissas $x_1 = a$ and $x_2 = b$, i.e. $m = 2$ and $r(x) = (x - a)(x - b)$. Instead of computing a polynomial $s(x)$ that satisfies condition 2 of Theorem 18.6, we construct the polynomial $\tilde{p}_{n+2}(x) = r(x)s(x)$ of degree $n + 2$. We will determine this polynomial from the following two requirements:

1. $\tilde{p}_{n+2}$ contains s as a factor:

$$\tilde{p}_{n+2}(x_1) = \tilde{p}_{n+2}(x_2) = 0. \tag{18.71}$$

2. The equation

$$\int_a^b \tilde{p}_{n+2}(x)p(x)\omega(x)\,dx = 0 \tag{18.72}$$

applies for all polynomials p up to degree $n - 1$.

We choose for $\tilde{p}_{n+2}$ the polynomial which is defined implicitly by the matrix equation (cf. [16])

$$x \begin{bmatrix} p_0 \\ \vdots \\ \vdots \\ p_n \\ \tilde{p}_{n+1} \end{bmatrix} = \begin{bmatrix} \alpha_1 & \beta_1 & & & \\ \beta_1 & \ddots & \ddots & & \\ & \ddots & \ddots & \beta_n & \\ & & \beta_n & \alpha_{n+1} & \tilde{\beta}_{n+1} \\ & & & \tilde{\beta}_{n+1} & \tilde{\alpha}_{n+2} \end{bmatrix} \begin{bmatrix} p_0 \\ \vdots \\ \vdots \\ p_n \\ \tilde{p}_{n+1} \end{bmatrix} + \begin{bmatrix} 0 \\ \vdots \\ \vdots \\ 0 \\ \tilde{p}_{n+2} \end{bmatrix}, \tag{18.73}$$

where $\tilde{\beta}_{n+1} \neq 0$ and $\tilde{\alpha}_{n+2}$ are yet unspecified.

We now show that this choice of $\tilde{p}_{n+2}$ will meet condition (18.72) for any $\tilde{\beta}_{n+1}$ and $\tilde{\alpha}_{n+2}$. In view of (18.21) the penultimate equation of (18.73)

$$xp_n = \beta_n p_{n-1} + \alpha_{n+1} p_n + \tilde{\beta}_{n+1} \tilde{p}_{n+1}$$

defines the polynomial $\tilde{p}_{n+1}$ as a multiple of p_{n+1}:

$$\tilde{p}_{n+1} = \frac{\beta_{n+1}}{\tilde{\beta}_{n+1}} p_{n+1}. \tag{18.74}$$

By substituting (18.74) into the last equation of (18.73)

$$x\tilde{p}_{n+1} = \tilde{\beta}_{n+1} p_n + \tilde{\alpha}_{n+2} \tilde{p}_{n+1} + \tilde{p}_{n+2}$$

we get

$$x\frac{\beta_{n+1}}{\tilde{\beta}_{n+1}} p_{n+1} = \tilde{\beta}_{n+1} p_n + \tilde{\alpha}_{n+2} \frac{\beta_{n+1}}{\tilde{\beta}_{n+1}} p_{n+1} + \tilde{p}_{n+2}.$$

Due to

$$xp_{n+1} = \beta_{n+1} p_n + \alpha_{n+2} p_{n+1} + \beta_{n+2} p_{n+2}$$

we can represent the polynomial $\tilde{p}_{n+2}$ by

$$\tilde{p}_{n+2} = (\frac{\beta_{n+1}^2}{\tilde{\beta}_{n+1}} - \tilde{\beta}_{n+1})p_n + \frac{\beta_{n+1}}{\tilde{\beta}_{n+1}}(\alpha_{n+2} - \tilde{\alpha}_{n+2})p_{n+1} + \frac{\beta_{n+1}\beta_{n+2}}{\tilde{\beta}_{n+1}}p_{n+2}. \quad (18.75)$$

The main observation here is that $\tilde{p}_{n+2}$ is given by a linear combination of p_n, p_{n+1}, and p_{n+2}, so that Equation (18.72) is always satisfied.

Now we will determine the values of $\tilde{\beta}_{n+1} \neq 0$ and $\tilde{\alpha}_{n+2}$ such that $\tilde{p}_{n+2}(a) = \tilde{p}_{n+2}(b) = 0$. But this is equivalent to requiring that the matrix in (18.73) possesses the two eigenvalues a and b. Consider the eigenvalue equations

$$a \left[\begin{array}{c} \mathbf{y} \\ \hline \eta \end{array} \right] = \left[\begin{array}{c|c} T_{n+1} & \\ & \tilde{\beta}_{n+1} \\ \hline \tilde{\beta}_{n+1} & \tilde{\alpha}_{n+2} \end{array} \right] \left[\begin{array}{c} \mathbf{y} \\ \hline \eta \end{array} \right]$$

and

$$b \left[\begin{array}{c} \mathbf{z} \\ \hline \zeta \end{array} \right] = \left[\begin{array}{c|c} T_{n+1} & \\ & \tilde{\beta}_{n+1} \\ \hline \tilde{\beta}_{n+1} & \tilde{\alpha}_{n+2} \end{array} \right] \left[\begin{array}{c} \mathbf{z} \\ \hline \zeta \end{array} \right],$$

or, equivalently,

$$a\mathbf{y} = T_{n+1}\mathbf{y} + \tilde{\beta}_{n+1}\eta\mathbf{e}_{n+1}, \quad (18.76)$$

$$a\eta = \tilde{\beta}_{n+1}y_{n+1} + \tilde{\alpha}_{n+2}\eta, \quad (18.77)$$

$$b\mathbf{z} = T_{n+1}\mathbf{z} + \tilde{\beta}_{n+1}\zeta\mathbf{e}_{n+1}, \quad (18.78)$$

$$b\zeta = \tilde{\beta}_{n+1}z_{n+1} + \tilde{\alpha}_{n+2}\zeta. \quad (18.79)$$

We can now see from the Equations (18.76) and (18.78) that the vectors $\mathbf{y}$ and $\mathbf{z}$ are multiples of $(T_{n+1} - aI)^{-1}\mathbf{e}_{n+1}$ and $(T_{n+1} - bI)^{-1}\mathbf{e}_{n+1}$, respectively. If we define $\mathbf{y}$ and $\mathbf{z}$ by

$$\mathbf{y} := (T_{n+1} - aI)^{-1}\mathbf{e}_{n+1}, \quad (18.80)$$

$$\mathbf{z} := (T_{n+1} - bI)^{-1}\mathbf{e}_{n+1}, \quad (18.81)$$

Equations (18.76) and (18.78) imply

$$\eta = \zeta = -\frac{1}{\tilde{\beta}_{n+1}}. \quad (18.82)$$

Furthermore $T_{n+1} - aI$ and $T_{n+1} - bI$ represent a positive and negative definite matrix, respectively. In particular we have the inequality

$$y_{n+1} > 0 > z_{n+1}. \quad (18.83)$$

After substituting (18.82) into (18.77) and (18.79), and after multiplying by $\tilde{\beta}_{n+1}$ we get the linear system

$$\left[\begin{array}{cc} 1 & -y_{n+1} \\ 1 & -z_{n+1} \end{array} \right] \left[\begin{array}{c} \tilde{\alpha}_{n+2} \\ \tilde{\beta}_{n+1}^2 \end{array} \right] = \left[\begin{array}{c} a \\ b \end{array} \right]. \quad (18.84)$$

ALGORITHM 18.6. *Gauss-Lobatto Quadrature Rule.*

```
Lobatto := proc (mu0, alpha, beta, n, x1, x2, x, w)
  local alphan2tilde, betan1tilde, D, e, i, Q, T, y, z;

  T := array (1..n+1, 1..n+1, symmetric, [[0$n+1]$n+1]);
  for i from 1 to n+1 do
    T[i,i] := alpha[i] - x1;
  od;
  for i from 1 to n do
    T[i,i+1] := beta[i];
    T[i+1,i] := beta[i];
  od;

  e := linalg[vector] (n+1, 0);  e[n+1] := 1;
  y := linalg[linsolve] (T, e);
  for i from 1 to n+1 do
    T[i,i] := alpha[i] - x2;
  od;
  z := linalg[linsolve] (T, e);
  alphan2tilde := simplify ((x2 * y[n+1] - x1*z[n+1]) /
                            (y[n+1] - z[n+1]));
  betan1tilde := simplify (sqrt ((x2-x1) /
                            (y[n+1] - z[n+1])));

  T := array (1..n+2, 1..n+2, symmetric, [[0$n+2]$n+2]);
  for i from 1 to n+1 do
    T[i,i] := alpha[i];
  od;
  T[n+2,n+2] := alphan2tilde;
  for i from 1 to n do
    T[i,i+1] := beta[i];
    T[i+1,i] := beta[i];
  od;
  T[n+1,n+2] := betan1tilde;
  T[n+2,n+1] := betan1tilde;

  D := evalf (Eigenvals (T, Q));
  x := array (1..n+2, [seq (D[i], i=1..n+2)]);
  w := array (1..n+2,
              [seq (evalf (mu0) * Q[1,i]^2, i=1..n+2)]);

  RETURN (NULL);
end:
```

Because of (18.83) the linear system (18.84) has the unique solution

$$\tilde{\alpha}_{n+2} = \frac{by_{n+1} - az_{n+1}}{y_{n+1} - z_{n+1}}, \tag{18.85}$$

$$\tilde{\beta}_{n+1} = \sqrt{\frac{b - a}{y_{n+1} - z_{n+1}}} > 0. \tag{18.86}$$

This means that we have eventually determined the polynomial $\tilde{p}_{n+2}(x)$ whose zeroes represent the abscissas of the Gauss-Lobatto quadrature rule.

18.7   Weights

The procedure for computing the weights is the same for each of the above mentioned quadrature rules. We have to determine the weights $w_1, \ldots, w_{m+n}$ such that condition 1 of Theorem 18.6 is met. Equivalently, we may require that the polynomials $p_0, \ldots, p_n$ and $\tilde{p}_{n+1}, \ldots, \tilde{p}_{m+n-1}$ from (18.64) and (18.73) are integrated exactly:

$$\int_a^b p_k(x)\omega(x)\,dx = \sum_{i=1}^{m+n} w_i p_k(x_i) = \frac{1}{p_0}\delta_{k0}, \qquad k = 0, \ldots, \min(n, m+n-1),$$

$$\int_a^b \tilde{p}_k(x)\omega(x)\,dx = \sum_{i=1}^{m+n} w_i \tilde{p}_k(x_i) = 0, \qquad k = n+1, \ldots, m+n-1.$$

In matrix terms this can be written as

$$\begin{bmatrix} p_0(x_1) & \cdots & \cdots & p_0(x_{m+n}) \\ \vdots & & & \vdots \\ \vdots & & & \vdots \\ \tilde{p}_{m+n-1}(x_1) & \cdots & \cdots & \tilde{p}_{m+n-1}(x_{m+n}) \end{bmatrix} \begin{bmatrix} w_1 \\ \vdots \\ \vdots \\ w_{m+n} \end{bmatrix} = \begin{bmatrix} 1/p_0 \\ 0 \\ \vdots \\ 0 \end{bmatrix}$$

or $P\mathbf{w} = \mathbf{e}_1/p_0$. Because of Theorem 18.5 we have $Q = PW$, such that

$$\mathbf{w} = \frac{1}{p_0} W Q^{\mathsf{T}} \mathbf{e}_1.$$

From (18.14) and (18.6) it follows that

$$\int_a^b p_0^2(x)\omega(x)\,dx = p_0^2\mu_0 = 1,$$

such that finally the weights w_i are given by

$$w_i = \mu_0 q_{1i}^2, \qquad i = 1, \ldots, m+n. \tag{18.87}$$

As an application we will now compute a Gauss-Lobatto quadrature rule corresponding to the interval $[1, 2]$ and the weight function

$$\omega(x) := \frac{1}{x}. \tag{18.88}$$

The following MAPLE statements compute the three-term recurrence relationship (18.21) of the orthogonal polynomials. Then, the quadrature rule is computed by a call of Algorithm 18.6:

```
> N := 11:
> Digits := 100:
> a := 1:
> b := 2:
> omega := x -> 1 / x:

> alphahat := array (1..2*N-1, [0$2*N-1]):
> betahat := array (1..2*N-1, [1$2*N-1]):
> gammahat := array (1..2*N-2, [0$2*N-2]):

> pi0 := 1:
> nu[0] := ln (b) - ln (a):
> for n from 1 to 2*N-1 do
>   nu[n] := (b^n - a^n) / n;
> od:
> mu0 := nu[0] / pi0:

> SackDonovan (pi0, alphahat, betahat, gammahat, nu, N,
>              'alpha', 'beta');
> Lobatto (mu0, evalf(op(alpha)), evalf(op(beta)), N-1, a, b,
              'x', 'w');
```

The abscissas and weights are shown as Table 18.3. The results have been rounded to twenty decimal digits.

18.8 Quadrature Error

We will now consider the question how to express the quadrature error

$$E[f] := \int_a^b f(x)\omega(x)\,dx - \sum_{i=1}^{m+n} w_i f(x_i). \tag{18.89}$$

We assume that the integrand f is a $(m + 2n)$-times continuously differentiable function on the interval $[a, b]$. Furthermore, let $H(x)$ denote the uniquely determined polynomial of degree $\leq m + 2n - 1$, which satisfies the Hermite interpolation conditions (cf. [29, p. 44])

$$H(x_i) = f(x_i), \qquad i = 1, \ldots, m + n,$$
$$H'(x_i) = f'(x_i), \qquad i = m + 1, \ldots, m + n.$$

First, we examine how well the polynomial $H(x)$ approximates the given function $f(x)$ on the interval $[a, b]$.

THEOREM 18.7. *Let*

$$Q(x) := \prod_{i=1}^{m}(x - x_i) \cdot \prod_{i=m+1}^{m+n}(x - x_i)^2 \tag{18.90}$$

TABLE 18.3.
Abscissas and Weights of a Gauss-Lobatto Quadrature
Rule.

k	x_k	w_k
1	1.0000000000000000000	0.0073101161434772987648
2	1.0266209022507358746	0.043205676918139196040
3	1.0875753284733094064	0.070685480638318232547
4	1.1786507622317286926	0.088433599898292821869
5	1.2936151328648399126	0.096108531378420052625
6	1.4243072396520977438	0.095097966102219702354
7	1.5611522948208446896	0.087445652257865527102
8	1.6938824042052787649	0.075118353999491464492
9	1.8123676368833335432	0.059695944656549921510
10	1.9074586992416619710	0.042332113537812345903
11	1.9717495114038218100	0.023827527632972240207
12	2.0000000000000000000	0.0038862173963865060038

denote a polynomial of degree $m + 2n$. Then for each x in the interval $[a, b]$ there exists a ξ with $a < \xi < b$ such that

$$f(x) = H(x) + \frac{f^{(m+2n)}(\xi)}{(m + 2n)!} Q(x). \tag{18.91}$$

PROOF. For $x = x_i$ equation (18.91) is trivially satisfied. Therefore we assume that $x \neq x_i$ and $x \in [a, b]$ in the following. Let us consider the function

$$F(t) := f(t) - H(t) - \frac{f(x) - H(x)}{Q(x)} Q(t). \tag{18.92}$$

Obviously, $F(t)$ possesses at least $m + 2n + 1$ zeroes in the interval $[a, b]$. As a consequence of Rolle's theorem $F^{(m+2n)}(t)$ has at least one zero ξ in the interior of $[a, b]$:

$$F^{(m+2n)}(\xi) = f^{(m+2n)}(\xi) - \frac{f(x) - H(x)}{Q(x)}(m + 2n)! = 0.$$

But this establishes Equation (18.91). $\square$

The reader is referred to [20, p. 49] and [29, p. 48] for similar proofs.

After these preparations we are in a position to give a quantitative expression for the quadrature error (18.89). See also [20, pp. 162–163] and [29, p. 134].

THEOREM 18.8. *The error of the quadrature rule (18.58) is given by*

$$E[f] = \frac{f^{(m+2n)}(\xi)}{(m+2n)!} \int_a^b Q(x)\omega(x)\, dx \tag{18.93}$$

with $a < \xi < b$.

PROOF. According to (18.91) we can represent the integrand $f(x)$ for $a \leq x \leq b$ by

$$f(x) = H(x) + \frac{f^{(m+2n)}(\xi(x))}{(m+2n)!}Q(x) \tag{18.94}$$

with $a < \xi(x) < b$. Because of

$$f^{(m+2n)}(\xi(x)) = (m+2n)!\frac{f(x) - H(x)}{Q(x)}$$

we have in $f^{(m+2n)}(\xi(x))$ a continuous function for $a \leq x \leq b$.

The integration of the representation (18.94) of $f(x)$ yields

$$\int_a^b f(x)\omega(x)\, dx = \int_a^b H(x)\omega(x)\, dx + \\ \frac{1}{(m+2n)!} \int_a^b f^{(m+2n)}(\xi(x))Q(x)\omega(x)\, dx. \tag{18.95}$$

The interpolating polynomial $H(x)$ is of degree $\leq m + 2n - 1$ and is therefore integrated exactly:

$$\int_a^b H(x)\omega(x)\, dx = \sum_{i=1}^{m+n} w_i H(x_i) = \sum_{i=1}^{m+n} w_i f(x_i). \tag{18.96}$$

On the other hand $Q(x)$ has a constant sign on $[a, b]$. This allows us to make use of the generalized mean value theorem of integral calculus [19, p. 477], giving

$$\int_a^b f^{(m+2n)}(\xi(x))Q(x)\omega(x)\, dx = f^{(m+2n)}(\xi) \int_a^b Q(x)\omega(x)\, dx \tag{18.97}$$

for $a < \xi < b$. After substituting (18.96) and (18.97) into (18.95) we get

$$\int_a^b f(x)\omega(x)\, dx = \sum_{i=1}^{m+n} w_i f(x_i) + \frac{f^{(m+2n)}(\xi)}{(m+2n)!} \int_a^b Q(x)\omega(x)\, dx$$

with $a < \xi < b$. This establishes Equation (18.93). $\square$

References

[1] N. I. AKHIEZER, *The Classical Moment Problem*, translated by N. Kemmer, Oliver & Boyd, Edinburgh and London, 1965.

[2] J. BOUZITAT, *Sur l'intégration numérique approchée par la méthode de Gauss généralisée et sur une extension de cette méthode*, C. R. Acad. Sci. Paris, 229 (1949), pp. 1201–1203.

[3] E. B. CHRISTOFFEL, *Über die Gaussische Quadratur und eine Verallgemeinerung derselben*, J. Reine Angew. Math., 55 (1858), pp. 61–82.

[4] P. J. DAVIS AND P. RABINOWITZ, *Methods of Numerical Integration*, Academic Press, Orlando, 1984.

[5] R. N. DESMARAIS, *Programs for computing abscissas and weights for classical and non-classical Gaussian quadrature formulas*, NASA Report TN D-7924, NASA Langley Research Center, Hampton VA, 1975.

[6] A. ERDÉLYI ET AL., *Higher Transcendental Functions*, Bateman Manuscript Project, McGraw-Hill, New York, 1953.

[7] C. F. GAUSS, *Methodus Nova Integralium Valores per Approximationem Inveniendi*, Werke, Vol. 3, Göttingen, 1866, pp. 163–196.

[8] W. GAUTSCHI, *Computational aspects of three-term recurrence relations*, SIAM Review, 9 (1967), pp. 24–82.

[9] W. GAUTSCHI, *Construction of Gauss-Christoffel Quadrature Formulas*, Math. Comp., 22 (1968), pp. 251–270.

[10] W. GAUTSCHI, *On the Construction of Gaussian Quadrature Rules from Modified Moments*, Math. Comp., 24 (1970), pp. 245–260.

[11] W. GAUTSCHI, *Minimal Solutions of Three-Term Recurrence Relations and Orthogonal Polynomials*, Math. Comp., 36 (1981), pp. 547–554.

[12] W. GAUTSCHI, *A Survey of Gauss-Christoffel Quadrature Formulae*, in E. B. Christoffel, The Influence of His Work on Mathematics and the Physical Sciences, ed. P. L. Butzer and F. Fehér, Birkhäuser, Basel, 1981, pp. 72–147.

[13] W. GAUTSCHI, *An algorithmic implementation of the generalized Christoffel theorem*, in Numerical Integration, Internat. Ser. Numer. Math., ed. G. Hämmerlin, Birkhäuser, Basel, 57 (1982), pp. 89–106.

[14] W. GAUTSCHI, *Questions of numerical condition related to polynomials*, in Studies in Mathematics, Volume 24: Studies in Numerical Analysis, ed. G. H. Golub, Math. Assoc. Amer., Washington, 1984, pp. 140–177.

[15] W. GAUTSCHI, *Orthogonal polynomials—constructive theory and applications*, J. Comput. Appl. Math., 12&13 (1985), pp. 61–76.

[16] G. H. GOLUB, *Some Modified Matrix Eigenvalue Problems*, SIAM Review, 15 (1973), pp. 318–334.

[17] G. H. GOLUB AND C. F. VAN LOAN, *Matrix Computations*, Second Edition, The Johns Hopkins University Press, Baltimore, 1989.

[18] G. H. GOLUB AND J. H. WELSCH, *Calculation of Gauss Quadrature Rules*, Math. Comp., 23 (1969), pp. 221–230.

[19] H. HEUSER, *Lehrbuch der Analysis, Teil 1*, Teubner, Stuttgart, 1986.

[20] V. I. KRYLOV, *Approximate Calculation of Integrals*, translated by A. H. Stroud, Macmillan, New York, 1962.

[21] A. MARKOFF, *Sur la méthode de Gauss pour le calcul approché des intégrales*, Math. Ann., 25 (1885), pp. 427–432.

[22] I. P. MYSOVSKIH, *On the Construction of Cubature Formulas with Fewest Nodes*, Soviet Math. Dokl., 9 (1968), pp. 277–280.

[23] B. N. PARLETT, *The Symmetric Eigenvalue Problem*, Prentice-Hall, Englewood Cliffs, 1980.

[24] W. H. PRESS AND S. A. TEUKOLSKY, *Orthogonal Polynomials and Gaussian Quadrature with Non-classical Weight Functions*, Computers in Physics, 4 (1990), pp. 423–426.

[25] P. RABINOWITZ, *Abscissas and Weights for Lobatto Quadrature of High Order*, Math. Comp., 14 (1960), pp. 47–52.

[26] R. RADAU, *Étude sur les formules d'approximation qui servent à calculer la valeur numérique d'une intégrale définie*, J. Math. Pures Appl., Ser. 3, 6 (1880), pp. 283–336.

[27] R. A. SACK AND A. F. DONOVAN, *An Algorithm for Gaussian Quadrature given Modified Moments*, Numer. Math., 18 (1972), pp. 465–478.

[28] J. A. SHOHAT AND J. D. TAMARKIN, *The Problem of Moments*, Second Edition, American Mathematical Society, New York, 1950.

[29] J. STOER, *Einführung in die Numerische Mathematik I*, Springer-Verlag, 1983.

[30] J. STOER AND R. BULIRSCH, *Einführung in die Numerische Mathematik II*, Springer-Verlag, 1978.

[31] A. H. STROUD AND D. SECREST, *Gaussian Quadrature Formulas*, Prentice-Hall, Englewood Cliffs, New Jersey, 1966.

[32] G. SZEGŐ, *Orthogonal Polynomials*, American Mathematical Society, New York, 1939.

[33] H. S. WALL, *Analytic Theory of Continued Fractions*, van Nostrand, New York, 1948.

[34] J. C. WHEELER, *Modified moments and Gaussian quadratures*, Rocky Mountain J. of Math., 4 (1974), pp. 287–296.

Chapter 19. Symbolic Computation of Explicit Runge-Kutta Formulas

D. Gruntz

19.1 Introduction

In this chapter we show how MAPLE can be used to derive explicit Runge-Kutta formulas which are used in numerical analysis to solve systems of differential equations of the first order. We show how the nonlinear system of equations for the coefficients of the Runge-Kutta formulas are constructed and how such a system can be solved. We close the chapter with an overall procedure to construct Runge-Kutta formulas for a given size and order. We will see up to which size such a general purpose program is capable of solving the equations obtained.

The solution of the initial value problem

$$y'(x) = f(x, y(x)), \qquad y(x_k) = y_k \tag{19.1}$$

can be approximated by a Taylor series around x_k, which is obtained from (19.1) by repeated differentiation and replacing $y'(x)$ by $f(x, y(x)))$ every time it appears.

$$
\begin{aligned}
y(x_k + h) &= \sum_{i=0}^{\infty} y^{(i)}(x_k)\tfrac{h^i}{i!} \\
&= y(x_k) + hf(x_k, y(x_k)) + \tfrac{h^2}{2}\left(\left.\tfrac{\partial}{\partial x}f(x, y(x))\right|_{x=x_k}\right) + \ldots \\
&= y(x_k) + h\Big(f(x_k, y(x_k)) \\
&\qquad + \underbrace{\tfrac{h}{2}\big(f_x(x_k, y(x_k)) + f(x_k, y(x_k))f_y(x_k, y(x_k))\big) + \ldots}_{\Phi(x_k, y(x_k), h)}\Big)
\end{aligned}
$$

$$\tag{19.2}$$

A numerical approximation to the solution of (19.1) can be computed by using only a few terms of (19.2) with h sufficiently small and by repeating this computation from the point $x_{k+1} = x_k + h$.

The idea of the Runge-Kutta methods is to approximate the Taylor series (19.2) up to order m by using only *values* of $f(x, y(x))$ and no derivatives of it. For example, the Taylor series of the improved Euler method,

$$y_{k+1} = y_k + h\, f\Big(x_k + \frac{h}{2}, y_k + \frac{h}{2} f(x_k, y_k)\Big), \qquad (19.3)$$

is the same as (19.2) up to the coefficient of h^2, i.e. up to the order $m = 2$. We will prove this statement later in this chapter.

The general scheme of a Runge-Kutta formula was formulated by Kutta [9] and has the following form: Let s be an integer (the "number of stages"), and let $a_{i,j}$, b_i and c_i be real coefficients. Then the method

$$
\begin{aligned}
k_1 &= f(x, y), \\
k_2 &= f(x + c_2\, h, y + h\, a_{2,1}\, k_1), \\
&\vdots \\
k_s &= f\Big(x + c_s\, h, y + h \sum_{j=1}^{s-1} a_{s,j}\, k_j\Big), \\
\Phi(x, y, h) &= \sum_{i=1}^{s} b_i k_i, \\
y_{k+1} &= y_k + h\, \Phi(x_k, y_k, h),
\end{aligned}
$$

is called an *s-stage explicit Runge-Kutta method*. s is also the number of evaluations of f used to compute y_{k+1}. Such a Runge-Kutta scheme is defined by the $(s^2 + 3s - 2)/2$ variables $a_{i,j}$, b_i and c_i. It has become customary in the literature to symbolize such a method with the following table of coefficients:

$$
\begin{array}{c|ccccc}
0 & & & & & \\
c_2 & a_{2,1} & & & & \\
c_3 & a_{3,1} & a_{3,2} & & & \\
\vdots & \vdots & \vdots & \ddots & & \\
c_s & a_{s,1} & a_{s,2} & \cdots & a_{s,s-1} & \\
\hline
 & b_1 & b_2 & \cdots & b_{s-1} & b_s
\end{array}
$$

The derivation of an s-stage Runge-Kutta method of order m consists of two steps:

1. From the condition that the Runge-Kutta formula and the Taylor series of the solution must agree up to order m, a set of nonlinear equations for the parameters $a_{i,j}$, b_i and c_i can be constructed, and

2. this system of equations must either be solved or its inconsistency must be proven.

Of these two steps the second is by far the more difficult (see [12, 4, 2]). The problem of the generation of Runge-Kutta equations was also asked in the problem section of the SIGSAM Bulletin [8], but an answer never appeared.

In the next section, we will solve Step 1 and Step 2 with MAPLE for $s = 3$ and $m = 3$.

19.2 Derivation of the Equations for the Parameters

For the derivation of the system of equations for the parameters of a Runge-Kutta formula, we must be able to compute the Taylor series of the solution $y(x + h)$ and of the Runge-Kutta formula. MAPLE knows how to differentiate a function with two parameters that both depend on x, i.e.

```
> diff(f(x, y(x)), x);
```

$$D_1(f)(x, \, y(x)) + D_2(f)(x, \, y(x)) \, (\frac{\partial}{\partial x} y(x)) \, ,$$

where $D_1(f)(x, \, y(x))$ stands for the derivative of f with respect to the first argument, i.e. $f_x(x, y(x))$. For an extended discussion of the D-Operator in MAPLE we refer the reader to [10, 11].

The next problem is to replace $y'(x)$ by $f(x, y(x))$ whenever it appears. There are several ways of doing that. One possibility is to use the interface offered by the procedure `diff` which allows the user to install his own differentiation functions. (Note: Since `diff` remembers every function call, this will work only after restarting MAPLE.)

```
> restart;
> 'diff/y' := (a, x) -> f(a, y(a))*diff(a, x):
> diff(y(x), x);
```

$$f(x, \, y(x))$$

```
> diff(y(x), x$2);
```

$$D_1(f)(x, \, y(x)) + D_2(f)(x, \, y(x)) \, f(x, \, y(x))$$

Another possibility is to overwrite the derivative of the operator y by our own definition. Whenever D is called with the argument y, the user-defined procedure is returned. With this definition it is possible to compute the Taylor series of $y(x + h)$ around $h = 0$, and that is exactly what we need. We feel that this is the most elegant solution:

```
> D(y) := x -> f(x, y(x)):
> taylor(y(x+h), h=0, 3);
```

$$y(x) + f(x, \, y(x)) \, h + (\frac{1}{2} D_1(f)(x, \, y(x)) + \frac{1}{2} D_2(f)(x, \, y(x)) \, f(x, \, y(x))) \, h^2$$
$$+O(h^3)$$

This series is correct but unfortunately not very easy to read, because of the complicated notation for the derivatives of $f(x, y(x))$. Therefore we introduce some `alias` definitions for the derivatives which will make the expressions readable. The first few derivatives of $y(x)$ will then have their well-known form:

```
> alias(F = f(x,y(x)),
>     F[x] = D[1](f)(x,y(x)), F[y] = D[2](f)(x,y(x)),
>     F[x,x] = D[1,1](f)(x,y(x)), F[x,y] = D[1,2](f)(x,y(x)),
>     F[y,y] = D[2,2](f)(x,y(x))):
```

```
> diff(F, x);
```

$$F_x + F_y\,F$$

```
> diff(F, x$2);
```

$$F_{x,x} + F_{x,y}\,F + (F_{x,y} + F_{y,y}\,F)\,F + F_y\,(F_x + F_y\,F)$$

```
> taylor(y(x+h), h=0, 3);
```

$$y(x) + F\,h + (\frac{1}{2}\,F_x + \frac{1}{2}\,F_y\,F)\,h^2 + O(h^3)$$

For computing the Taylor series of the function Φ, which we defined in the general Runge-Kutta scheme, we must be able to compute the series of the expressions $k_i = f(x + c_i\,h, y + h\sum_{j=1}^{i-1} a_{i,j}\,k_j)$ around $h = 0$. This can be performed by simply using the `taylor` command as well. For multivariate Taylor series expansions, the function `mtaylor` could be used, but in our case, this is not necessary.

```
> taylor(f(x+h, y(x)+h), h=0, 3);
```

$$F + (F_x + F_y)\,h + (\frac{1}{2}\,F_{x,x} + F_{x,y} + \frac{1}{2}\,F_{y,y})\,h^2 + O(h^3)$$

Now we have prepared the tools for computing the Taylor series of the solution $y(x)$ and of the Runge-Kutta formula. In the following we try to compute the coefficients of a Runge-Kutta scheme of order three $(m = 3)$ with three stages $(s = 3)$.

```
> m := 3:
> taylor(y(x+h), h=0, m+1);
```

$$y(x) + F\,h + (\frac{1}{2}\,F_x + \frac{1}{2}\,F_y\,F)\,h^2 +$$
$$(\frac{1}{6}\,F_{x,x} + \frac{1}{3}\,F_{x,y}\,F + \frac{1}{6}\,F_{y,y}\,F^2 + \frac{1}{6}\,F_y\,F_x + \frac{1}{6}\,F_y{}^2\,F)\,h^3 + O(h^4)$$

```
> TaylorPhi := normal((convert(", polynom) - y(x))/h);
```

$$TaylorPhi := F + \frac{1}{2}\,h\,F_x + \frac{1}{2}\,h\,F_y\,F + \frac{1}{6}\,h^2\,F_{x,x} + \frac{1}{3}\,h^2\,F_{x,y}\,F$$
$$+ \frac{1}{6}\,h^2\,F_{y,y}\,F^2 + \frac{1}{6}\,h^2\,F_y\,F_x + \frac{1}{6}\,h^2\,F_y{}^2\,F$$

The `convert` command converts the Taylor series into a polynomial, i.e. it cuts off the O-term. The variable *TaylorPhi* corresponds to Φ in Equation (19.2).

For the Runge-Kutta scheme we get the following Taylor series. Note that we keep the parameters $a_{i,j}$, b_i and c_i as symbolic objects.

```
> k[1]:=taylor(f(x,         y(x)),                              h=0,m):
> k[2]:=taylor(f(x+c[2]*h,y(x)+h*(a[2,1]*k[1])),               h=0,m):
> k[3]:=taylor(f(x+c[3]*h,y(x)+h*(a[3,1]*k[1]+a[3,2]*k[2])),h=0,m):
> RungeKuttaPhi := series(b[1]*k[1]+b[2]*k[2]+b[3]*k[3], h, m):
```

```
> RungeKuttaPhi := convert(RungeKuttaPhi, polynom);
```

$$RungeKuttaPhi := b_1\, F + b_2\, F + b_3\, F$$
$$+ \left(b_2\,(F_x\, c_2 + F_y\, a_{2,1}\, F) + b_3\,(F_x\, c_3 + F_y\, a_{3,1}\, F + F_y\, a_{3,2}\, F)\right) h +$$
$$\left(b_2\,(\tfrac{1}{2}\, F_{x,x}\, c_2{}^2 + c_2\, F_{x,y}\, a_{2,1}\, F + \tfrac{1}{2}\, a_{2,1}{}^2\, F^2\, F_{y,y}) + b_3(\tfrac{1}{2}\, F_{x,x}\, c_3{}^2\right.$$
$$+\, c_3\, F_{x,y}\, a_{3,1}\, F + c_3\, F_{x,y}\, a_{3,2}\, F + \tfrac{1}{2}\, a_{3,1}{}^2\, F^2\, F_{y,y}$$
$$+\, a_{3,1}\, F^2\, F_{y,y}\, a_{3,2} + \tfrac{1}{2}\, a_{3,2}{}^2\, F^2\, F_{y,y} + F_y\, a_{3,2}\, F_x\, c_2$$
$$\left. +\, F_y{}^2\, a_{3,2}\, a_{2,1}\, F)\right)h^2$$

The difference d of the two polynomials *TaylorPhi* and *RungeKuttaPhi* should be zero. We consider d as a polynomial in the unknowns h, F, F_x, F_y, $F_{x,x}$, etc. and we set the coefficients of that polynomial to zero. This gives us the nonlinear system of equations (19.4) which must be solved.

```
> d := expand(TaylorPhi - RungeKuttaPhi):
> eqns := {coeffs(d, [h,F,F[x],F[y],F[x,x],F[x,y],F[y,y]])};
```

$$eqns := \{\tfrac{1}{2} - b_3\, c_3 - b_2\, c_2,\ -\tfrac{1}{2}\, b_2\, c_2{}^2 - \tfrac{1}{2}\, b_3\, c_3{}^2 + \tfrac{1}{6},$$
$$\tfrac{1}{6} - b_3\, a_{3,2}\, a_{2,1},\ -b_3\, a_{3,2}\, c_2 + \tfrac{1}{6},$$
$$-b_3\, a_{3,1}\, a_{3,2} - \tfrac{1}{2}\, b_3\, a_{3,2}{}^2 - \tfrac{1}{2}\, b_2\, a_{2,1}{}^2 + \tfrac{1}{6} - \tfrac{1}{2}\, b_3\, a_{3,1}{}^2, \qquad (19.4)$$
$$1 - b_3 - b_2 - b_1,\ \tfrac{1}{3} - b_3\, c_3\, a_{3,1} - b_2\, c_2\, a_{2,1} - b_3\, c_3\, a_{3,2},$$
$$-b_3\, a_{3,2} - b_2\, a_{2,1} + \tfrac{1}{2} - b_3\, a_{3,1}\}$$

```
> vars := indets(eqns);
```

$$vars := \{a_{2,1},\, c_2,\, a_{3,1},\, a_{3,2},\, c_3,\, b_1,\, b_2,\, b_3\}$$

19.3   Solving the System of Equations

In this section we discuss the second step of the derivation of Runge-Kutta formulas, i.e. the question how to solve the system (19.4) of nonlinear equations . We note that we have to deal with a system of *polynomial* equations in the unknowns. Special algorithms exist for this type of problem.

    We describe two algorithms which are used in computer algebra systems to solve systems of polynomial equations. The first is based on the theory of Gröbner bases for polynomial ideals, and the second uses polynomial resultants for performing nonlinear elimination steps. For a good introduction to both methods we refer the reader to [5].

But first, we give MAPLE's `solve` command a try. This command tries to solve the system of equations in a natural way using a substitution approach. For system (19.4) this approach is quite good, because the equations are almost in a triangular form as we will see later. The substitution algorithm used by MAPLE's `solve` command is described in detail in [6].

```
> sols := solve(eqns, vars);
```

$$
\begin{aligned}
sols := \{&b_3 = b_3,\ a_{3,2} = \frac{1}{4}\,\frac{2\,b_3\,c_3 - 1}{b_3\,(3\,b_3\,c_3{}^2 - 1)},\ c_3 = c_3,\\
&b_1 = \frac{1}{4}\,\frac{12\,b_3\,c_3{}^2 - 1 + 4\,b_3 - 12\,b_3\,c_3}{3\,b_3\,c_3{}^2 - 1},\ a_{2,1} = \frac{2}{3}\,\frac{3\,b_3\,c_3{}^2 - 1}{2\,b_3\,c_3 - 1},\\
&a_{3,1} = \frac{1}{4}\,\frac{-6\,b_3\,c_3 + 1 + 12\,b_3{}^2\,c_3{}^3}{b_3\,(3\,b_3\,c_3{}^2 - 1)},\ b_2 = -\frac{3}{4}\,\frac{4\,b_3{}^2\,c_3{}^2 - 4\,b_3\,c_3 + 1}{3\,b_3\,c_3{}^2 - 1},\\
&c_2 = \frac{2}{3}\,\frac{3\,b_3\,c_3{}^2 - 1}{2\,b_3\,c_3 - 1}\},\ \{c_3 = \frac{2}{3},\ a_{2,1} = \frac{2}{9}\,\frac{1}{a_{3,2}},\ b_2 = 0,\ a_{3,2} = a_{3,2},\\
&b_1 = \frac{1}{4},\ b_3 = \frac{3}{4},\ a_{3,1} = \frac{2}{3} - a_{3,2},\ c_2 = \frac{2}{9}\,\frac{1}{a_{3,2}}\}
\end{aligned}
$$

Thus, for $s = 3$, we found two (parameterized) solutions which can be represented by the following coefficient schemes. Note that in the first solution, the unknowns c_3 and b_3 are free parameters, i.e. they can take on any value. This is indicated by the entries $c_3 = c_3$ and $b_3 = b_3$ in the solution set. For the second solution, $a_{3,2}$ is a free parameter (see also Figure 19.1). From the first solution we get the method of Heun of third order if we set $a_{3,2} = 2/3$.

FIGURE 19.1.
Three Step Runge-Kutta Methods of Order 3.

$$
\begin{array}{c|ccc}
0 & & & \\
\frac{2}{3}\,\frac{3\,b_3\,c_3{}^2-1}{2\,b_3\,c_3-1} & \frac{2}{3}\,\frac{3\,b_3\,c_3{}^2-1}{2\,b_3\,c_3-1} & & \\
c_3 & \frac{1-6\,b_3\,c_3+12\,b_3{}^2c_3{}^3}{4\,b_3\,(3\,b_3\,c_3{}^2-1)} & \frac{2\,b_3\,c_3-1}{4\,b_3\,(3\,b_3\,c_3{}^2-1)} & \\
\hline
 & \frac{12\,b_3\,c_3{}^2-1+4\,b_3-12\,b_3\,c_3}{12\,b_3\,c_3{}^2-4} & \frac{3}{4}\,\frac{(2b_3\,c_3-1)^2}{1-3b_3\,c_3{}^2} & b_3
\end{array}
\qquad
\begin{array}{c|ccc}
0 & & & \\
\frac{2}{9a_{3,2}} & \frac{2}{9a_{3,2}} & & \\
\frac{2}{3} & \frac{2}{3} - a_{3,2} & a_{3,2} & \\
\hline
 & \frac{1}{4} & 0 & \frac{3}{4}
\end{array}
$$

We will now present two other methods to solve the system of equations.

19.3.1 Gröbner Bases

A wide variety of problems in computer algebra involving polynomials may be formulated in terms of *polynomial ideals*. Examples of such problems are simplifications with respect to (polynomial) side relations or solutions of systems of (polynomial) equations.

We recall the definition of an ideal here. Given a (finite) set $F = \{f_1, f_2, \ldots, f_n\}$ of multivariate polynomials over a field $\mathbf{K}$, then the ideal $\langle F \rangle$ generated by the set F is defined as

$$\langle F \rangle = \langle f_1, f_2, \ldots, f_n \rangle = \left\{ \sum_{i=1}^{m} h_i f_i \ \middle| \ h_i \in \mathbf{K}[x_1, \ldots, x_n] \right\},$$

that is, $\langle F \rangle$ is the set of all the polynomials that can be constructed from combinations of the polynomials in F. The elements in F are called a *basis* for the ideal $\langle F \rangle$.

The method of Gröbner bases consists in transforming the given set F into a canonical basis G (called a Gröbner basis) first, such that $\langle G \rangle = \langle F \rangle$, and then in solving the problems with $\langle G \rangle$ rather than with $\langle F \rangle$. This transformation process is done by eliminating terms of the set of polynomials similar to the process of Gaussian elimination. New polynomials are formed from pairs of old ones f_1, f_2 as $p = \alpha f_1 + \beta f_2$ where α and β are properly chosen polynomials. By this process, variable by variable will be eliminated (according to a specified ordering). If the system has finitely many solutions, eventually a polynomial in one variable remains which can then be solved and substituted into the other elements of the Gröbner basis. This way one gets a generalization of the triangularization process of Gaussian elimination. We will not go into any details of the theory and the algorithm to compute Gröbner bases and refer the reader to [1] for an introduction. However, we will use the Gröbner bases algorithm implemented in the MAPLE system [3] to solve our system of equations:

```
> G := grobner[gbasis](eqns,
>          [b[1],b[2],c[2],a[2,1],a[3,1],a[3,2],b[3],c[3]], plex);
```

$$
\begin{aligned}
G := [\, &b_1 - 6\,a_{3,2}\,b_3{}^2\,c_3 - 1 + b_3 + 3\,b_3\,a_{3,2}, \\
&b_2 + 6\,a_{3,2}\,b_3{}^2\,c_3 - 3\,b_3\,a_{3,2},\ c_2 - a_{2,1}, \\
&3\,a_{3,2}\,a_{2,1} + 6\,a_{3,2}\,b_3\,c_3{}^2 - c_3 - 2\,a_{3,2}, \\
&6\,a_{2,1}\,b_3\,c_3 - 3\,a_{2,1} - 6\,b_3\,c_3{}^2 + 2,\ a_{3,1} + a_{3,2} - c_3, \\
&1 - 2\,b_3\,c_3 + 12\,b_3{}^2\,c_3{}^2\,a_{3,2} - 4\,b_3\,a_{3,2}\,]
\end{aligned}
$$

The last equation contains three unknowns which means, that the system has two free parameters. We can solve this last equation with respect to $a_{3,2}$.

```
> a[3,2] := normal(solve(G[7], a[3,2]));
```

$$a_{3,2} := \frac{1}{4}\,\frac{2\,b_3\,c_3 - 1}{b_3\,(3\,b_3\,c_3{}^2 - 1)}$$

This expression is only a solution, if the denominator is not zero. Let us first assume this case. The values for the other unknowns can be obtained by a back substitution process over the Gröbner bases G.

```
> a[3,1] := normal(solve(G[6], a[3,1]));
```

$$a_{3,1} := \frac{1}{4} \frac{-6\,b_3\,c_3 + 1 + 12\,b_3{}^2\,c_3{}^3}{b_3\,(3\,b_3\,c_3{}^2 - 1)}$$

```
> a[2,1] := normal(solve(G[5], a[2,1]));
```

$$a_{2,1} := \frac{2}{3} \frac{3\,b_3\,c_3{}^2 - 1}{2\,b_3\,c_3 - 1}$$

```
> normal(G[4]);
```

$$0$$

```
> c[2] := normal(solve(G[3], c[2]));
```

$$c_2 := \frac{2}{3} \frac{3\,b_3\,c_3{}^2 - 1}{2\,b_3\,c_3 - 1}$$

```
> b[2] := factor(solve(G[2], b[2]));
```

$$b_2 := -\frac{3}{4} \frac{(2\,b_3\,c_3 - 1)^2}{3\,b_3\,c_3{}^2 - 1}$$

```
> b[1] := normal(solve(G[1], b[1]));
```

$$b_1 := \frac{1}{4} \frac{12\,b_3\,c_3{}^2 - 1 + 4\,b_3 - 12\,b_3\,c_3}{3\,b_3\,c_3{}^2 - 1}$$

and we get the same solution as with the `solve` command directly.

For $b_3\,(3\,b_3\,c_3{}^2 - 1) = 0$ we arrive at another solution by adding the polynomial $3\,b_3{}^2\,c_3{}^2 - b_3$ to the ideal we obtained from the latter Gröbner basis computation. First we must clear the parameters which we assigned above.

```
> a := 'a': b := 'b': c := 'c':
> grobner[gbasis]([G[], 3*b[3]^2*c[3]^2-b[3]],
>           [b[1],b[2],c[2],a[2,1],a[3,1],a[3,2],b[3],c[3]], plex);
```

$$[4\,b_1 - 1,\ b_2,\ c_2 - a_{2,1},\ 9\,a_{3,2}\,a_{2,1} - 2,\ 3\,a_{3,1} + 3\,a_{3,2} - 2,$$
$$-3 + 4\,b_3,\ 3\,c_3 - 2]$$

This corresponds to the second solution computed above with the `solve` command.

19.3.2 Resultants

The solution process using resultants is also similar to Gaussian elimination, because the resultant of two polynomials $f, g \in \mathbf{R}[x]$ is an eliminant, that is, $\mathrm{res}_x(f, g) \in \mathbf{R}$. The resultant of two polynomials $f, g \in \mathbf{R}[x]$ (written $\mathrm{res}_x(f, g)$) is defined to be the determinant of the Sylvester matrix of f, g (see any introductory algebra book, e.g. [13]). The following theorem (taken from [5]) shows how resultants may be used to solve systems of polynomial equations.

Let $\overline{F}$ be an algebraically closed field, and let

$$f = \sum_{i=0}^{m} a_i(x_2, \ldots, x_r)\, x_1^i, \quad g = \sum_{i=0}^{n} b_i(x_2, \ldots, x_r)\, x_1^i,$$

be elements of $\overline{F}[x_1, \ldots, x_r]$ of positive degrees in x_1. Then, if $(\alpha_1, \ldots, \alpha_r)$ is a common zero of f and g, their resultant with respect to x_1 satisfies

$$\mathrm{res}_{x_1}(\alpha_2, \ldots, \alpha_r) = 0.$$

Conversely, if $\mathrm{res}_{x_1}(\alpha_2, \ldots, \alpha_r) = 0$ then one of the following conditions holds:

$$a_m(\alpha_2, \ldots, \alpha_r) = b_n(\alpha_2, \ldots, \alpha_r) = 0,$$
$$\forall\, x \in \overline{F} \;:\; f(x, \alpha_2, \ldots, \alpha_r) = 0,$$
$$\forall\, x \in \overline{F} \;:\; g(x, \alpha_2, \ldots, \alpha_r) = 0,$$
$$\exists\, \alpha_1 \in \overline{F} \quad \text{such that} \quad (\alpha_1, \alpha_2, \ldots, \alpha_r) \;\; \text{is a common zero of } f \text{ and } g,$$

where the last case is the interesting one for us.

We now try to use MAPLE's `resultant` function to transform the system of equations into triangular form. We first collect the equations in different sets B_j such that

$$B_j = \left\{ p \in \overline{F}[x_j, \ldots, x_r] - \overline{F}[x_{j+1}, \ldots, x_r] \right\}.$$

We consider in the sequel the equations as elements in $\mathbf{Q}[b_1, a_{3,1}, b_2, c_2, a_{2,1}, a_{3,2}, b_3, c_3]$ and define the sets B_j.

```
> X := [b[1],a[3,1],b[2],c[2],a[2,1],a[3,2],b[3],c[3]]:
> for i to nops(X) do B[i] := {} od:
> for p in eqns do
>     for i while not has(p, X[i]) do od;
>     B[i] := B[i] union {primpart(p)}
> od:
```

Let us look at the sets B_i. Consider especially the number of elements in each of these sets.

```
> seq('B['.i.']' = B[i], i=1..nops(X));
```

$$B[1] = \{1 - b_3 - b_2 - b_1\},\; B[2] = \{$$
$$-6\, b_3\, a_{3,1}\, a_{3,2} - 3\, b_3\, a_{3,2}^{\,2} - 3\, b_2\, a_{2,1}^{\,2} + 1 - 3\, b_3\, a_{3,1}^{\,2},$$
$$-2\, b_3\, a_{3,2} - 2\, b_2\, a_{2,1} + 1 - 2\, b_3\, a_{3,1},$$
$$1 - 3\, b_3\, c_3\, a_{3,1} - 3\, b_2\, c_2\, a_{2,1} - 3\, b_3\, c_3\, a_{3,2}\},$$
$$B[3] = \{1 - 2\, b_3\, c_3 - 2\, b_2\, c_2,\; -3\, b_2\, c_2^{\,2} - 3\, b_3\, c_3^{\,2} + 1\},$$
$$B[4] = \{-6\, b_3\, a_{3,2}\, c_2 + 1\},\; B[5] = \{1 - 6\, b_3\, a_{3,2}\, a_{2,1}\},$$
$$B[6] = \{\},\; B[7] = \{\},\; B[8] = \{\}$$

Thus we can eliminate b_2 from the two elements in B_3 giving an additional element for the set B_4, which then has two elements too. Eliminating c_2 from these two equations yields a resultant in $\mathbf{Q}[a_{3,2}, b_3, c_3]$ which goes into B_6.

```
> primpart(resultant(B[3][1], B[3][2], b[2]));
```

$$3\,c_2^2 - 6\,c_2^2\,b_3\,c_3 - 2\,c_2 + 6\,c_2\,b_3\,c_3^2$$

```
> B[4] := B[4] union {"}:
> primpart(resultant(B[4][1], B[4][2], c[2]));
```

$$1 - 2\,b_3\,c_3 + 12\,b_3^2\,c_3^2\,a_{3,2} - 4\,b_3\,a_{3,2}$$

```
> B[6] := B[6] union {"}:
```

Since we know from the above computation that the whole system has two free parameters, we can stop the process and start back-substituting the solution. We will only construct the second solution, which is valid in the case $4\,b_3(3b_3c_3^2 - 1) \neq 0$.

```
> a[3,2] := normal(solve(B[6][1], a[3,2]));
```

$$a_{3,2} := \frac{1}{4}\,\frac{2\,b_3\,c_3 - 1}{b_3\,(3\,b_3\,c_3^2 - 1)}$$

```
> a[2,1] := normal(solve(B[5][1], a[2,1]));
```

$$a_{2,1} := \frac{2}{3}\,\frac{3\,b_3\,c_3^2 - 1}{2\,b_3\,c_3 - 1}$$

```
> c[2] := normal(solve(B[4][1], c[2]));
```

$$c_2 := \frac{2}{3}\,\frac{3\,b_3\,c_3^2 - 1}{2\,b_3\,c_3 - 1}$$

```
> b[2] := normal(solve(B[3][1], b[2]));
```

$$b_2 := -\frac{3}{4}\,\frac{(2\,b_3\,c_3 - 1)^2}{3\,b_3\,c_3^2 - 1}$$

```
> a[3,1] := normal(solve(B[2][2], a[3,1]));
```

$$a_{3,1} := \frac{1}{4}\,\frac{-6\,b_3\,c_3 + 1 + 12\,b_3^2\,c_3^3}{b_3\,(3\,b_3\,c_3^2 - 1)}$$

```
> b[1] := normal(solve(B[1][1], b[1]));
```

$$b_1 := \frac{1}{4}\,\frac{12\,b_3\,c_3^2 - 1 + 4\,b_3 - 12\,b_3\,c_3}{3\,b_3\,c_3^2 - 1}$$

and we obtain the same results as when using Gröbner basis.

19.4   The Complete Algorithm

In this section we compile all the MAPLE statements that we have used to derive the Runge-Kutta scheme for $s = 3$ into a single MAPLE procedure called RungeKutta(s,m). It computes the coefficients for an arbitrary Runge-Kutta method with s stages of order m. Unfortunately, for values of s greater than 3, the equations can no longer be solved by MAPLE and other Computer Algebra Systems directly. The system of polynomial equations has already become too complex. It can be simplified by adding the so-called symmetry conditions,

$$c_i = \sum_{j=1}^{i-1} a_{i,j}, \quad i = 2..s,$$

which we obtain by requiring that for the differential equation $y' = 1, y(0) = 0$ all the predictor values $y_i^* = y + h \sum_{j=1}^{i-1} a_{i,j}$ at position $x + c_i h$ are the same as the values obtained by inserting the exact solution $y(x) = x$. With these additional conditions, Runge-Kutta formulas can be computed for the order $m = 1, \cdots, 4$. The complete MAPLE code is shown in Algorithm 19.1.

19.4.1   Example 1:

We first test this procedure for the parameters $s = 2$ and $m = 2$

> RungeKutta(2, 2);

$$\{b_1 = 1 - b_2, \ a_{2,1} = \frac{1}{2} \frac{1}{b_2}, \ b_2 = b_2, \ c_2 = \frac{1}{2} \frac{1}{b_2}\}$$

Again we have one free parameter, namely b_2. In Figure 19.2 this result is shown in the table notation.

FIGURE 19.2. General two Stage Runge-Kutta Schema.

$$
\begin{array}{c|cc}
0 & & \\
\frac{1}{2b_2} & \frac{1}{2b_2} & \\
\hline
& 1 - b_2 & b_2
\end{array}
$$

For $b_2 = 1$ we obtain the improved Euler method (see Equation (19.3)), and for the choice $b_2 = 1/2$ we get the rule

$$y_{k+1} = y_k + \frac{h}{2} \left(f(x_k, y_k) + f(x_k + h, y_k + h\, f(x_k, y_k)) \right)$$

which is known as Heun's method.

ALGORITHM 19.1. *Procedure* RungeKutta.

```
RungeKutta := proc(s, m)
  local TaylorPhi, RungeKuttaPhi, d, vars, eqns, k, i, j;
  global a, b, c, h;
  # Taylor series
  D(y) := x -> f(x,y(x)):
  TaylorPhi := convert(taylor(y(x+h),h=0,m+1), polynom):
  TaylorPhi := normal((TaylorPhi - y(x))/h);
  # RK-Ansatz:
  c[1] := 0;
  for i from 1 to s do
     k[i] := taylor(f(x+c[i]*h, y(x) +
                sum(a[i,j]*k[j], j=1..i-1)*h), h=0, m):
  od:
  RungeKuttaPhi := 0:
  for i from 1 to s do
     RungeKuttaPhi := RungeKuttaPhi + b[i] * k[i]:
  od:
  RungeKuttaPhi := series (RungeKuttaPhi, h, m):
  RungeKuttaPhi := convert(RungeKuttaPhi, polynom);
  d := expand(TaylorPhi - RungeKuttaPhi):
  vars := {seq(c[i], i=2..s),
              seq(b[i], i=1..s),
              seq((seq(a[i,j], j=1..i-1)), i = 2..s)};
  eqns := {coeffs(d, indets(d) minus vars)};
  # symmetry condition:
  eqns := eqns union {seq(sum(a[i,'j'], 'j'=1..i-1) -
                      c[i], i=2..s)};
  solve(eqns, vars);
end:
```

19.4.2   Example 2:

In this example we compute the 4-stage Runge-Kutta formulas of order 4.

```
> RK4 := RungeKutta(4, 4):
> RK4[2];
```

$$\{b_4 = b_4,\; b_1 = \frac{1}{6},\; a_{4,2} = -\frac{1}{12}\frac{1}{b_4},\; b_2 = -b_4 + \frac{1}{6},\; a_{3,1} = \frac{3}{8},\; a_{4,3} = \frac{1}{3}\frac{1}{b_4},$$
$$a_{3,2} = \frac{1}{8},\; b_3 = \frac{2}{3},\; c_4 = 1,\; c_2 = 1,\; a_{4,1} = \frac{1}{4}\frac{-1 + 4\,b_4}{b_4},\; a_{2,1} = 1,\; c_3 = \frac{1}{2}\}$$

We get two solutions, where one has b_4 as a free parameter and can be represented by the scheme shown in Figure 19.3.

The other solution has the two free parameters c_2 and $a_{3,2}$. The % terms stand for common subexpressions and are displayed underneath the solution set, and the *RootOf* expression stands for the two roots of the quadratic equation

$$576\,(1 - 1\,c_2 - 2\,a_{3,2}\,c_2 + 4\,a_{3,2}\,c_2{}^2)\,x^2 + 24\,(c_2 - 2)\,x + 1.$$

FIGURE 19.3. *Simple Solution for $s = 4$ and $m = 4$.*

$$
\begin{array}{c|cccc}
0 & & & & \\
1 & 1 & & & \\
\frac{1}{2} & \frac{3}{8} & \frac{1}{8} & & \\
1 & 1 - \frac{1}{4b_4} & -\frac{1}{12b_4} & \frac{1}{3b_4} & \\
\hline
 & \frac{1}{6} & \frac{1}{6} - b_4 & \frac{2}{3} & b_4
\end{array}
$$

```
> RK4[1];
```

$$
\left\{ a_{2,1} = c_2,\ a_{3,2} = a_{3,2},\ c_2 = c_2, a_{4,2} = \frac{1}{2}\left(-1 - 5\,a_{3,2}\,c_2^{\ 2} + 24\,\%1\,c_2^{\ 2} \right.\right.
$$

$$
- 168\,\%4 + a_{3,2}\,c_2 + 2\,c_2 + 384\,\%1\,c_2^{\ 4}\,a_{3,2}^{\ 2} - 576\,\%1\,c_2^{\ 3}\,a_{3,2}^{\ 2}
$$
$$
+ 192\,\%1\,c_2^{\ 2}\,a_{3,2}^{\ 2} + 4\,a_{3,2}\,c_2^{\ 3} - c_2^{\ 2} + 24\,\%1 - 48\,\%1\,c_2 + 264\,\%3
$$
$$
\left. - 96\,\%2\right)\Big/\!\left(a_{3,2}\,\%5\,c_2^{\ 2}\right),\ a_{3,1} = -\frac{1}{24}\,\frac{1 - 24\,\%1 + 24\,\%1\,a_{3,2}}{\%1},
$$
$$
b_2 = -\frac{1}{576}\,\frac{96\,\%2 - 24\,\%1 + 1}{(c_2 - 1)\,\%1\,c_2^{\ 2}\,a_{3,2}},\quad b_3 = \frac{\%1}{c_2\,a_{3,2}},
$$
$$
a_{4,3} = 12\,\frac{\%1\left(2\,c_2 - 1 - 6\,a_{3,2}\,c_2^{\ 2} + 2\,a_{3,2}\,c_2 + 4\,a_{3,2}\,c_2^{\ 3} - c_2^{\ 2}\right)}{c_2\,a_{3,2}\,\%5},
$$
$$
b_4 = \frac{1}{288}\,\frac{\%5}{\left(1 - 2\,a_{3,2}\,c_2 + 4\,a_{3,2}\,c_2^{\ 2} - c_2\right)\%1\,(c_2 - 1)},\quad c_4 = 1, a_{4,1} = \frac{1}{2}\big(
$$
$$
1 + 3\,a_{3,2}\,c_2^{\ 2} - a_{3,2}\,c_2 - 2\,c_2 - 24\,\%1 - 192\,\%1\,c_2^{\ 2}\,a_{3,2}^{\ 2} - 408\,\%3
$$
$$
+ 96\,\%2 + 24\,\%1\,c_2^{\ 3} - 72\,\%1\,c_2^{\ 2} + 72\,\%1\,c_2 + 552\,\%4
$$
$$
- 1728\,\%1\,c_2^{\ 4}\,a_{3,2}^{\ 2} + 960\,\%1\,c_2^{\ 3}\,a_{3,2}^{\ 2} - 288\,\%1\,c_2^{\ 4}\,a_{3,2}
$$
$$
\left. + 1152\,a_{3,2}^{\ 2}\,c_2^{\ 5}\,\%1 + c_2^{\ 2}\right)\Big/\!\left(a_{3,2}\,\%5\,c_2^{\ 2}\right),\ c_3 = \frac{1}{24}\,\frac{-1 + 24\,\%1}{\%1},
$$
$$
\left. b_1 = \frac{1}{576}\,\frac{288\,\%3 - 96\,\%2 - 24\,\%1\,c_2 - 1 + 24\,\%1}{\%1\,c_2^{\ 2}\,a_{3,2}}\right\}
$$

$$
\%1 := \mathrm{RootOf}\big(1 + (-48 + 24\,c_2)\,\_Z
$$
$$
+ (576 - 1152\,a_{3,2}\,c_2 + 2304\,a_{3,2}\,c_2^{\ 2} - 576\,c_2)\,\_Z^2\big)
$$
$$
\%2 := \%1\,c_2\,a_{3,2}
$$
$$
\%3 := \%1\,c_2^{\ 2}\,a_{3,2}
$$
$$
\%4 := \%1\,c_2^{\ 3}\,a_{3,2}
$$
$$
\%5 := -48\,\%1 + 192\,\%2 - 672\,\%3 + 120\,\%1\,c_2 + 576\,\%4
$$
$$
- 96\,\%1\,c_2^{\ 2} - 1 + 2\,c_2
$$

From this solution we obtain the classical Runge-Kutta method of order 4 if we require that the three parameters $a_{3,1}$, $a_{4,1}$ and $a_{4,2}$ are all zero. This is obtained if we set $c_2 = 1/2$ and $a_{3,2} = 1/2$ and if we take the root $x = 1/12$ of the polynomial $288\,x^2 - 36\,x + 1$.

```
> solve(subs(RK4[1], {a[3,1]=0, a[4,1]=0, a[4,2]=0}));
```

$$\left\{c_2 = \frac{1}{2},\ a_{3,2} = \frac{1}{2}\right\}$$

```
> subs(", RK4[1]);
```

$$\left\{c_4 = 1,\ a_{4,2} = -\frac{1}{3}\frac{-\frac{3}{8}+\frac{9}{2}\%1}{\%1},\ a_{3,1} = -\frac{1}{24}\frac{1-12\%1}{\%1},\ a_{2,1} = \frac{1}{2},\ \frac{1}{2} = \frac{1}{2},\right.$$

$$c_3 = \frac{1}{24}\frac{-1+24\%1}{\%1},\ b_1 = \frac{1}{72}\frac{-1+24\%1}{\%1},\ a_{4,1} = -\frac{1}{3}\frac{\frac{3}{8}-\frac{9}{2}\%1}{\%1},$$

$$\left. a_{4,3} = 1,\ b_4 = \frac{1}{6},\ b_2 = \frac{1}{36}\frac{1}{\%1},\ b_3 = 4\%1\right\}$$

$$\%1 := \text{RootOf}(1 - 36\,\_Z + 288\,\_Z^2)$$

```
> allvalues(%1);
```

$$\frac{1}{24},\ \frac{1}{12}$$

```
> subs(%1=1/12, "");
```

$$\left\{a_{3,1} = 0,\ a_{4,1} = 0,\ b_2 = \frac{1}{3},\ b_3 = \frac{1}{3},\ b_1 = \frac{1}{6},\ c_4 = 1,\ a_{2,1} = \frac{1}{2},\ \frac{1}{2} = \frac{1}{2},\right.$$

$$\left. a_{4,3} = 1,\ b_4 = \frac{1}{6},\ c_3 = \frac{1}{2},\ a_{4,2} = 0\right\}$$

If we set $c_2 = 1/3$ and $a_{3,2} = 1$ we get two other Runge-Kutta methods, where the one which corresponds to the root $1/8$ of the polynomial $256\,x^2 - 40\,x + 1$ is called the 3/8-Rule, according to its weights b_i.

```
> subs(c[2]=1/3, a[3,2]=1, RK4[1]);
```

$$\left\{c_3 = \frac{1}{24}\frac{-1+24\%1}{\%1},\ b_1 = \frac{1}{64}\frac{16\%1-1}{\%1},\ b_3 = 3\%1,\right.$$

$$a_{4,3} = -\frac{32}{3}\frac{\%1}{-8\%1-\frac{1}{3}},\ b_4 = -\frac{3}{256}\frac{-8\%1-\frac{1}{3}}{\%1},\ a_{4,2} = \frac{9}{2}\frac{-\frac{14}{27}+\frac{176}{27}\%1}{-8\%1-\frac{1}{3}},$$

$$a_{3,1} = -\frac{1}{24}\frac{1}{\%1},\ b_2 = \frac{3}{128}\frac{8\%1+1}{\%1},\ a_{4,1} = \frac{9}{2}\frac{\frac{4}{9}-\frac{160}{27}\%1}{-8\%1-\frac{1}{3}},\ c_4 = 1,\ 1 = 1,$$

$$\left. a_{2,1} = \frac{1}{3},\ \frac{1}{3} = \frac{1}{3}\right\}$$

$$\%1 := \text{RootOf}(1 - 40\,\_Z + 256\,\_Z^2)$$

```
> allvalues(%1);
```

$$\frac{1}{32}, \frac{1}{8}$$

```
> subs(%1=1/8, "");
```

$$\{a_{4,1} = 1, a_{3,1} = \frac{-1}{3}, b_2 = \frac{3}{8}, a_{4,2} = -1, b_4 = \frac{1}{8}, b_1 = \frac{1}{8}, b_3 = \frac{3}{8}, c_3 = \frac{2}{3},$$
$$c_4 = 1, 1 = 1, a_{4,3} = 1, a_{2,1} = \frac{1}{3}, \frac{1}{3} = \frac{1}{3}\}$$

The coefficient schemes of these two methods are shown below:

FIGURE 19.4. *classical Runge-Kutta method*   FIGURE 19.5. *3/8-Rule*

0				
$\frac{1}{2}$	$\frac{1}{2}$			
$\frac{1}{2}$	0	$\frac{1}{2}$		
1	0	0	1	
	$\frac{1}{6}$	$\frac{1}{3}$	$\frac{1}{3}$	$\frac{1}{6}$

0				
$\frac{1}{3}$	$\frac{1}{3}$			
$\frac{2}{3}$	$-\frac{1}{3}$	1		
1	1	-1	1	
	$\frac{1}{8}$	$\frac{3}{8}$	$\frac{3}{8}$	$\frac{1}{8}$

19.5 Conclusions

We have demonstrated that MAPLE is a great help in deriving the equations which define an explicit Runge-Kutta formula. We note that we have made no attempt to simplify these equations. Techniques for doing so are well-known. The equations we constructed are the result of brute force formula manipulations, and, consequently, only Runge-Kutta formulas of up to order 4 can be derived. The equations corresponding to higher order formulas are still too big to be solved by today's computer algebra systems.

As early as 1966 Moses conjectured in [12] that the known inconsistency of the system of equations which corresponds to the five-stage Runge-Kutta method of order five could not be established using a computer algebra system. This is still valid for today's computer algebra systems and hence the algorithms to solve systems of polynomial equations must still be improved. One research direction is to take advantage of symmetries in the equations.

This chapter also justifies the mathematicians' efforts to simplify the systems in order to construct Runge-Kutta formulas of up to order 10 (17 stages) [7].

References

[1] B. BUCHBERGER, *Gröbner Bases: An Algorithmic Method in Polynomial Ideal Theory*, in Progress, directions and open problems in multidimensional systems theory, ed. N.K. Bose, D. Reidel Publishing Co, 1985, pp. 189-232.

[2] J.C. BUTCHER, *The non-existence of ten Stage eight Order Explicit Runge-Kutta Methods*, BIT, 25, 1985, pp. 521-540.

[3] S. CZAPOR and K. GEDDES, *On Implementing Buchbergers's Algorithm for Gröbner Bases*, ISSAC86, 1986, pp. 233-238.

[4] G.E. COLLINS, *The Calculation of Multivariate Polynomial Resultants*, Journal of the ACM,18, No. 4, 1971, pp. 512-532.

[5] K.O. GEDDES, S.R. CZAPOR, and G. LABAHN, *Algorithms for Computer Algebra*, Kluwer, 1992.

[6] G.H. GONNET and M.B. MONAGAN, *Solving systems of Algebraic Equations, or the Interface between Software and Mathematics*, Computers in mathematics, Conference at Stanford University, 1986.

[7] E. HAIRER, S.P. NØRSETT and G. WANNER, *Solving Ordinary Differential Equations I*, Springer-Verlag Berlin Heidelberg, 1987.

[8] R.J. JENKS, *Problem #11: Generation of Runge-Kutta Equations*, SIGSAM Bulletin,10, No. 1, 1976, p. 6.

[9] W. KUTTA, *Beitrag zur näherungsweisen Integration totaler Differentialgleichungen*, Zeitschrift für Math. u. Phys., Vol. 46, 1901, pp. 435-453.

[10] M. MONAGAN and J.S. DEVITT, *The D Operator and Algorithmic Differentiation*, Maple Technical Newsletter, No. 7, 1992.

[11] M. MONAGAN and R.R. RODINI, *An Implementation of the Forward and Reverse Modes of Automatic Differentiation in Maple*, Proceedings of the Santa Fe conference on Computational Differentiation, SIAM, 1996.

[12] J. MOSES, *Solution of Systems of Polynomial Equations by Elimination*, Comm. of the ACM,9, No. 8, 1966, pp. 634-637.

[13] B.L. VAN DER WAERDEN, *Algebra I*, Springer-Verlag, Berlin, 1971.

Chapter 20. Transient Response of a Two-Phase Half-Wave Rectifier

H.J. Halin and R. Strebel

20.1 Introduction

Electronic circuits are typically governed by linear differential equations with constant or time-dependent coefficients. The numerical simulation of such systems in the time-domain can be quite demanding, especially if the systems are very large and if they feature widely distributed eigenvalues. Methods for the analysis of electronic circuits have been presented in several textbooks, e.g. [2]. There are also a number of different packages for computer-aided analyses on the market with programs which are offsprings of codes such as ECAP [3] or SPICE2 [5]. These programs employ fully numerical methods and have therefore well-known pros and cons.

In what follows we would like to take advantage of the analytical capabilities and the accuracy of MAPLE V [1] in order to elegantly solve a small but tricky sample problem from the area of electronic circuits. It will be outlined why the problem, the simulation of the transient response of a two-phase half-wave rectifier, is a demanding problem in many ways. This will be explained below in more detail.

For the numerical solution by means of conventional programs a straightforward implementation of the mathematical model would not be sufficient. Instead some 'tricks' would have to be used to overcome the several numerical difficulties to be discussed later on. This is why an unexperienced analyst most likely would not immediately succeed in performing this simulation study.

20.2 Problem Outline

The problem to be investigated is the simulation of the electrical transient in a two-phase half-wave rectifier after two alternating voltage sources which act as external driving functions have been put in the circuit. This problem has originally been described in [4]. The structure of the model is illustrated in figure 20.1 The system is composed of two ideal diodes D_1 and D_2, two resistances and two impedances R_1, R_2, L_1, and L_2, respectively, the two open-circuit voltages $u_1(t)$, $u_2(t)$, and a load expressed by the resistance R_3 and the impedance L_3. The currents flowing through the diodes are $i_1(t)$ and $i_2(t)$,

FIGURE 20.1. *Structure of the Rectifier System*

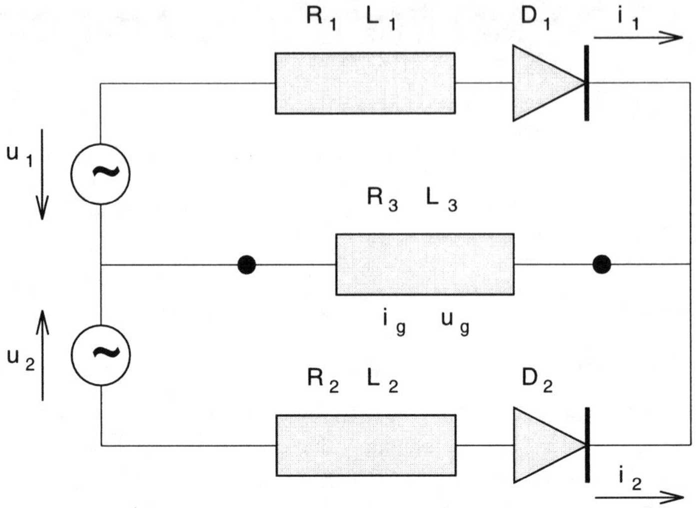

respectively. $i_g(t)$ is the rectified current, while $u_g(t)$ is the rectified voltage. Time is represented by t.

Depending on which diode is conducting three different cases can be distinguished. Which of the three cases applies will depend on the logical values of two time-dependent variables $diod_1$ and $diod_2$. We will define these variables later on but we would like to emphasize already that we use them only for didactic reasons. In the listing of the program that follows these variables do not show up explicitly.

1. Only diode D_1 is conducting. This case applies if $diod_1 = true$ and $diod_2 = false$:

$$\frac{d}{dt}i_1(t) = \frac{u_1 - i_1 R_{13}}{a_2} \tag{20.1}$$

$$\frac{d}{dt}i_2(t) = 0 \tag{20.2}$$

2. Only diode D_2 is conducting. This case applies if $diod_1 = false$ and $diod_2 = true$:

$$\frac{d}{dt}i_1(t) = 0 \tag{20.3}$$

$$\frac{d}{dt}i_2(t) = \frac{u_2 - i_2 R_{23}}{a_1} \tag{20.4}$$

3. Diode D_1 and diode D_2 are conducting. This case applies if $diod_1 = true$ and $diod_2 = true$:

$$\frac{d}{dt}i_1(t) = \frac{a_1u_1 - L_3u_2 - z_1i_1 - z_2i_2}{b} \qquad (20.5)$$

$$\frac{d}{dt}i_2(t) = \frac{a_2u_2 - L_3u_1 - z_3i_2 - z_4i_1}{b} \qquad (20.6)$$

where

$$
\begin{aligned}
a_1 &= L_2 + L_3 \\
a_2 &= L_1 + L_3 \\
b &= L_1L_2 + L_1L_3 + L_2L_3 \\
R_{13} &= R_1 + R_3 \\
R_{23} &= R_2 + R_3 \\
z_1 &= a_1R_1 + L_2R_3 \\
z_2 &= L_2R_3 - L_3R_2 \\
z_3 &= a_2R_2 + L_1R_3 \\
z_4 &= L_1R_3 - L_3R_1
\end{aligned}
$$

Note that only one of these three cases will hold at any time and that the fourth case where both variables are $false$ is not meaningful except in the steady state situation when the voltage sources are disconnected. The necessary conditions on which one of the three cases is relevant at a given time will be discussed later on when introducing the conditions for the values of $diod_1$ and $diod_2$.

The rectified voltage $u_g(t)$ and the rectified current $i_g(t)$ are given by

$$u_g(t) = R_3(i_1 + i_2) + L_3\left(\frac{d}{dt}i_1(t) + \frac{d}{dt}i_2(t)\right) \qquad (20.7)$$

$$i_g(t) = i_1 + i_2, \qquad (20.8)$$

respectively.
The voltages $u_1(t)$ and $u_2(t)$ read

$$u_1(t) = U\sin(\omega t) \qquad (20.9)$$

$$u_2(t) = -u_1(t). \qquad (20.10)$$

where

$$U = \sqrt{2}U_{eff} \qquad (20.11)$$

and

$$\omega = 2\Pi\nu. \qquad (20.12)$$

In this U_{eff} is the effective voltage while ν is the frequency.

In order to specify which of the diodes is conductive, we had already introduced the two logical variables $diod_1$ and $diod_2$. At any time t the logical values of these variables are governed by the following relations:

$$diod_1 = \begin{cases} true & \text{if } i_1 > 0 \quad \text{or} \quad u_1 > u_g, \\ false & \text{otherwise} \end{cases} \qquad (20.13)$$

$$diod_2 = \begin{cases} true & \text{if } i_2 > 0 \quad \text{or} \quad u_2 > u_g, \\ false & \text{otherwise} \end{cases} \qquad (20.14)$$

The transient will now be studied during a time interval $t_{start} \leq t \leq t_{final}$, where t_{start} and t_{final} denote the lower and the upper value of simulation time.

For the study the following numerical values will be used:

$$R_1 = 2[\Omega] \qquad (20.15)$$
$$R_2 = 2[\Omega] \qquad (20.16)$$
$$R_3 = 10[\Omega] \qquad (20.17)$$
$$L_1 = 0.04[H] \qquad (20.18)$$
$$L_2 = 0.04[H] \qquad (20.19)$$
$$L_3 = 0.20[H] \qquad (20.20)$$
$$U_{eff} = 100[V] \qquad (20.21)$$
$$\nu = 50[Hz]. \qquad (20.22)$$

As initial conditions we choose

$$i_1(0) = 0 \qquad (20.23)$$
$$i_2(0) = 0. \qquad (20.24)$$

20.3   Difficulties in Applying Conventional Codes and Software Packages

When using an integration algorithm such as Runge-Kutta for solving the problem outlined above the user needs to write a main program or a driver routine from which an integrator is invoked. These calls are embedded in a do-loop and are such that the program runs under control of the integrator until the integration over a specified interval of the independent variable is completed. When program control is returned to the main program the solution at the end of the last interval is available and can be used for output. After this the integrator is called again for doing the integration over the next interval while using the last values of the solution as initial conditions.

When performing an integration step the integrator will call a routine that needs to be provided by the user for doing a so-called 'function evaluation'. In this the present value of the independent variable t, the number of first order ODEs and the solution vector at t are transferred to the referencing routine. This routine would essentially contain the code of the differential equations, i.e.

(20.1-20.5) together with mechanisms for making constants, such as (20.15), locally available, which were introduced in the main program. It is then possible to evaluate the right-hand sides of the differential equations and to return a vector of the time-derivatives of the solution to the integrator.

Prior to doing a 'function evaluation' of our problem it is necessary to determine the logical values of the variables $diod_1$ and $diod_2$ from (20.13) in order to decide which of the three forms of our system of two ODEs applies. Since the value of u_g, (20.13), is needed for doing the evaluation of $diod_1$ and $diod_2$ we obviously have an implicit relation. For evaluating $\frac{d}{dt}i_1$ and $\frac{d}{dt}i_2$ the rectified voltage u_g must be known. Thereafter $diod_1$ and $diod_2$ can be found and subsequently $\frac{d}{dt}i_1$ and $\frac{d}{dt}i_2$ from the case that applies momentarily.

In order to deal with this difficulty, Jentsch proposes in [4] the usage of a value of u_g which is delayed by τ seconds, where τ is some small amount of time. The delayed value u_{gd} to be used in (20.13) is given by

$$u_{gd}(t) = \begin{cases} u_g(t - \tau) & \text{if } t > \tau \\ u_g(0) & \text{if } t \le \tau \end{cases}$$

Note that this is equivalent to

$$u_{gd}(s) = e^{(-\tau s)} u_g(s)$$

when using a standard engineering technique for linear problems by applying a Laplace transform.

In a numerical computation $u_{gd}(t)$ is found by interpolation between some values $u_g(t_i)$ and $u_g(t_{i+1})$, $(i = 1, 2, \ldots)$ where $(t_i \le t - \tau \le t_{i+1})$. $u_g(t_i)$ and $u_g(t_{i+1})$ are the values of u_g at the end of the $(i - 1)$-th and i-th integration step, respectively.

Yet another difficulty arises at the beginning of the very first integration step at time $t_0 = t_{start}$. Since the initial conditions (20.23), i.e. $i_1(t_0)$, and $i_2(t_0)$ and the two voltages, which act as excitations, (20.9), i.e. $u_1(t_0)$, $u_2(t_0)$, $u_g(t_0)$ are all zero it follows $diod_1 = diod_2 = false$, so that it cannot be decided which case applies.

Even when arbitrarily assuming one of the three cases to be valid it turns out that $\frac{d}{dt}i_1(t_0) = \frac{d}{dt}i_2(t_0) = 0$. Clearly, from a physical point of view at least one of the diodes must become conductive, since $u_1(t)$ has a positive derivative and likewise $u_2(t)$ a negative derivative at $t = t_0$. This implies that one of the two logical variables must be $true$. Therefore at $t = t_0$ the evaluation of $diod_1$ and $diod_2$, respectively, has to be done on the basis of derivatives of $u_1(t)$, $u_2(t)$, and $u_g(t)$.

The decision can also be made when doing the initial evaluation of $diod_1$ and $diod_2$ at some location $t_0 + \epsilon$, where ϵ is a small positive number, rather than at t_0.

20.4   Solution by Means of MAPLE

In solving the sample problem by means of MAPLE V [1] the numerical difficulties mentioned above can be avoided elegantly.

In the following a program is presented which is suited for computing the transient response of the rectifier. The program is composed of three files halfwave.map, data.map, and procs.map, respectively.

The program starts by referencing the file halfwave.map. After a restart we read and load from within halfwave.map a library routine for the later usage of an unassign-statement. After this the files data.map and procs.map will be read. This is followed by assigning values to t_{start} and t_{final} so as to specify the range of the independent variable t.

Example calling sequence in halfwave.map

```
readlib (unassign):
read ('data.map');
read ('procs.map');
tstart := 0:
tfinal := 0.050:
i := rectify_solve (array(1..2,[0,0]), tstart..tfinal):
ig := unapply(i[1](t)+i[2](t),t):
ug := unapply(R.3*ig(t) + L.3*diff(ig(t),t),t):

plot (i[1], tstart..tfinal, title='Plot 1: i1(t)');
plot (i[2], tstart..tfinal, title='Plot 2: i2(t)');
plot (ig, tstart..tfinal, title='Plot 3: ig(t)');
plot (ug, tstart..tfinal, title='Plot 4: ug(t)');
```

The integration of our differential equations is initiated by referencing procedure rectify_solve. This procedure which is listed at the very end of file procs.map serves as "main program". For the moment it is sufficient to mention that rectify_solve returns the analytic solutions $i_1(t)$ and $i_2(t)$ as elements of the two-dimensional vector ddata. Since analytic solutions can be found in piecewise form only each element itself is representing a list. Note that the initial conditions of the two state variables and the range of the simulation time are also entered when referencing rectify_solve. The plots that follow at the end of file halfwave.map will be described at the end of this chapter.

Note that subscripts in equations (20.1-20.24) are represented throughout the program by a period followed by the subscript, e.g. R_3 in one of the equations will be denoted as R.3 in the program.

In file data.map the external forcing functions $u_1(t)$ and $u_2(t)$ are given first. After this some necessary parameters for our problem are introduced. In this nof_cases and nof_states represent the number of different cases to be encountered during the operation of the rectifier and the number of state variables, respectively.

Solution of the differential equations in `data.map`

```
#------- input

u[1] := t ->  U*sin(Omega*t):
u[2] := t -> -U*sin(Omega*t):

#------- parameters

U := sqrt(2)*100:
nu := 50:
Omega := 2*Pi*nu:

R.1 := 2:
R.2 := 2:
R.3 := 10:
L.1 := convert(0.04,rational):
L.2 := convert(0.04,rational):
L.3 := convert(0.2,rational):
a.1 := L.2 + L.3:
a.2 := L.1 + L.3:
b := L.1*L.2 + L.1*L.3 + L.2*L.3:
R.13 := R.1 + R.3:
R.23 := R.2 + R.3:
z.1 := a.1*R.1 + L.2*R.3:
z.2 := L.2*R.3 - L.3*R.2:
z.3 := a.2*R.2 + L.1*R.3:
z.4 := L.1*R.3 - L.3*R.1:

#------- equations

nof_cases := 3:
nof_states := 2:

vars := {seq(j[k](t), k=1..nof_states)}:
deqn := array (1..3, [
  { diff(j[1](t),t) = j[1](t)*(-R.13/a.2) + u[1](t)/a.2,
    diff(j[2](t),t) = 0,
    j[1](t0) = j10,
    j[2](t0) = j20
  },
  { diff(j[1](t),t) = 0,
    diff(j[2](t),t) = j[2](t)*(-R.23/a.1) + u[2](t)/a.1,
    j[1](t0) = j10,
    j[2](t0) = j20
  },
  { diff(j[1](t),t) = j[1](t)*(-z.1/b) + j[2](t)*(-z.2/b)
                + 1/b*(a.1*u[1](t) - L.3*u[2](t)),
    diff(j[2](t),t) = j[1](t)*(-z.4/b) + j[2](t)*(-z.3/b)
                + 1/b*(a.2*u[2](t) - L.3*u[1](t)),
    j[1](t0) = j10,
    j[2](t0) = j20
  }]):

#------- global params

eps := 1/nu*1.0e-6; # get some 'relative' gap
Digits := 14;
```

```
_Envsignum0 := 0;

#------- solve the system for all cases

unassign ('k','t0','j10','j20'):
for k from 1 to nof_cases do
  dsol[k] := simplify (dsolve (deqn[k], vars)):
  assign (dsol[k]);
  for n from 1 to nof_states do
    i[k][n] := unapply (j[n](t), t, t0, j10, j20):
  od:
  unassign ('j[1](t)','j[2](t)'):
od:
```

This is followed by the formulation of the three sets of differential equations for the three cases possible and by their subsequent formal solution. Note that neither the starting time t0 nor the initial conditions j10 and j20 have been specified numerically so far.

For finding roots of the state variables $i_1(t)$ and $i_2(t)$, respectively, at some location t it will be necessary to provide some tolerance e.g. $\epsilon = 1.0e - 6$. Since the length of time intervals in which each of the three cases applies depends in some way on the wavelength $1/\nu$ of the external forcing functions $u_1(t)$ and $u_2(t)$ it is more appropriate to introduce a "relative" tolerance $\epsilon = 1.0e - 6/\nu$.

The last file procs.map contains the "main program" and some procedures and functions for doing the numerical computations.

Procedure definitions in procs.map

```
#------- Utilities

'diff/piecewise' := proc ()
  description 'cheap diff on piecewise expression.',
              'Maple V4's library function considers\
               discontinuities, but is _much_ slower.';
  local g, s, n, k, t;
  g := [args[1..nargs-1]];
  t := args[nargs];  n := nops(g);
  s := seq(op([g[2*k-1],diff(g[2*k],t)]),k=1..floor(n/2));
  if (type(n,odd)) then s := s,diff(g[n],t); fi;
  RETURN (piecewise(s));
end:

signum_rightof := proc (f, t::numeric)
  description 'return signum of f(t+)';
  global eps;
  local r, s, x;
  if (type (t, rational)) then
    s := simplify (series (f(x), x=t));
    r := op(1,s);
    RETURN (signum(r));
  fi;
  RETURN (signum(evalf (f(t+eps))));
end:
```

```
fsolve_smallest := proc (f, r::range)
  description 'return smallest zero in [r]. NULL if none.';
  global eps;
  local tfrom, tto, s, sprev, t;
  tfrom := op(1,r); tto := op(2,r);
  sprev := NULL;
  while (tfrom <= tto) do
    s := fsolve (f(t), t, t=tfrom..tto);
    if (whattype(s) <> float) then RETURN(sprev); fi;
    if ((s < tfrom) or (tto < s)) then RETURN(sprev); fi;
    tto := s-eps;
    sprev := s;
  od;
  RETURN(sprev);
end:

#------- get next interval

case[1,0] := 1:  case[0,1] := 2:  case[1,1] := 3:
# i[2] = 0:  i[1] = 0:

next_case := proc (t0::numeric, j::array)
  description 'returns case that follows point t0.';
  global eps, case;
  RETURN (case[signum_rightof(j[1],t0), signum_rightof(j[2],t0)]);
end:

next_interval := proc (t0::numeric, i0::array, VAR_case)
  description 'returns currents for next interval.',
              'In VAR_case next case.';
  local t, j, s, case;
  for s from nof_cases by -1 to 1 do
    j := array(1..2);
    j[1] := unapply (i[s][1](t,t0,i0[1],i0[2]), t);
    j[2] := unapply (i[s][2](t,t0,i0[1],i0[2]), t);
    case := next_case (t0, j);
    if (s = case) then VAR_case := s; RETURN (j); fi;
  od;
  ERROR ('Continuation failed');
end:

crit_smallest := proc (f, r::range(numeric))
  description 'smallest zero of f in ]r] (without left border).',
              'right(r) if (f == 0) or (f(t) <> 0) in r.';
  global eps;
  local s, tfrom, tto;
  tfrom := op(1,r); tto := op(2,r);
  if (f = 0) then RETURN (tto); fi;
  s := fsolve_smallest (f, tfrom+eps..tto);
  if (whattype(s) <> float) then RETURN (tto); fi;
  RETURN (s);
end:

next_crit := proc (j::array, r::range(numeric),
                   case::integer, VAR_i)
  description 'returns next critical point.',
              'In VAR_i current i[] at this point.';
  local t, ug, ud, crit, tcrit, i;
```

```
ug := unapply (evalf(R.3*(j[1](t) + j[2](t)) +
      L.3*(diff(j[1](t),t)+diff(j[2](t),t))), t);
ud[1] := unapply (u[1](t) - ug(t), t);
ud[2] := unapply (u[2](t) - ug(t), t);
if (case = 1) then
  crit[1] := crit_smallest (j[1], r);
  crit[4] := crit_smallest (ud[2], r);
  tcrit := min(crit[1], crit[4]);
elif (case = 2) then
  crit[2] := crit_smallest (j[2], r);
  crit[3] := crit_smallest (ud[1], r);
  tcrit := min(crit[2], crit[3]);
elif (case = 3) then
  crit[1] := crit_smallest (j[1], r);
  crit[2] := crit_smallest (j[2], r);
  crit[3] := crit_smallest (ud[1], r);
  crit[4] := crit_smallest (ud[2], r);
  tcrit := min(crit[1], crit[2], crit[3], crit[4]);
fi;
i := array (1..2, [j[1](tcrit), j[2](tcrit)]);
if (crit[1] = tcrit) then i[1] := 0; fi;
if (crit[2] = tcrit) then i[2] := 0; fi;
VAR_i := i;
RETURN (tcrit);
end:


#------- main program

rectify_solve := proc (i0::array(numeric), r::range(numeric))
  description 'returns i[1..2] as piecewise functions';
  global nof_states, eps;
  local tfrom, tto, tcur, scur, res, i, j, k, ires, n, nof, d, s, t;
  tfrom := op(1,r); tto := op(2,r);
  res := array (1..nof_states);
  for k from 1 to nof_states do res[k] := NULL; od;
  tcur := tfrom; i := i0;
  while (tcur < tto) do
    unassign ('scur');
    j := next_interval (tcur, i, scur);
    for k from 1 to nof_states do
      res[k] := res[k], [tcur, j[k], scur];
    od;
    unassign ('i');
    tcur := next_crit (j, tcur..tto, scur, i);
  od;
  ires := array (1..nof_states);
  for k from 1 to nof_states do
    nof := nops([res[k]]);
    d := res[k],[tto+eps];
    s := seq(op([d[n][1] <= t and t < d[n+1][1],d[n][2](t)]),n=1..nof);
    ires[k] := unapply(piecewise(s),t);
  od;
  RETURN (ires);
end:
```

The "main program" in file procs.map is the procedure that governs most of the execution. The procedure uses the initial values of the state variables and the value of the independent variable at the beginning of each interval for which one of the three possible states holds. By making successive references to procedure `next_interval` it will be determined first which of the three cases applies. In doing this, procedure `next_case` needs to be invoked which itself will call procedure `sign_rightof`. Of course the solutions in analytic form found already in file `halfwave.map` will be used throughout. As can be seen in `sign_rightof`, the determination of the case that applies is done in analytical form at $t = t_{start}$ by evaluation of higher derivatives of the state variables $i_1(t)$ and $i_2(t)$ up to an order where they do not vanish any more. Otherwise the determination is done by looking for the states at some location $t + \epsilon$, where t denotes the last location of a change of the case of operation. Once the case is determined information regarding the results will be stored in lists in the "main program". After this the next interval will be considered until the whole range of the independent variable is covered.

The solution is illustrated by a number of plots for displaying the currents $i_1(t)$, $i_2(t)$, the rectified current $i_g(t)$, and the rectified voltage $u_g(t)$ over time.

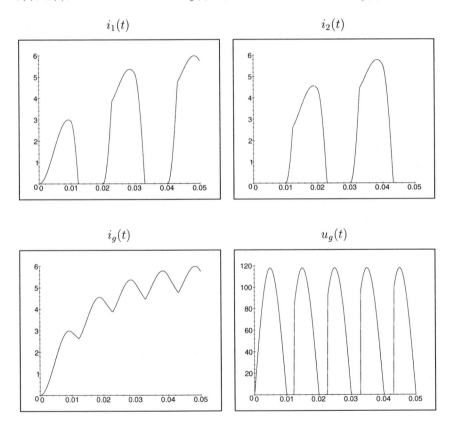

References

[1] B.W. CHAR, K.O. GEDDES, G.H. GONNET, B.L. LEONG, M.B. MONAGAN, AND S.M. WATT, MAPLE *Language/Reference Manual*, Springer Verlag, New York, 1991.

[2] L.O. CHUA AND P.-M. LIN, *Computer-Aided Analysis of Electronic Circuits: Algorithms and Computational Techniques*, Prentice-Hall, Englewood Cliffs, N.J., 1975.

[3] R.W. JENSEN AND M.D. LIEBERMAN, *IBM Electronic Circuit Analysis Program*, Prentice-Hall, Englewood Cliffs, N.J., 1968.

[4] W. JENTSCH, *Digitale Simulation kontinuierlicher Systeme*, Oldenbourg Verlag, München und Wien, 1969.

[5] L.W. NAGEL, *SPICE2: A Computer Program to Simulate Semiconductor Circuits*, University of California, Ph.D. Dissertation, Berkeley, California, 1975.

Chapter 21. Circuits in Power Electronics

J. Waldvogel

21.1 Introduction

Over the last few years high-power semi conductor devices with intrinsic turn-off capability have become available. These devices, called gate turn-off (GTO) thyristors, consist of several layers of silicon with appropriate dotations; they are able to turn off currents of 1000 Amperes at thousands of volts within microseconds. In circuits used in power electronics the usual resistive, inductive and capacitive circuit elements are combined with thyristors which may simply be considered as switches. This technology is still an active field of research, and it has many important applications such as AC/DC conversion (both ways), speed control of locomotives and electric cars, control of power stations and power networks, etc.

For every fixed state of the thyristor switches Kirchhoff's laws must be satisfied, and therefore the dynamical behavior of such a circuit is described by a system of linear ordinary differential equations with constant coefficients, assuming linearity of the circuit elements. If the switches change their positions the structure of the circuit changes, but the final state of the currents in the circuit before switching determines the initial conditions after the switching.

Therefore the mathematical model of a circuit in power electronics is a system of linear differential equations with piecewise constant coefficients if the switching times are neglected. We assume the dynamics of the circuit to be described by n continuous functions of time t, which are represented by the vector $\mathbf{x}(t) \in \mathbf{R}^n$ of dependent variables. By using matrix notation the model may be written as

$$\dot{\mathbf{x}} = A(t)\mathbf{x} + \mathbf{p}(t) \tag{21.1}$$

where dots denote derivatives with respect to time, $A(t)$ is a step function, i.e., a given piecewise constant $n \times n$ matrix, and $\mathbf{p}(t) \in \mathbf{R}^n$ is a given forcing function. In the environment of AC (alternating current) circuits $\mathbf{p}(t)$ and $A(t)$ are often periodic functions. With no loss of generality the period will be normalized to 2π, and in view of the Fourier decomposition of $\mathbf{p}(t)$ we will use the first harmonic as a model case. Usually the solution $\mathbf{x}(t)$ is specified by initial conditions $\mathbf{x}(0) = \mathbf{x}_0$, but other specifications, e.g. periodicity of $\mathbf{x}(t)$, will be considered in Section 21.3.

The initial value problem of systems of linear differential equations with a

FIGURE 21.1. *Simplified SVC circuit*

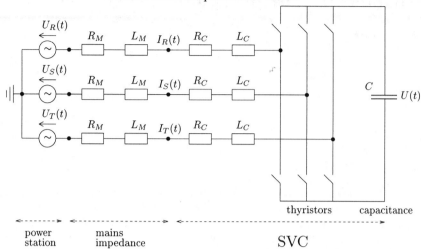

Description of the SVC circuit

Variable	Description
$U_R(t), U_S(t), U_T(t)$	AC voltages generated by the power station
R_M, L_M	Resistance and inductance of the mains
R_C, L_C, C	Resistance, inductance and capacitance of the SVC
$I_R(t), I_S(t), I_T(t)$	Currents injected in the mains
$U(t)$	Voltage across the DC capacitance of the SVC

constant matrix A is a topic of elementary calculus (cf. [1]) and may be handled via the eigenvalues and eigenvectors of A or via the matrix exponential e^{At}. Even if the matrix $A(t)$ is piecewise constant the explicit solution of the initial value problem is straight-forward, although quite laborious if $A(t)$ has many discontinuities.

It turns out that the features of MATLAB allow a very elegant construction of the solution $\mathbf{x}(t)$ of this initial value problem. Also, periodic solutions may be easily calculated and plotted. In this article we will use a specific device from the field of electric power network control, the so-called Static Var Compensator (SVC) to explain the use of MATLAB in power electronics. Var stands for Volt-Ampere reactive.

SVCs are used in electric power networks to compensate for the voltage drop due to the losses in the power lines and due to a variable user load. In the usual three-phase AC system six switching operations per period are needed, in order to transfer an impulse of reactive power from one phase to another during an appropriate time interval in each period.

In Figure 21.1 a simplified SVC circuit including the mains is shown, the thyristors being represented by switches. For simplicity, no loads of the electric network are taken into consideration. In each circuit corresponding to a single phase only one switch may be closed at any time. For the circuit of Figure 21.1 it is sufficient to choose $n = 3$ independent variables, e.g.

$$x_1(t) = U(t), \quad x_2(t) = I_R(t), \quad x_3(t) = \frac{1}{\sqrt{3}}(I_S(t) - I_T(t)) \tag{21.2}$$

Then the dynamics of the circuit is described by Equation (21.1) with

$$\mathbf{p}(t) = \frac{1}{L}(0, \cos t, \sin t)^T, \quad L = L_M + L_C, \quad R = R_M + R_C \tag{21.3}$$

and

$$A(t) = B(\varphi(t)), \quad B(\varphi) = \begin{bmatrix} 0 & \frac{1}{C}\cos\varphi & \frac{1}{C}\sin\varphi \\ -\frac{2}{3L}\cos\varphi & -\frac{R}{L} & 0 \\ -\frac{2}{3L}\sin\varphi & 0 & -\frac{R}{L} \end{bmatrix}. \tag{21.4}$$

The switching angle $\varphi(t)$ is a given piecewise constant function that controls the operation of the thyristors. In the 6-pulse SVC it is chosen as

$$\varphi(t) = \frac{\pi}{3} \text{round}\left(\frac{3}{\pi}(t - \tau)\right), \tag{21.5}$$

where the shift τ is a parameter of the SVC to be chosen in $|\tau| \leq R$. Typical values of the parameters C, L, R, expressed in normalized units such that the period of the AC (often 0.02 sec) becomes 2π are

$$C = 0.2, \quad L = 0.15, \quad R = 0.005. \tag{21.6}$$

In practice, the orders of magnitude of C, L and R are 150nF, 2Hy and 20Ω respectively. For further technical details the reader is referred to the textbooks [2, 3].

21.2 Linear Differential Equations with Piecewise Constant Coefficients

We consider the differential equation (21.1),

$$\dot{\mathbf{x}} = A(t)\mathbf{x} + \mathbf{p}(t),$$

for the unknown function $\mathbf{x}(t) \in \mathbf{R}^n$. For simplicity the given 2π-periodic forcing function $\mathbf{p}(t)$ is assumed to contain the first harmonic only,

$$\mathbf{p}(t) = \mathbf{b}e^{it} + \bar{\mathbf{b}}e^{-it}, \quad \mathbf{b} \in \mathbf{C}^n. \tag{21.7}$$

The use of complex notation turns out to be advantageous since it greatly simplifies the equations, and it is fully supported by MATLAB. The piecewise constant real matrix $A(t)$ is assumed to be 2π-periodic as well. Therefore the $m+1$ (possible) discontinuities (jumps) t_k of $A(t)$ will be introduced as

$$0 = t_0 < t_1 < t_2 < \ldots < t_{m-1} < t_m = 2\pi,$$

and the discrete values of the matrix are denoted by

$$A(t) = A_k \text{ in } t_k \le t < t_{k+1}, \quad k = 0, \ldots, m-1. \tag{21.8}$$

Given initial conditions $\mathbf{x}(0) = \mathbf{x}_0$, Equation (21.1) has a unique solution $\mathbf{x}(t)$; its values at the jumps are denoted by

$$\mathbf{x}_k := \mathbf{x}(t_k), \quad k = 0, \ldots, m.$$

First, we construct the explicit solution $\mathbf{x}(t)$ of (21.1) in the k-th subinterval $t \in [t_k, t_{k+1}]$ satisfying the appropriate differential equation and the initial condition:

$$\left.\begin{array}{l} \dot{\mathbf{x}}(t) = A_k\mathbf{x}(t) + \mathbf{p}(t), \quad t \in [t_k, t_{k+1}] \\ \mathbf{x}(t_k) = \mathbf{x}_k \end{array}\right\} \quad k = 0, \ldots, m-1. \tag{21.9}$$

As usual we use the decomposition

$$\mathbf{x}(t) = \mathbf{y}(t) + \mathbf{z}(t) \tag{21.10}$$

into a conveniently chosen particular solution $\mathbf{z}(t)$ and the solution $\mathbf{y}(t)$ of the homogeneous problem satisfying

$$\dot{\mathbf{y}}(t) = A_k\mathbf{y}(t), \quad \mathbf{y}(t_k) = \mathbf{x}_k - \mathbf{z}(t_k). \tag{21.11}$$

Using the well-known matrix exponential we obtain

$$\mathbf{y}(t) = e^{A_k t}\mathbf{c}_k \tag{21.12}$$

with $\mathbf{c}_k$ determined from the second equation of (21.11).

The computation of the matrix exponential is a non trivial problem with a long history, as is seen from the title of the survey paper [4], "Nineteen Dubious Ways to Compute the Exponential of a Matrix". In the MATLAB command expm the "least dubious" way, the method of Padé approximation, is implemented and works reliably, accurately and quickly in almost all cases. It is an expensive operation, however, requiring up to $30n^3$ flops for an $n \times n$ matrix.

If the matrix B is diagonalizable e^B may easily be computed via the eigenvalue factorization $B = TDT^{-1}$, $D = \operatorname{diag}(\lambda_1, \ldots, \lambda_n)$ as

$$e^B = Te^D T^{-1}, \quad e^D = \operatorname{diag}(e^{\lambda_1}, \ldots, e^{\lambda_n}), \tag{21.13}$$

where λ_j is the jth eigenvalue of B.

In the case of the matrix $A(t) = B(\varphi(t))$ defined in (21.4) the eigenvalues turn out to be independent of φ and may be written explicitly as

$$\lambda_1 = -r, \quad \lambda_{2,3} = -r/2 \pm i\omega \tag{21.14}$$

where

$$r = -R/L, \quad \omega = \sqrt{\frac{2/3}{LC} - \frac{R^2}{4L^2}}.$$

Therefore, $\lambda_1, \lambda_2, \lambda_3$ are also the eigenvalues of all the matrices A_k for any choice of the switching times t_k. This is seen by means of the similarity relation

$$B(\varphi) = S(\varphi)B(0)S(\varphi)^{-1}$$

with the orthogonal matrix

$$S(\varphi) = \begin{bmatrix} 1 & 0 & 0 \\ 0 & \cos(\varphi) & -\sin(\varphi) \\ 0 & \sin(\varphi) & \cos(\varphi) \end{bmatrix}.$$

We then have

$$e^{B(\varphi)} = S(\varphi)e^{B(0)}S(\varphi)^{-1},$$

with the great advantage that $e^{B(0)}$ has a simple explicit representation due to the block diagonal structure of B(0):

$$e^{B(0)} = e^{r/2} \begin{bmatrix} \cos(\omega)I + \sin(\omega)/\omega\, B_0 & 0 \\ 0 & e^{r/2} \end{bmatrix}.$$

Here I is the 2×2 unit matrix, B_0 is the upper left 2×2 block of $B(0)$, and r, ω are defined above. The reader is invited to derive or verify these relations by means of MAPLE. An implementation of $e^{B(\varphi)}$ is possible with about 30 elementary operations and 6 function calls (such as exp or sin).

To complete the construction of the solution in the k–th time interval a particular solution $\mathbf{z}(t)$ of (21.9) has to be chosen. The simplest choice is a harmonic oscillation of the same frequency as $\mathbf{p}(t)$, i.e.

$$\mathbf{z}(t) = -\mathbf{u}_k e^{it} - \bar{\mathbf{u}}_k e^{-it} \tag{21.15}$$

where the complex vector $\mathbf{u}_k \in \mathbf{C}^n$ must be determined such that (21.9) is satisfied. Inserting (21.15) into (21.9) yields the condition

$$(A_k - iI)\mathbf{u}_k = \mathbf{b}, \tag{21.16}$$

a system of linear equations in $\mathbf{C}$ for $\mathbf{u}_k$, where I is the $n \times n$ unit matrix. Therefore, the necessary and sufficient condition for the existence of a solution of the form (21.15) is $\det(A_k - iI) \neq 0$, i.e. $\pm i$ must not be an eigenvalue of A_k. From (21.14) there follows that this is satisfied as long as

$$R \neq 0 \quad \text{or} \quad \frac{2L}{3C} - \frac{R^2}{4} \neq L^2, \tag{21.17}$$

which is true for the specific data given in (21.6). The resonant case (Condition (21.17) violated) may be handled by augmenting (21.15) with terms such as $\mathbf{v}_k t e^{it}$, but this case will not be pursued further.

Finally, combining (21.10), (21.12) and (21.15) yields the explicit solution

$$\mathbf{x}(t) = e^{A_k t}\mathbf{c}_k - 2\,\mathrm{Re}\,(\mathbf{u}_k e^{it}), \quad t \in [t_k, t_{k+1}] \tag{21.18}$$

with

$$\mathbf{u}_k = (A_k - iI)^{-1}\mathbf{b} \tag{21.19}$$

and

$$\mathbf{c}_k = e^{-A_k t_k}(\mathbf{x}_k + 2\,\mathrm{Re}\,(\mathbf{u}_k e^{it_k})), \tag{21.20}$$

as follows from (21.18) with $t = t_k$. Putting $t = t_{k+1}$ in (21.18) yields the value of $\mathbf{x}(t)$ at the next jump,

$$\mathbf{x}_{k+1} = e^{A_k t_{k+1}}\mathbf{c}_k - 2\,\mathrm{Re}\,(\mathbf{u}_k e^{it_{k+1}}). \tag{21.21}$$

For evaluating $\mathbf{x}(t)$ at many points it is best to pre-compute and store $\mathbf{u}_k$, $\mathbf{c}_k$, $\mathbf{x}_{k+1}$ according to Equations (21.19), (21.20) and (21.21) in a loop running over $k = 0, \ldots, m-1$. Then (21.18) yields $\mathbf{x}(t)$ involving at most one matrix exponential.

21.3   Periodic Solutions

In technical applications such as SVCs one is often interested in periodic solutions of the corresponding differential equations. However, periodic solutions are of practical significance only if they are attractive; then they arise naturally after a long time from an arbitrary initial state in their basin of attraction.

In linear problems such as (21.1) the principle of superimposition holds; therefore the stability of periodic solutions is determined by the corresponding homogeneous problem defined by $\mathbf{p}(t) = 0$ or $\mathbf{b} = 0$ (see Equation (21.7)). From Equations (21.19), (21.20), (21.21) with $\mathbf{b} = 0$ and $\mathbf{x}_k$ replaced by $\mathbf{y}_k$ we obtain

$$\mathbf{y}_{k+1} = e^{A_k(t_{k+1}-t_k)}\mathbf{y}_k \tag{21.22}$$

since the matrices $A_k t_{k+1}$ and $A_k t_k$ commute. Here $\mathbf{y}_k$ denotes the value of a solution of the homogeneous equation at the jump $t_k (k = 0, \ldots, m)$. Therefore, after a full period $t_m = 2\pi$, the value of $\mathbf{y}_m$ is given by the linear map

$$\mathbf{y}_m = M\mathbf{y}_0, \tag{21.23}$$

where

$$M = \prod_{k=0}^{m-1} e^{A_k(t_{k+1}-t_k)} \tag{21.24}$$

(the product taken from right to left) is the so-called monodromy matrix. There follows that a periodic solution of (21.1) is globally attractive if $|\mu_j| < 1$ holds for all eigenvalues μ_j of M.

It turns out that for the 6-pulse SVC given by the matrix (21.4) and the switching function (21.5) the eigenvalues μ_j are independent of the shift τ. In the example (21.6) the values

$$\mu_1 = 0.84311362558494 \tag{21.25}$$
$$\mu_{2,3} = -0.77497080502505 \pm 0.42379742896324i \tag{21.26}$$

are obtained; hence if a periodic solution exists it is globally attractive.

To construct such a solution $\mathbf{x}(t) = \mathbf{x}_P(t)$, an initial value vector $\mathbf{x}_0 \in \mathbf{R}^n$ has to be found such that

$$\mathbf{x}_m = \mathbf{x}_0 \quad \text{or} \quad \mathbf{f}(\mathbf{x}_0) := \mathbf{x}_m - \mathbf{x}_0 = 0 \tag{21.27}$$

in the notation of (21.21). Due to the linearity of the problem the vector valued function $\mathbf{f}$ defined in (21.27) is itself linear. Hence it suffices to compute $n + 1$ values of $\mathbf{f}$ in order to define the system of linear equations (21.27). This is necessary because $\mathbf{f}$ is defined only indirectly by means of the rather complicated algorithm described at the end of Section 21.2.

The $n+1$ points of evaluation are conveniently chosen as the origin and the n unit points $\mathbf{e}_j = (0, \ldots, 0, 1, 0, \ldots, 0)^{\mathrm{T}}$, where the non-vanishing component is in position j, $(j = 1, \ldots, n)$. If we denote

$$\mathbf{f}_0 := \mathbf{f}(0), \quad \mathbf{f}_j := \mathbf{f}(\mathbf{e}_j), \quad (j = 1, \ldots, n) \tag{21.28}$$

the linear function $\mathbf{f}(\mathbf{x})$ is explicitly given by

$$\mathbf{f}(\mathbf{x}) = \mathbf{f}_0 + \sum_{j=1}^n (\mathbf{f}_j - \mathbf{f}_0)\, x_j \tag{21.29}$$

where $\mathbf{x} = (x_1, \ldots, x_n)^{\mathrm{T}}$. The initial value $\mathbf{x}_0$ satisfying $\mathbf{f}(\mathbf{x}_0) = 0$ is therefore obtained from the linear system

$$F\mathbf{x}_0 = \mathbf{f}_0 \tag{21.30}$$

where F is the matrix

$$F = [\mathbf{f}_0 - \mathbf{f}_1, \ldots, \mathbf{f}_0 - \mathbf{f}_n]. \tag{21.31}$$

A unique periodic solution exists if the matrix F is regular. In the numerical example (21.6) we obtain for all $\tau \in [-R, R]$ $\operatorname{cond}(F) \doteq 11.74$, hence in this case F is far away from a singular matrix.

21.4 A MATLAB Implementation

In this section we will present a complete MATLAB program capable of carrying out the tasks listed below for the example of the 6-pulse SVC. It is based on the explicit solution of $n = 3$ linear differential equations with piecewise constant coefficients as described in the previous sections.

(a) The periodic solution $\mathbf{x}_P$ discussed in Section 21.3 is generated, and its values at the $m + 1$ jumps of $A(t)$ are stored. Possible near-degeneracies may be detected by means of $\operatorname{cond}(F)$.

ALGORITHM 21.1. *Function* matrix.

```
function [A] = matrix(k)
% generates the matrix A[k], k=0,...,m-1
%
  global m C L R

  phi = k*2*pi/(m-1);
  A   = [
                    0,  cos(phi)/C,  sin(phi)/C
        -cos(phi)/1.5/L,     -R/L,          0
        -sin(phi)/1.5/L,       0,         -R/L
      ];

  end % matrix
```

(b) Computation of the monodromy matrix M associated with x_P together with its eigenvalues for discussing the stability of x_P.

(c) Efficient tabulation and plotting of x_P (dense output).

(d) Fourier analysis of the periodic solution x_P.

The program is kept general as much as possible although some particular features of the specific example necessarily appear. This will enable the reader to adapt the program to any other problem involving linear ODEs with piecewise constant coefficients. The main objectives of the code are efficiency and simplicity, not luxury of the input and output. However, compared to the shortness of the program a fair amount of luxury and a high degree of reliability is achieved.

The core of the program is the function f = solution(x0) given in Algorithm 21.2. It solves (21.1) with the initial vector x0 = x_0 and returns the value f = $f(x_0)$ of the function defined in (21.27). According to the number nargout of output arguments in the actual call (a permanent variable of MATLAB) the matrices xx, uu, M are also computed. The matrices A_k specific to this example (see Equations (21.4), (21.5)) are generated by the function A = matrix(k) in Algorithm 21.1. Table 21.1 describes the variables which are passed as global parameters for convenience.

TABLE 21.1. *Description of Global Variables*

Variable	Description
m, n	Dimensional parameters
C, L, R	System parameters
b	Inhomogeneity **b** from Equations (21.7),(21.3)
tt(1:m+1)	Array of jumps, tt(1+k)= t_k, $(k = 0 \ldots m)$

ALGORITHM 21.2. *Function* solution.

```
function [f, xx, uu, M] = solution(x0)
%SOLUTION          Solves the SVC problem over a period.
%                  f == 0 <=> periodic solution
%
  global m n tt b

  if (nargout > 1), uu = []; xx = x0; end;
  if (nargout > 3), M  = eye(n); end;
  x = x0;
  for k = 1:m,
    A = matrix(k-1);
    E = expm(A*(tt(k+1)-tt(k)));
    u = (A-i*eye(n))\b;
    x = E*(x+2*real(u*exp(i*tt(k)))) ...
            - 2*real(u*exp(i*tt(k+1)));
    if (nargout > 1), uu = [uu,u]; xx = [xx,x]; end;
    if (nargout > 3), M  = E*M; end;
  end;
  f = x - x0;

end % solution
```

On exit from solution, xx will contain the values x_k, uu the intermediate results u_k and M the monodromy matrix. These variables are initialized in the first lines of solution. In the subsequent loop over k we first generate the matrix A_{k-1} by calling the function matrix. Then Equations (21.19), (21.20) and (21.21) are evaluated as described at the end of Section 21.2. The only modification is that c_k in (21.21) has been substituted by the expression in (21.20). This halves the number of matrix exponentials to be computed. Furthermore, it turns out that storing the vectors c_k can be avoided at almost no cost.

Since the MATLAB indices are ≥ 1 the loop index k has been shifted by 1. The newly computed vectors $u = u_k$, $x = x_{k+1}$ are stored by appending them to the matrices uu and xx, respectively. In the same loop the partial product M is updated according to (21.24). The final statement defines the function value (21.27).

In Algorithm 21.3 the main program per.m performing the tasks (a) through (d) by means of calls to the function solution is given. After initializing some global variables, the input of the shift τ is done via a program request. In this way the sensitive dependence of the circuit's behavior on τ may easily be studied. In the next statement the array of the jumps tt(1:m+1) (with the index shifted by 1) is defined:

$$tt(1) = 0$$
$$tt(1+k) = t_k = \tau + (k - \tfrac{1}{2})\frac{\pi}{3}, \quad (k = 1, \dots, 6)$$
$$tt(8) = 2\pi$$

ALGORITHM 21.3. *Script* per.

```
global m n C L R tt b

m = 7; n = 3;
C = 0.2; L = 0.15; R = 0.005;
b = [0;1;-i]/2/L;
N = 512;

tau = input('tau = ');
tt  = [0, 2*pi/(m-1)*[1/2:(m-3/2)]+tau, 2*pi];

f0 = solution(zeros(n,1));
F = []; for x = eye(n),
  F = [F, f0 - solution(x)];
end;
x0 = F\f0;
[ff, xx, uu, M] = solution(x0);

delta = 2*pi/N;
k = 0; xtab = [];
for j = 0:N-1,
  t = j*delta;
  if (tt(k+1) <= t),
    k = k+1;
    A = matrix(k-1);
    aux = expm(A*(t-tt(k))) ...
            *(xx(:,k)+2*real(uu(:,k)*exp(i*tt(k))));
    E = expm(A*delta);
  else
    aux = E*aux;
  end;
  xtab = [xtab, aux-2*real(uu(:,k)*exp(i*t))];
end;

figure(1); plot(xtab(1,:)');
figure(2); plot(xtab(2:3,:)');

cc = fft(1/N*xtab'); cc(1:32,:)
```

(see (21.5)). Then the periodic solution $\mathbf{x}_P$ together with its initial value x0 = $\mathbf{x}_0$ is computed. The program closely follows Equations (21.28) through (21.31) and is self-explanatory.

In the next section of the program $\mathbf{x}_P(t)$ is tabulated with step $\Delta = 2\pi/N$:

$$\text{xtab}(:,\mathrm{j}) = \mathbf{x}_P((j-1)\Delta), \quad j = 1,\ldots,N.$$

In view of the subsequent fast-Fourier analysis N must be a power of 2. The algorithm is organized to work efficiently if $N \gg m$ as follows. Given a value of t, the index k is such that $t \in [t_k, t_{k+1})$. If $t \in [t_k, t_{k+1})$ is the first evaluation

FIGURE 21.2. *Voltage and currents in an SVC*

Parameters: $C = 0.2$, $L = 0.15$, $R = 0.005$ according to Equation (21.6). The three cases of the shifts $\tau = -R$, $-0.375R$, R are shown. In the left-hand figures the voltage $x_1(t) = U(t)$ is plotted versus time t, whereas the currents $x_2(t) = I_S(t)$ (solid line) and $x_3(t) = (I_S(t) - I_T(t))/\sqrt{3}$ (dashed line) are plotted in the right-hand figures.

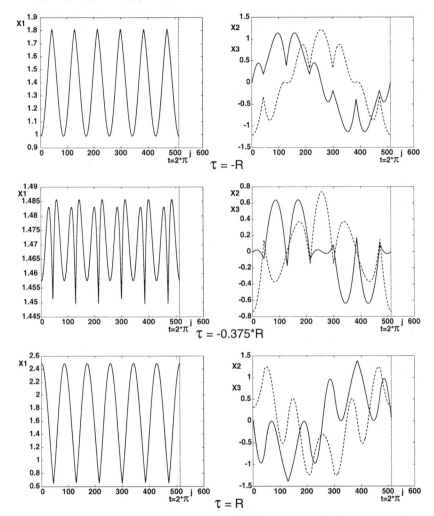

point in this interval, $\mathbf{x}_P(t)$ must be calculated according to Equation (21.18). Hence we begin by computing the first term of (21.18) as

$$\mathtt{aux} := \mathrm{e}^{A_k t}\mathbf{c}_k = \mathrm{e}^{A_k(t-t_k)}(\mathbf{x}_k + 2\,\mathrm{Re}\,(\mathbf{u}_k\mathrm{e}^{it_k})),$$

where the vectors $\mathbf{x}_k$ and $\mathbf{u}_k$ are taken from the arrays **xx** and **uu**, respectively. Furthermore, the matrix $\mathtt{E} := \mathrm{e}^{A_k\Delta}$ is computed and stored at this point, thus avoiding its frequent re-computation at later points in the same interval. Otherwise, it suffices to update aux as aux := E*aux.

In the final sections of the program the first component of $\mathbf{x}_P(t)$, i.e. $U(t)$ (the voltage across the capacitance C) is plotted versus $j = 1 + t/\Delta$ as MATLAB Figure 1. The second and third components, which are both variable currents, are simultaneously plotted as MATLAB Figure 2. Finally, the 3 components of $\mathbf{x}_P$ are separately Fourier-analyzed by means of the MATLAB command fft that requires a column vector as its argument. The matrix cc serves for printing the first 32 complex Fourier coefficients of the three components of $\mathbf{x}(t)$.

The data (21.6) in the three cases $\tau = -R$, $\tau = -0.375R$ and $\tau = R$ produce the plots presented in Figure 21.2. The sensitivity to small changes in τ is obvious. Ideally, the currents $x_2(t)$, $x_3(t)$ should be sinusoidal. One goal of research in this field is to reduce the disturbances due to the higher harmonics present in the periodic solution $\mathbf{x}_P$.

21.5   Conclusions

A computer simulation of a technical process is a research tool useful in designing and optimizing the process. The above specialized circuit simulator is more than ten times faster than a general-purpose simulator on the same problem, and it produces highly accurate approximations (14 decimals) to the exact solutions of the mathematical model. This enables the user to obtain reliable spectra of periodic solutions to high order, which, in turn, enable the designer to eliminate some of the unwanted harmonics.

Clearly, the success of this simulation is largely due to the high standards of the MATLAB software. Best results are obtained, however, if good software is combined with a careful mathematical analysis.

The author is indebted to Gerald Scheuer of the Institute of Power Electronics (ETH Zürich) for providing the differential equations of the SVC. Helpful comments by Rolf Strebel are gratefully acknowledged.

References

[1] M. BRAUN, *Differential Equations and their Applications, Fourth ed.*, Springer, New York, 1993, 578 pp.

[2] M. MEYER, *Leistungselektronik*, Springer, Berlin, 1990, 349 pp.

[3] T. J. E. MILLER, *Reactive Power Control in Electric Systems*, J. Wiley and Sons, New York, 1982, 381 pp.

[4] C.B. MOLER AND C.F. VAN LOAN, *Nineteen Dubious Ways to Compute the Exponential of a Matrix*, SIAM Review 20, 1978, pp. 801–836.

Chapter 22. Newton's and Kepler's laws

S. Bartoň

22.1 Introduction

The goal of this article is to demonstrate the use of computer algebra in physics teaching. Seven practical examples from Newton's theory of gravity will be solved with support of MAPLE. All the examples will use no more than then celebrated Newton's law for the gravitational force

$$F = G\frac{m_1 m_2}{r^2}$$

or the formula for the corresponding potential energy.

22.2 Equilibrium of Two Forces

Let's start with a simple problem: Find the point of equilibrium located between two points with masses m_1 and m_2, the distance of the points being r, (see Figure 22.1).

FIGURE 22.1. *Definition of the problem*

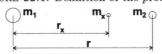

```
> R1 := G*m[1]*m[x]/rx^2 = G*m[x]*m[2]/(r - rx)^2;
```

$$R1 := \frac{G\, m_1\, m_x}{rx^2} = \frac{G\, m_x\, m_2}{(r - rx)^2}$$

```
> Sr := solve(R1, rx);
```

$$Sr := \frac{1}{2}\frac{(2\, m_1 + 2\,\sqrt{m_1\, m_2})\, r}{m_1 - m_2},\ \frac{1}{2}\frac{(2\, m_1 - 2\,\sqrt{m_1\, m_2})\, r}{m_1 - m_2}$$

```
> rx := factor(Sr[2]):
```

```
> rnx := simplify(subs(r = 1, m[2] = N^2*m[1], rx),symbolic):
> rnx := subs(N = sqrt(n), rnx);
```

$$rnx := \frac{1}{\sqrt{n}+1}$$

```
> plot(rnx, n=0..10, labels=['n', 'rx']);
```

FIGURE 22.2. $\dfrac{r_x}{r} = \mathrm{f}\left(\dfrac{m_1}{m_2}\right)$

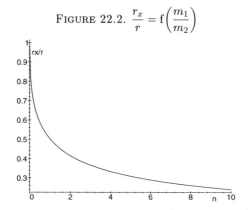

Figure 22.2 shows the coordinate of the equilibrium point as a function of m_2/m_1, using $r = 1$.

22.3   Equilibrium of Three Forces

Now we would like to solve the analogous problem with three points of mass m_1, m_2 and m_3. Their coordinates are given as indicated in Figure 22.3. Consider a point m at (x, y). We split the gravity forces $\vec{F_i}$ generated by m_i as sum of the forces $F_i x$ and $F_i y$ as shown in Figure 22.3.

FIGURE 22.3.
Three forces in the equilibrium, problem description

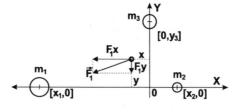

```
> r1 := sqrt((x1 - x)^2 + y^2):
> r2 := sqrt((x2 - x)^2 + y^2):
> r3 := sqrt(x^2 + (y3 - y)^2):
```

```
> F1  := G*m1*m/r1^2:
> F2  := G*m2*m/r2^2:
> F3  := G*m3*m/r3^2:
> F1x := F1*(x1 - x)/r1: F1y := -F1*y/r1:
> F2x := F2*(x2 - x)/r2: F2y:= -F2*y/r2:
> F3x := -F3*x/r3: F3y := F3*(y3 - y)/r3:
```

The mass m will be at the equilibrium point if the sum of the forces acting on it is zero.

```
> Rx := F1x + F2x + F3x=0;  Ry := F1y + F2y + F3y=0;
```

$$Rx := \frac{G\, m1\, m\, (x1 - x)}{(x1^2 - 2\, x1\, x + x^2 + y^2)^{3/2}} + \frac{G\, m2\, m\, (x2 - x)}{(x2^2 - 2\, x2\, x + x^2 + y^2)^{3/2}}$$
$$- \frac{G\, m3\, m\, x}{(x^2 + y3^2 - 2\, y3\, y + y^2)^{3/2}} = 0$$

$$Ry := -\frac{G\, m1\, m\, y}{(x1^2 - 2\, x1\, x + x^2 + y^2)^{3/2}} - \frac{G\, m2\, m\, y}{(x2^2 - 2\, x2\, x + x^2 + y^2)^{3/2}}$$
$$+ \frac{G\, m3\, m\, (y3 - y)}{(x^2 + y3^2 - 2\, y3\, y + y^2)^{3/2}} = 0$$

We obtain a nonlinear system of equations for x and y, which has no analytical solution in general. The approximate solution for given values $x_1 = -5$, $x_2 = 3$, $y_3 = 8$, $m_1 = 12$, $m_2 = 9$ and $m_3 = 6$ can be found graphically. By plotting $R_x(x, y) = 0$ and $R_y(x, y) = 0$ with `implicitplot` (see Figure 22.4), we can determine intervals where the solutions are. These intervals are then used by `fsolve` for the numerical computation. The equations contain the variables G and m as common factors, which can be cancelled by setting $G = 1$ and $m = 1$.

```
> System := subs(G = 1, m = 1, m1 = 12, x1 = -5, m2 = 9,
>                 x2 = 3, m3 = 6, y3 = 8, {Rx, Ry});
```

$$System := \{12\, \frac{-5 - x}{(25 + 10\, x + x^2 + y^2)^{3-2}} + 9\, \frac{3 - x}{(9 - 6\, x + x^2 + y^2)^{3/2}}$$
$$- 6\, \frac{x}{(x^2 + 64 - 16\, y + y^2)^{3/2}} = 0, -12\, \frac{y}{(25 + 10\, x + x^2 + y^2)^{3/2}}$$
$$- 9\, \frac{y}{(9 - 6\, x + x^2 + y^2)^{3/2}} + 6\, \frac{8 - y}{(x^2 + 64 - 16\, y + y^2)^{3/2}} = 0\}$$

```
> with(plots):
> implicitplot(System, x = -4..2, y = 0..6, numpoints = 800,
>                 labels = ['x','y'], color = black);
```

FIGURE 22.4. *Implicit graph of* Rx *and* Ry

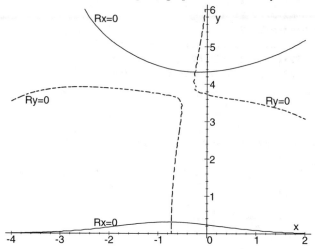

```
> Digits:=20:
> NS1 := fsolve(System, {x, y}, {x = -1..0, y = 0..1});
> NS2 := fsolve(System, {x, y}, {x = -1..0, y = 4..5});
```

$$NS1 := \{x = -.72487727980894034937, y = .30815197495485312203\}$$

$$NS2 := \{x = -.25295953560065612935, y = 4.3264261442756129585\}$$

```
> evalf(subs(NS1, System)); evalf(subs(NS2, System));
```

$$\{-.1\,10^{-19} = 0, .346\,10^{-19} = 0\}$$

$$\{.2\,10^{-19} = 0, -.11\,10^{-19} = 0\}$$

The Figure 22.4 shows two intersection points, i.e we have two solutions and therefore two points of equilibrium. For the numerical calculation, we use an accuracy of 20 digits and verify the solution by back-substitution.

22.4   Equilibrium of Three Forces, Computed from the Potential Energy

The same problem as in the previous section can be solved by finding critical points of the potential energy, so we need to find the coordinates x and y satisfying $\operatorname{grad} U(x, y) = 0$.

```
> Digits:=10:
> U := - G*m1/r1 - G*m2/r2 - G*m3/r3:
> with(linalg):
> g := grad(U,[x,y]);
```

$$g := \left[\frac{1}{2}\frac{G\,m1\,(-2\,x1 + 2\,x)}{(x1^2 - 2\,x1\,x + x^2 + y^2)^{3/2}} + \frac{1}{2}\frac{G\,m2\,(-2\,x2 + 2\,x)}{(x2^2 - 2\,x2\,x + x^2 + y^2)^{3/2}}\right.$$
$$+ \frac{G\,m3\,x}{(x^2 + y3^2 - 2\,y3\,y + y^2)^{3/2}}, \frac{G\,m1\,y}{(x1^2 - 2\,x1\,x + x^2 + y^2)^{3/2}}$$
$$\left.+ \frac{G\,m2\,y}{(x2^2 - 2\,x2\,x + x^2 + y^2)^{3/2}} + \frac{1}{2}\frac{G\,m3\,(-2\,y3 + 2\,y)}{(x^2 + y3^2 - 2\,y3\,y + y^2)^{3/2}}\right]$$

Naturally, the solution should be the same as before; in our case, even the resulting system of equations is identical to the one in Section 22.3. This time we will generate a plot of isopotential energy curves and a 3-dimensional plot of the potential energy.

```
> Un := subs(G = 1, m = 1, m1 = 12, x1 = -5,
>              m2 = 9, x2 = 3, m3 = 6, y3 = 8, U);
```

$$Un :=$$
$$-\frac{12}{\sqrt{25 + 10\,x + x^2 + y^2}} - \frac{9}{\sqrt{9 - 6\,x + x^2 + y^2}} - \frac{6}{\sqrt{x^2 + 64 - 16\,y + y^2}}$$

```
> gn := grad(Un, [x, y]);
```

$$gn := \left[6\frac{10 + 2\,x}{(25 + 10\,x + x^2 + y^2)^{3/2}} + \frac{9}{2}\frac{-6 + 2\,x}{(9 - 6\,x + x^2 + y^2)^{3/2}}\right.$$
$$+ 6\frac{x}{(x^2 + 64 - 16\,y + y^2)^{3/2}}, 12\frac{y}{(25 + 10\,x + x^2 + y^2)^{3/2}}$$
$$\left.+ 9\frac{y}{(9 - 6\,x + x^2 + y^2)^{3/2}} + 3\frac{-16 + 2\,y}{(x^2 + 64 - 16\,y + y^2)^{3/2}}\right]$$

```
> implicitplot({seq(Un = -i/6, i = 24..50)}, x=-6..3, y=-1..9,
>     scaling = constrained, labels = ['x', 'y'], color = black,
>     numpoints = 800);
> plot3d(Un, x = -6..3, y = -1..9, view = -12..-2,
>     orientation = [-100, 75], style = hidden, color = black,
>     numpoints = 40^2, axes = boxed, labels = ['x', 'y', 'U']);
```

We now recompute the stationary points of the energy and use the Hessian to determine what kind of critical points we found.

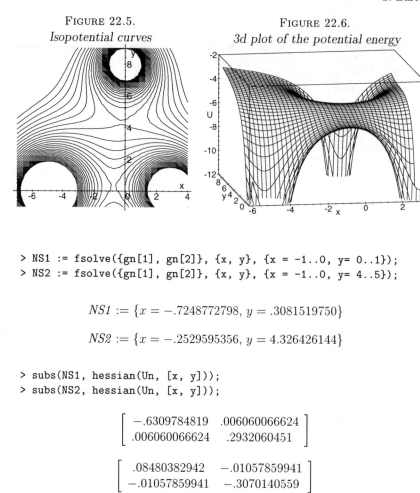

FIGURE 22.5.   FIGURE 22.6.

Isopotential curves   3d plot of the potential energy

```
> NS1 := fsolve({gn[1], gn[2]}, {x, y}, {x = -1..0, y= 0..1});
> NS2 := fsolve({gn[1], gn[2]}, {x, y}, {x = -1..0, y= 4..5});
```

$$NS1 := \{x = -.7248772798, y = .3081519750\}$$

$$NS2 := \{x = -.2529595356, y = 4.326426144\}$$

```
> subs(NS1, hessian(Un, [x, y]));
> subs(NS2, hessian(Un, [x, y]));
```

$$\begin{bmatrix} -.6309784819 & .006060066624 \\ .006060066624 & .2932060451 \end{bmatrix}$$

$$\begin{bmatrix} .08480382942 & -.01057859941 \\ -.01057859941 & -.3070140559 \end{bmatrix}$$

The opposite signs of the diagonal elements of the Hessian matrices imply that
the matrices are indefinite, and that both stationary points are saddle points.
Of course, this can be concluded as well from Figures 22.5 and 22.6. Usually,
deriving the system of equations via potential energy is simpler and faster than
with some force decomposition.

22.5   Gravitation of the Massive Line Segment

22.5.1   Potential and Intensity

We consider the problem of finding the gravitational potential and the intensity
of the gravitational field of a line segment with mass m and length l.

```
> restart;
> with(plots): with(linalg):
```

FIGURE 22.7. Computation of the potential energy

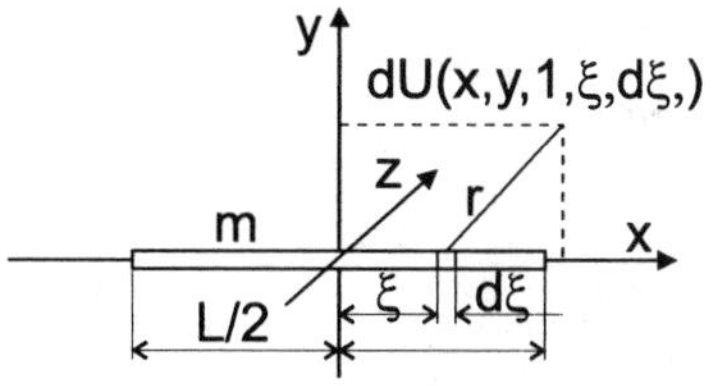

```
> r := sqrt((xi - x)^2 +y^2 +z^2):
> sigma := m/L: # linear mass density
> U := Int(G*sigma/r, xi = -L/2..L/2);
```

$$U := \int_{-1/2\,L}^{1/2\,L} \frac{G\,m}{L\,\sqrt{\xi^2 - 2\,\xi\,x + x^2 + y^2 + z^2}}\,d\xi$$

```
> U := normal(value(U));
```

$$U := G\,m(\ln(L - 2\,x + \sqrt{L^2 + 4\,x^2 + 4\,y^2 - 4\,L\,x + 4\,z^2})$$
$$- \ln(-L - 2\,x + \sqrt{L^2 + 4\,x^2 + 4\,y^2 + 4\,L\,x + 4\,z^2}) - \Big/L$$

We plot the potential energy and the isopotential curves and the vector field of intensity in a plane parallel to the xy-plane at a distance of $z = 1/5$.

```
> Us := subs(G=1,m=1,L=1, z=1/5, U);
```

$$Us := \ln(1 - 2\,x + \sqrt{\frac{29}{25} + 4\,x^2 + 4\,y^2 - 4\,x})$$
$$- \ln(-1 - 2\,x + \sqrt{\frac{29}{25} + 4\,x^2 + 4\,y^2 + 4\,x})$$

```
> plot3d(Us, x = -2..2, y=-2..2, axes = boxed, style = hidden,
>            color = black, orientation = [-80, -130],
>            numpoints = 35^2, labels = ['x', 'y', 'U']);
> p1 := implicitplot({seq(Us = i/10, i = 5..30)}, x = -1..1,
>            y = -1..1, color = black, scaling = constrained):
> p2 := gradplot(Us, x = -1..1, y = -1..1, color = black,
>            scaling = constrained, arrows = SLIM, numpoints = 40^2):
> display({p1,p2},labels = ['x','y']);
```

Figure 22.9 can be used as a good example to show that the field vectors and isopotential curves are always perpendicular. Let's check the correctness of the result by computing $\lim_{L \to 0} U$ and $\lim_{L \to 0} g$.

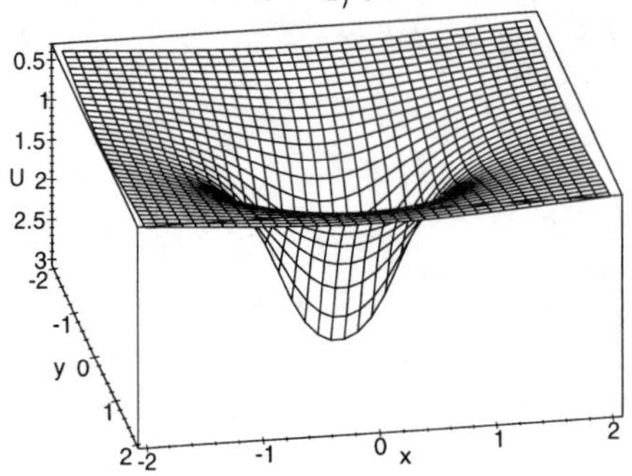

FIGURE 22.8.
Potential energy,
$z = 1/5$

FIGURE 22.9.
Isopotential curves
+ field vectors

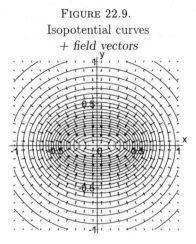

```
> Uinf := Limit(U, L=0);
```

$$Uinf := \lim_{L \to 0} G\,m(\ln(L - 2\,x + \sqrt{L^2 + 4\,x^2 + 4\,y^2 - 4\,L\,x + 4\,z^2})$$
$$- \ln(-L - 2\,x + \sqrt{L^2 + 4\,x^2 + 4\,y^2 + 4\,L\,x + 4\,z^2}))\Big/L$$

```
> Uinf := value(Uinf);
```

$$Uinf := \frac{G\,m}{\sqrt{z^2 + x^2 + y^2}}$$

```
> ginf := grad(Uinf, [x, y, z]);
```

$$ginf := \left[-\frac{x\,G\,m}{(z^2 + x^2 + y^2)^{3/2}}, -\frac{G\,m\,y}{(z^2 + x^2 + y^2)^{3/2}}, -\frac{G\,m\,z}{(z^2 + x^2 + y^2)^{3/2}} \right]$$

```
> abs(ginf) = simplify(norm(ginf, 2), symbolic);
```

$$|ginf| = \frac{m\,G}{z^2 + x^2 + y^2}$$

For long distances, $r \gg L$, the line segment looks like a point and the gravitational field is nearly spherically symmetric. The limit computations of U and g for $L \to 0$ return the classical Newton's law for a point mass. Thus our solution is consistent so far.

ALGORITHM 22.1. *Procedure* `dsnumsort`

```
dsnumsort := proc(numpr::list, Coor::list)
  local i, j, n;
  global C1, C2, C3, V1, V2, V3;
  n := nops(Coor):
  print('Order of the variables:');
  for i from 2 to 2*n + 1 do;
    for j from 1 to n do
      if [numpr[i]] =
              select(has, numpr, diff(Coor[j](t),t)) then
        C.j := i - 1; V.j := i;
        print(Coor[j], C.j,'  ', diff(Coor[j](t),t), V.j);
      fi;
    od;
  od;
end:
```

22.5.2 The Particle Trajectory

We will look at a moving particle in the gravitational field of the mass line segment. By the way, we will find out that Kepler's second law–the principle of constant "surface" velocity is not applicable in this case.

We use Newton's equations of motion and solve them numerically:

```
> Us := subs(G = 1, m = 1, L = 1, U):
> D2r := [diff(x(t), t, t), diff(y(t), t, t), diff(z(t), t, t)]:
> g := subs(x = x(t), y = y(t), z = z(t), grad(Us, [x, y, z])):
> IniC:= x(0) = 1, D(x)(0) = 0, y(0) = 0, D(y)(0) = 1,
>        z(0) = 3/4, D(z)(0) = 0:
> Ns := dsolve({seq(D2r[i] = g[i], i = 1..3), IniC},
>              {x(t), y(t), z(t)}, numeric);
```

$$Ns := \mathbf{proc}(rkf45\_x) \ldots \mathbf{end}$$

It is not possible to know in advance in which order MAPLE returns the functions. Therefore we need to use Algorithm 22.1 to detect the ordering of the variables in the numerical solution of the differential equation. Algorithm 22.1 delivers the indices of the functions in C1,C2, C3 and the indices of their derivatives in V1,V2, V3.

```
> read 'dsnumsort.map';
> dsnumsort(Ns(0), [x,y,z]):
```

Order of the variables :

$z, 2, \quad , \frac{\partial}{\partial t} z(t), 3$

$x, 4, \quad , \frac{\partial}{\partial t} x(t), 5$

$y, 6, \quad , \frac{\partial}{\partial t} y(t), 7$

The MAPLE commands below plots the trajectory and verifies Kepler's second law.

```
> for i from 0 to 1000 do;
>    T := i/25;
>    NsT := Ns(T):
>    X[i] := rhs(NsT[C1]); Vx[i] := rhs(NsT[V1]);
>    Y[i] := rhs(NsT[C2]); Vy[i] := rhs(NsT[V2]);
>    Z[i] := rhs(NsT[C3]); Vz[i] := rhs(NsT[V3]);
>    KepVec[i] := convert(crossprod([X[i], Y[i] ,Z[i]],
>       [Vx[i], Vy[i], Vz[i]]), list);
>    KepAbs[i] := norm(KepVec[i], 2);
> od:
> spacecurve({[seq([X[i], Y[i], Z[i]], i = 0..1000)],
>    [[-1/2, 0, 0], [1/2, 0, 0]]},  labels=['x', 'y', 'z']);
> spacecurve([seq(KepVec[i], i = 0..1000)],
>    orientation=[0,90], labels=['x', 'y', 'z']);
> plot([seq([i/25, KepAbs[i]], i = 0..1000)],
>    labels = ['t', 'MofI']);
```

The results are given in Figures 22.10, 22.12 and 22.14 for 40 [time units], with time step = 1/25 [time unit]. Figures, 22.11, 22.13 and 22.15 illustrate the motion of the particle with initial condition:

```
> IniC := x(0) = 1, D(x)(0) = 0, y(0) = 0, D(y)(0) = 1,
>            z(0) = 1/2, D(z)(0) = 1/2:
```

To prepare the data for graphical output, we have to compute the radius vector of the trajectory and the velocity vector. These vectors allow us to compute the "surface-velocity" which is the momentum if we assume that the mass of the particle is 1. Figures 22.10 and 22.11 show the trajectories, which are no more closed. Kepler's second law is not valid as we see from Figures 22.12 – 22.15. For a spherically symmetric gravity field, momentum is constant. In our case only M_x is constant due to the rotational symmetry of the problem along the x-axis. Of course the momentum is non constant for short distances because the gravitational field of the line segment has no spherical symmetry, see Figure 22.9.

22.6   The Earth Satellite

Now we will put a satellite into orbit around the earth. The velocity of the satellite at the perigee (the closest point to the earth) is 9000 m/s. The height of the perigee is 622 km. We find the satellite's trajectory and check the validity of Kepler's second law. In this case the orbit of the satellite is planar, so we will solve the problem in the plane using $x(t)$ and $y(t)$ coordinates.

The Trajectory

FIGURE 22.10.

FIGURE 22.11.

The Momentum of the Impulse, $\vec{M} = [\text{const. } M_y, M_z]$

FIGURE 22.12.

FIGURE 22.13.

$|\vec{M}| = |\vec{M}(t)|$

FIGURE 22.14.

FIGURE 22.15.

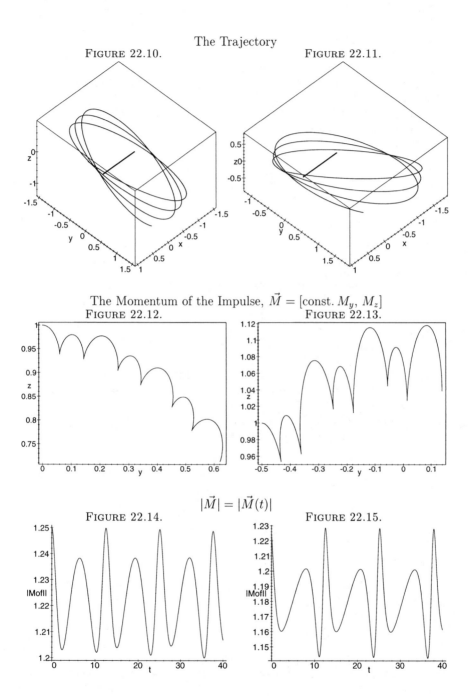

```
> restart;
> dx := diff(x(t), t, t) = -G*Mz*x(t)/(x(t)^2 + y(t)^2)^(3/2):
> dy := diff(y(t), t, t) = -G*Mz*y(t)/(x(t)^2 + y(t)^2)^(3/2):
> G:=6.67*10^(-11): Mz:=6*10^24:
> Inic := x(0) = 7*10^(6), D(x)(0)=0, y(0)=0, D(y)(0)=9*10^3:
> Digits := 15:
> Ns := dsolve({dx, dy, Inic}, {x(t),y(t)}, numeric):
> read 'dsnumsort.map';
> dsnumsort(Ns(0), [x, y]);
```

$$Order\ of\ the\ variables:$$
$$x, 2, \quad , \frac{\partial}{\partial t} x(t), 3$$
$$y, 4, \quad , \frac{\partial}{\partial t} y(t), 5$$

```
> for i from 0 to 400 do;
>     T := i*40;
>     NsT := Ns(T):
>     X[i] := rhs(NsT[C1]);  Vx[i] := rhs(NsT[V1]);
>     Y[i] := rhs(NsT[C2]);  Vy[i] := rhs(NsT[V2]);
>     MofI[i] := X[i]*Vy[i]-Y[i]*Vx[i];
> od:
> with(plots):
> p1 := polarplot(6378*10^3, phi = 0..2*Pi):
> p2 := plot([seq([X[i], Y[i]], i = 0..327)], thickness=2):
> display({p1, p2}, labels = ['x', 'y'], scaling = constrained);
> plot([seq([i*40,MofI[i] - 0.63*10^11], i = 0..400)],
>     labels=['t','Wp']);
```

FIGURE 22.16.
Trajectory, $1 : 10^3$ km

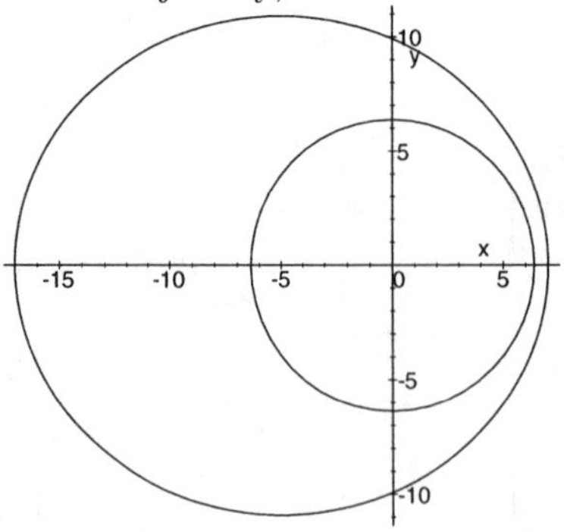

FIGURE 22.17.
The Relative Deviation of the Impulse

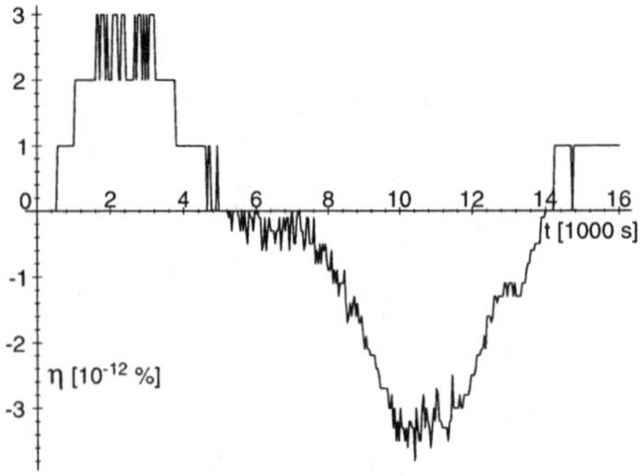

The MAPLE commands for the solution are similar as before. Kepler's second law is satisfied, as we can verify from Figure 22.17, which shows the relative deviation of the momentum.

We can observe that the period of the orbit ends between the 326th and the 327th time step. Let us to determine the period of the orbit by the method of bisection.

```
> tx := evalf((t1+t2)/2):
> t1 := 326*40: t2 := 327*40:
> y1 := rhs(Ns(t1)[C2]):
> y2 := rhs(Ns(t2)[C2]):

> while (t1 < tx) and (tx < t2) do:
>   yx := rhs(Ns(tx)[C2]):
>   if yx > 0 then
>     y2 := yx: t2 := tx:
>   else
>     y1 := yx: t1 := tx:
>   fi;
> od:
> Tx = evalf(tx);
```

$$Tx = 13060.28746$$

```
> Hx := floor(tx/3600):
> Mx := floor((tx - Hx*3600)/60):
> Sx := tx - Hx*3600 - Mx*60:
> Hx, Mx, Sx;
```

$$3, \ 37, \ 40.2874$$

The orbit period of the satellite is 3 hours, 37 minutes and 40.28746 seconds.

We save the main variables for use in further sections. For that purpose it is best to convert the indexed variables into a list.

```
> XS := [seq(X[i], i = 0..328)]:   YS := [seq(Y[i], i = 0..328)]:
> VxS := [seq(Vx[i], i = 0..328)]: VyS := [seq(Vy[i], i = 0..328)]:
> save(G, Mz, XS, YS, VxS, VyS,'orbit.sav');
```

22.7 Earth Satellite, Second Solution

Problem: Try to solve Example 22.6 using the law of motion in a homogeneous gravity field. Make the assumption that in the displacement of the satellite in a time step is so small that the gravitational acceleration can be assumed constant. Proceeding this way we integrate the equations of motion using the well known method of Euler.

```
> restart;  read('orbit.sav'): with(plots):
> ax := -G*Mz*x/(x^2 + y^2)^(3/2):
> ay := -G*Mz*y/(x^2 + y^2)^(3/2):
> i := 'i': j := i + 1:
> for k from 0 to 3 do:
>    x := 7*10^6: Vx := 0:
>    y := 0: Vy := 9000:
>    dt := evalf(1/2^k);
>    for i from 0 to 328 do:
>        X[i] := evalf(x); Y[i] := evalf(y):
>        for n from 1 to 40*2^k do:
>            x := evalf(ax*dt^2/2 + Vx*dt + x);
>            y := evalf(ay*dt^2/2 + Vy*dt + y);
>            Vx := evalf(ax*dt + Vx);
>            Vy := evalf(ay*dt + Vy);
>        od:
>        if i mod 41 = 0 then
>            dX[k, i] := X[i] - XS[j]; dY[k, i] := Y[i] - YS[j];
>        fi;
>    od:
>    p[k] := plot([seq([(X[i] - XS[j])/1000,
>            (Y[i] - YS[j])/1000], i = 0..328)], color = black):
> od:
> p1 := display({seq(p[k], k = 0..3)}, thickness = 3):
> SI := [seq(i*41, i = 0..8)]:
> p2 := plot({seq([seq([dX[k, i]/1000, dY[k, i]/1000],
>        k = 0..1), [0, 0]], i = SI)}, color = black):
> display({p1, p2}, scaling = constrained, labels = ['dx', 'dy']);
> display({p1, p2}, view = [-0.1..0.5, -0.4..0.2],
>        scaling = constrained, labels = ['dx', 'dy']);
```

This algorithm contains a systematic discretisation error, which can be reduced by choosing a smaller time step. We choose time steps from $\mathtt{dt} = 1s$ to $1/8s$.

Figure 22.18 shows the difference Δ between the exact position (as computed in the former section) and the approximate position after N time steps. If we connect these differences we see spiral like curves which are smaller if the time step is smaller. Studying the Figure 22.18, we see that the error is reduced by a factor of 2 when the step size is halved. So $\Delta \sim C\,\mathtt{dt}$, where C is nearly constant. More points are necessary for smaller time steps. If we look at the errors for corresponding times then it is interesting to note that these points are located on a straight line. Furthermore, the error does not increases uniformly, (see Figure 22.18); for the first 205 coarsest time steps the error is about $130km$, but in the next 100 steps it increases to more than $500km$.

FIGURE 22.18. *The Numerical–Approximate Error in km*

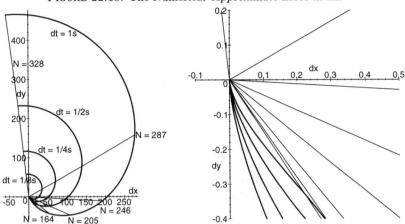

22.8   The Lost Screw

Consider an astronaut repairing something outside the spaceship. At time $t = 0$ he loses a screw. Assume that the velocity of the screw relative to the ship is $1m/s$. We consider the same initial conditions for the space ship as in the previous example, and let the relative velocity of the screw be perpendicular to the orbit of the ship: $\vec{v} = (1,\ 0)$ m/s.

We will compute the trajectory and velocity of the screw in the coordinate system of the spaceship, i.e., the spaceship will always be at the origin of the coordinate systems. First we use the translated coordinate system from Example 22.7, which keeps its orientation relative to the stars.

```
> restart; read('orbit.sav'):
> dx := diff(x(t), t, t) = -G*Mz*x(t)/(x(t)^2 + y(t)^2)^(3/2):
> dy := diff(y(t), t, t) = -G*Mz*y(t)/(x(t)^2 + y(t)^2)^(3/2):
> Inic := x(0) = 7*10^(6), D(x)(0)=1, y(0)=0, D(y)(0)=9*10^3:
> Ns := dsolve({dx, dy, Inic}, {x(t),y(t)}, numeric):
> read 'dsnumsort.map';  dsnumsort(Ns(0), [x, y]);
```

$$\textit{Order of the variables}:$$
$$y,\ 2,\quad,\ \tfrac{\partial}{\partial t}\,y(t),\ 3$$
$$x,\ 4,\quad,\ \tfrac{\partial}{\partial t}\,x(t),\ 5$$

```
> for i from 0 to 328 do;
>     T := i*40;  NsT := Ns(T);
>     X[i] := rhs(NsT[C1]); Vx[i] := rhs(NsT[V1]);
>     Y[i] := rhs(NsT[C2]); Vy[i] := rhs(NsT[V2]);
> od:

> i := 'i': j := i + 1:
```

```
> plot([seq([X[i] - XS[j], Y[i] - YS[j]], i = 0..328)],
>        labels = ['d x', 'd y']);
> plot([seq([Vx[i] - VxS[j], Vy[i] - VyS[j]], i = 0..328)],
>        labels = ['d Vx', 'd Vy']);
```

Second, we use a rotating coordinate system. The space ship is again at its origin, but the system rotates in such a way that the negative x-axis always points to the center of the earth. The astronaut now has a notion of *up, down, ahead, rear* in space: *down* means along the negative x-axis, *ahead* means along the y-axis.

```
> R := sqrt(XS[j]^2 + YS[j]^2):
> Sin := YS[j]/R:   Cos:=X[j]/R:
> GCX := (X[i] - XS[j])*Cos - (Y[i] - YS[j])*Sin:
> GCY := (X[i] - XS[j])*Sin + (Y[i] - YS[j])*Cos:
> GCVx := (Vx[i] - VxS[j])*Cos - (Vy[i] - VyS[j])*Sin:
> GCVy := (Vx[i] - VxS[j])*Sin + (Vy[i] - VyS[j])*Cos:
```

The variables beginning with GC describe the position and the velocity vector of the screw relative to the spaceship in the rotating coordinate system. This system is like a GeoCentric system.

```
> plot([seq([GCX, GCY], i = 0..328)], labels = ['d x', 'd y']);
> plot([seq([GCVx, GCVy], i = 0..327)], labels = ['d Vx','d Vy']);
```

The Figures 22.19 and 22.21 show that the screw is not lost! The astronaut has a chance to catch the lost screw if he is willing to wait for one orbit period.

22.9   Conclusions

1. MAPLE simplifies the use of mathematics in physics enormously. Consider the tools used in this chapter: symbolic computation, linear algebra, numerical solution of equations, numerical solution of differential equations, numerical integration and MAPLE graphics. Graphical interpretation of the results is very important in education.

2. Since MAPLE is a powerful tool for solving problems in classroom, the emphasis can be put on the physical problem and on its description in the mathematical language.

3. For an appropriate use of MAPLE, it is not sufficient to know MAPLE. It is also necessary to have a good mathematical knowledge.

4. Moral: If computer algebra systems will be used as teaching instrument, many changes in the current style of education will be necessary.

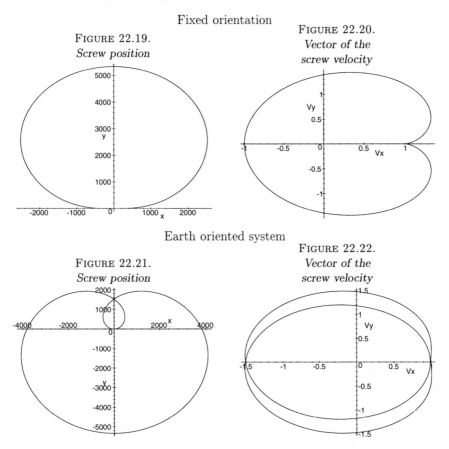

Fixed orientation

FIGURE 22.19.
Screw position

FIGURE 22.20.
*Vector of the
screw velocity*

Earth oriented system

FIGURE 22.21.
Screw position

FIGURE 22.22.
*Vector of the
screw velocity*

Acknowledgment

The author would like to thank Mgr. Miroslava Procházková, general manager
of Mendel Secondary School, Čechyňská 19, 602 00 Brno, for her continuous
interest and support. This chapter is based on lectures held by the author in
physics and mathematics with MAPLE support at this school.

References

[1] JOURNAL OF CELESTIAL MECHANICS: *J. Henrard* editor, Belgium

[2] GANDER W.: *Computer – Mathematik.* Birkhäuser, 1992

[3] GETTYS E., KELLER F., SKOVE M.: *Classical and Modern Physics*, McGraw–
Hill, 1989

[4] YOUNG H., FREEDMAN R.: *University Physics*, Addison Wesley, 1996

Chapter 23. Least Squares Fit of Point Clouds

W. Gander

23.1 Introduction

We consider a least squares problem in *coordinate metrology* (see [2], [1]): m points of a workpiece, so-called *nominal points* are given by their exact coordinates from construction plans when the workpiece is in *nominal position* in a reference frame. We denote the coordinate vectors of the nominal points in this position by

$$\mathbf{x}_1, \ldots, \mathbf{x}_m, \quad \mathbf{x}_i \in R^n, \quad 1 \le n \le 3.$$

Suppose now that a coordinate measuring machine gathers the same points of another workpiece. The machine records the coordinates of the *measured points*

$$\boldsymbol{\xi}_1, \ldots, \boldsymbol{\xi}_m, \quad \boldsymbol{\xi}_i \in R^n, \quad 1 \le n \le 3$$

which will be in a different frame than the frame of reference. The problem we want to solve is to find a frame transformation which maps the given nominal points onto the measured points.

We need to find a translation vector $\mathbf{t}$ and an orthogonal matrix Q with $\det(Q) = 1$ i.e., $Q^T Q = I$ such that

$$\boldsymbol{\xi}_i = Q\mathbf{x}_i + \mathbf{t}, \quad \text{for } i = 1, \ldots, m. \tag{23.1}$$

For $m > 6$ in 3D-space, Equation (23.1) is an over-determined system of equations and is only consistent if the measurements have no errors. This is not the case for a real machine and therefore our problem is to determine the unknowns Q and $\mathbf{t}$ of the least squares problem

$$\boldsymbol{\xi}_i \approx Q\mathbf{x}_i + \mathbf{t}. \tag{23.2}$$

This problem has been studied and solved in a nice paper by Hanson and Norris [4].

23.2 Computing the Translation

In the one-dimensional case we are given two sets of points on the line. The matrix Q is just the constant 1 and we have to determine a scalar t such that

$$\xi_i \approx x_i + t, \quad i = 1, \ldots, m.$$

With the notation $A = (1,\ldots,1)^T$, $\mathbf{a} = (\xi_1,,\ldots,\xi_m)^T$ and $\mathbf{b} = (x_1,\ldots,x_m)^T$ the problem becomes

$$At \approx \mathbf{a} - \mathbf{b}. \tag{23.3}$$

Using the normal equations $A^T A t = A^T(\mathbf{a} - \mathbf{b})$ we obtain $mt = \sum_{i=1}^{m}(\xi_i - x_i)$ and therefore

$$t = \bar{\xi} - \bar{x}, \quad \text{with} \quad \bar{\xi} = \frac{1}{m}\sum_{i=1}^{m}\xi_i \quad \text{and} \quad \bar{x} = \frac{1}{m}\sum_{i=1}^{m}x_i. \tag{23.4}$$

We can generalize this result for $n > 1$. Consider

$$\boldsymbol{\xi}_i \approx \mathbf{x}_i + \mathbf{t}, \quad i = 1,\ldots,m.$$

In matrix notation this least squares problem becomes (I is the $n \times n$ identity matrix):

$$\begin{pmatrix} I \\ I \\ \vdots \\ I \end{pmatrix} \mathbf{t} \approx \begin{pmatrix} \boldsymbol{\xi}_1 - \mathbf{x}_1 \\ \boldsymbol{\xi}_2 - \mathbf{x}_2 \\ \vdots \\ \boldsymbol{\xi}_m - \mathbf{x}_m \end{pmatrix}$$

The normal equations are $m\mathbf{t} = \sum_{i=1}^{m}(\boldsymbol{\xi}_i - \mathbf{x}_i)$ an thus again

$$\mathbf{t} = \bar{\boldsymbol{\xi}} - \bar{\mathbf{x}}, \quad \text{with} \quad \bar{\boldsymbol{\xi}} = \frac{1}{m}\sum_{i=1}^{m}\boldsymbol{\xi}_i \quad \text{and} \quad \bar{\mathbf{x}} = \frac{1}{m}\sum_{i=1}^{m}\mathbf{x}_i.$$

We therefore have shown that *the translation* $\mathbf{t}$ *is the vector connecting the two centers of gravity of the corresponding sets of points.*

23.3   Computing the Orthogonal Matrix

Applying the result from the previous subsection to the least squares problem

$$\boldsymbol{\xi}_i \approx Q\mathbf{x}_i + \mathbf{t} \iff \sum_{i=1}^{m} \|Q\mathbf{x}_i + \mathbf{t} - \boldsymbol{\xi}_i\|^2 = \min \tag{23.5}$$

we conclude that $\mathbf{t}$ is the vector connecting the two centers of gravity of the point sets $\boldsymbol{\xi}_i$ and $Q\mathbf{x}_i$, i.e.,

$$\mathbf{t} = \bar{\boldsymbol{\xi}} - Q\bar{\mathbf{x}}. \tag{23.6}$$

Using equation (23.6), we can eliminate $\mathbf{t}$ in (23.5) and so the problem becomes

$$G(Q) = \sum_{i=1}^{m} \|Q(\mathbf{x}_i - \bar{\mathbf{x}}) - (\boldsymbol{\xi}_i - \bar{\boldsymbol{\xi}})\|^2 = \min. \tag{23.7}$$

Introducing the new coordinates

$$\mathbf{a}_i := \mathbf{x}_i - \bar{\mathbf{x}} \quad \text{and} \quad \mathbf{b}_i = \boldsymbol{\xi}_i - \bar{\boldsymbol{\xi}}$$

the problem is:

$$G(Q) = \sum_{i=1}^{m} \|Q\mathbf{a}_i - \mathbf{b}_i\|^2 = \min. \tag{23.8}$$

We can collect the vectors in matrices

$$A := (\mathbf{a}_1, \ldots, \mathbf{a}_m), \quad \text{and} \quad B := (\mathbf{b}_1, \ldots, \mathbf{b}_m),$$

and rewrite the function G using the Frobenius norm

$$G(Q) = \|QA - B\|_F^2.$$

Since the Frobenius norm of a matrix is the same for the transposed matrix we finally obtain a *Procrustes problem* [5]: find an orthogonal matrix Q such that

$$\|B^T - A^T Q^T\|_F^2 = \min. \tag{23.9}$$

23.4   Solution of the Procrustes Problem

We consider the problem: given the matrices C and D, both $m \times n$ with $m \geq n$, find an orthogonal matrix P, $n \times n$, such that

$$\|C - DP\|_F^2 = \min.$$

We need some properties of the Frobenius norm. It is defined by

$$\|A\|_F^2 = \sum_{i=1}^{m} \sum_{j=1}^{n} a_{i,j}^2 = \sum_{j=1}^{n} \|\mathbf{a}_j\|_2^2, \quad \text{where } \mathbf{a}_j \text{ is the } j\text{th column of } A. \tag{23.10}$$

Note that

$$\|A\|_F^2 = \text{trace}(A^T A) = \sum_{i=1}^{n} \lambda_i(A^T A). \tag{23.11}$$

Remember that the trace of a matrix is defined as the sum of the diagonal elements and that the trace equals the sum of the eigenvalues. Equation (23.11) gives us some useful relations:

1. If P is orthogonal then $\|PA\|_F = \|A\|_F$.
   Additionally, since $\|A\|_F = \|A^T\|_F$, we have $\|AP\|_F = \|A\|_F$.

2.

$$\begin{aligned}
\|A + B\|_F^2 &= \text{trace}((A + B)^T (A + B)) \\
&= \text{trace}(A^T A + B^T A + A^T B + B^T B) \\
&= \text{trace}(A^T A) + 2\,\text{trace}(A^T B) + \text{trace}(B^T B) \\
\|A + B\|_F^2 &= \|A\|_F^2 + \|B\|_F^2 + 2\,\text{trace}(A^T B)
\end{aligned}$$

We now apply the last relation to the Procrustes problem:

$$\|C - DP\|_F^2 = \|C\|_F^2 + \|D\|_F^2 - 2\operatorname{trace}(P^T D^T C) = \min.$$

Computing the minimum is equivalent to maximizing

$$\operatorname{trace}(P^T D^T C) = \max.$$

Using the singular value decomposition $U\Sigma V^T$ of $D^T C$, we obtain

$$\operatorname{trace}(P^T D^T C) = \operatorname{trace}(P^T U\Sigma V^T).$$

Since U, V are orthogonal, we may write the unknown matrix P in the following form

$$P = U Z^T V^T, \quad \text{with } Z \text{ orthogonal.}$$

It follows that

$$
\begin{aligned}
\operatorname{trace}(P^T D^T C) &= \operatorname{trace}(V Z U^T U\Sigma V^T) = \operatorname{trace}(V Z\Sigma V^T) \\
&= \operatorname{trace}(Z\Sigma V V^T) = \operatorname{trace}(Z\Sigma) \\
&= \sum_{i=1}^{n} z_{ii}\sigma_i \le \sum_{i=1}^{n} \sigma_i,
\end{aligned}
$$

where the inequality follows from $z_{ii} \le 1$ for any orthogonal matrix Z. Furthermore, the bound is attained for $Z = I$. Notice that if $D^T C$ is rank deficient the solution is not unique (cf. [4]). So we have proved the following theorem:

Theorem

The Procrustes problem $\|C - DP\|_F^2 = \min$ is solved by the orthogonal polar factor of $D^T C$, i.e. $P = UV^T$ where $U\Sigma V^T$ is the singular value decomposition of $D^T C$.

The polar decomposition of a matrix is the generalization of the polar representation of complex numbers. The matrix is decomposed into the product of an orthogonal times a symmetric positive (semi-)definite matrix. The decomposition can be computed by the singular value decomposition or by other algorithms [3]. In our case we have

$$D^T C = U\Sigma V^T = \underbrace{UV^T}_{\text{orthogonal}} \underbrace{V\Sigma V^T}_{\substack{\text{positive} \\ \text{semidefinite}}}.$$

23.5 Algorithm

Given measured points $\boldsymbol{\xi}_i$ and corresponding nominal points $\mathbf{x}_i$ for $i = 1, \ldots, m$. We want to determine $\mathbf{t}$ and Q orthogonal such that $\boldsymbol{\xi}_i \approx Q\mathbf{x}_i + \mathbf{t}$.

1. Compute the centers of gravity:

$$\bar{\boldsymbol{\xi}} = \frac{1}{m}\sum_{i=1}^{m} \boldsymbol{\xi}_i \quad \text{and} \quad \bar{\mathbf{x}} = \frac{1}{m}\sum_{i=1}^{m} \mathbf{x}_i.$$

2. Compute the *relative coordinates*:

$$A := (\mathbf{a}_1, \ldots, \mathbf{a}_m), \qquad \mathbf{a}_i := \mathbf{x}_i - \bar{\mathbf{x}}$$
$$B := (\mathbf{b}_1, \ldots, \mathbf{b}_m), \qquad \mathbf{b}_i := \boldsymbol{\xi}_i - \bar{\boldsymbol{\xi}}$$

3. Solve the Procrustes problem $\|C - DP\|_F^2 = \min$ with $C = B^T$, $D = A^T$ and $P = Q^T$. Compute first the singular value decomposition

$$AB^T = U\Sigma V^T.$$

4. $Q^T = UV^T$ or $Q = VU^T$.

5. $\mathbf{t} = \bar{\boldsymbol{\xi}} - Q\bar{\mathbf{x}}$

For technical reasons it may be important to decompose the orthogonal matrix Q into elementary rotations. The algorithm that we developed so far computes an orthogonal matrix but there is no guarantee that Q can be represented as a product of rotations and that no reflection occurs. Q can be represented as a product of rotations if $\det(Q) = 1$. However, if $\det(Q) = -1$ then a reflection is necessary and this may be of no practical use. Therefore one would like to find the best orthogonal matrix with $\det(Q) = 1$.

It is shown in [4] that the constrained Procrustes problem

$$\|C - DP\|_F^2 = \min, \quad \text{subject to} \quad \det(P) = 1$$

has the solution

$$P = U \operatorname{diag}(1, \ldots, 1, \mu)V^T,$$

where $D^T C = U\Sigma V^T$ is the singular value decomposition and $\mu = \det(UV^T)$.

The proof is based on the fact, that for a real orthogonal $n \times n$ matrix Z with $\det(Z) < 1$ the trace is bounded by

$$\text{trace}(Z) \le n - 2 \quad \text{and} \quad \text{trace}(Z) = n - 2 \iff \lambda_i(Z) = \{1, \ldots, 1, -1\}.$$

This can be seen by considering the real Schur form [5] of Z. The maximum of $\text{trace}(Z\Sigma)$ is therefore $\sum_{i=1}^{n-1} \sigma_i - \sigma_n$ and is achieved for $Z = \operatorname{diag}(1, \ldots, 1, -1)$. Thus we obtain the MATLAB function `pointfit` shown in Algorithm 23.1:

ALGORITHM 23.1.

```
function [t, Q] = pointfit(xi,x);
%
xiq = sum(xi')/length(xi); xiq = xiq';
xq = sum(x')/length(x); xq = xq';
A = x - xq*ones(1,length(x));
B = xi - xiq*ones(1,length(xi));
[u sigma v] = svd(A*B');
Q = v*diag([ones(1,size(v,1)-1) det(v*u')])*u';
t = xiq - Q*xq;
```

23.6   Decomposing the Orthogonal Matrix

As mentioned before it may be useful to decompose the orthogonal matrix Q into elementary rotations. In [1] the affine transformation which maps the given nominal points onto the measured points is written as

$$\boldsymbol{\xi} = R_3 R_2 R_1 U_0 (\mathbf{x} - \mathbf{x}_0)$$

Here

$$R_1 = \begin{pmatrix} 1 & 0 & 0 \\ 0 & c_1 & s_1 \\ 0 & -s_1 & c_1 \end{pmatrix}, \quad R_2 = \begin{pmatrix} c_2 & 0 & s_2 \\ 0 & 1 & 0 \\ -s_2 & 0 & c_2 \end{pmatrix}, \quad \text{and } R_3 = \begin{pmatrix} c_3 & s_3 & 0 \\ -s_3 & c_3 & 0 \\ 0 & 0 & 1 \end{pmatrix},$$

$$(23.12)$$

are plane rotation matrices specified by $c_k = \cos\theta_k$ and $s_k = \sin\theta_k$, $k = 1, 2, 3$, defining rotations about the $x-$, $y-$ and $z-$axes, respectively, and U_0 is a fixed given orthogonal matrix.

We can easily compute the matrices R_k and the vector $\mathbf{x}_0$ from the known transformation $\boldsymbol{\xi} = Q\mathbf{x} + \mathbf{t}$. The vector $\mathbf{x}_0$ is related to $\mathbf{t}$ by $-Q\mathbf{x}_0 = \mathbf{t}$ thus

$$\mathbf{x}_0 = -Q^T \mathbf{t}.$$

To compute the matrices R_k we first note that

$$Q = R_3 R_2 R_1 U_0 \iff R_1^T R_2^T R_3^T Q U_0^T = I.$$

Then we can proceed as follows:

1. Form $H = Q U_0^T$.

2. Determine the rotation angle θ_3 such that in $H := R_3^T H$ the element h_{21} becomes zero:

$$\begin{pmatrix} c_3 & -s_3 & 0 \\ s_3 & c_3 & 0 \\ 0 & 0 & 1 \end{pmatrix} H = \begin{pmatrix} * & * & * \\ 0 & * & * \\ * & * & * \end{pmatrix} \iff \begin{matrix} \tan\theta_3 = -\frac{h_{21}}{h_{11}}, \\ c_3 = \cos\theta_3, \\ s_3 = \sin\theta_3. \end{matrix}$$

3. Annihilate h_{31} similarly in $H := R_2^T H$:

$$\tan\theta_2 = -\frac{h_{31}}{h_{11}}, \quad c_2 = \cos\theta_2, \quad s_2 = \sin\theta_2.$$

4. Finally rotate h_{32} to zero in $H := R_1^T H$:

$$\tan\theta_1 = -\frac{h_{32}}{h_{22}}, \quad c_1 = \cos\theta_1, \quad s_1 = \sin\theta_1.$$

ALGORITHM 23.2.

```
function [theta] = rotangle(H)
%
if det(H)<0
  error('The matrix is not a product of rotations')
end
n = size(H,1);
theta = [];
for i = 1 : n - 1
  for j = i + 1 : n
    theta_k = atan2(-H(j,i),H(i,i));
    theta = [theta_k, theta];
    c = cos(theta_k); s = sin(theta_k);
    R = eye(n);
    R(i,i) = c; R(j,j) = c;
    R(i,j) = -s; R(j,i) = s;
    H = R * H;
  end
end
```

Thus we obtain the MATLAB function `rotangle` shown in Algorithm 23.2.

23.7   Numerical Examples

We conclude with two three-dimensional examples and a test program for the functions `pointfit` and `rotangle`.

23.7.1   First example

First we define the nodes of a pyramid

```
>> A = [ 1    0    0    0;
>>       0    2    0    0;
>>       0    0    3    0 ];
```

To draw the pyramid we need to define a vector which indicates the edges to be plotted.

```
>> v = [ 1    2    3    4    1    3    4    2];
>> plot3(A(1,v),A(2,v),A(3,v),'-.');
>> hold on;
>> view(115,20);
>> axis([-2 2 0 5 0 4]);
```

Then we choose some nominal points on the pyramid

```
>> x = [ 0    0.5  0.5  0    0    0    1;
>>       0    0    1    1.5  0.5  0    0;
>>       0    1.5  0    0.75 2.25 2    0 ];
>> plot3(x(1,:),x(2,:),x(3,:),'*');
```

ALGORITHM 23.3.

```
function [Qr] = ang2orth(theta)
% generate orthogonal matrix from given angles
Qr = eye(3);
n = 1;
for i = 2 : -1 : 1
  for j = 3 : -1 : i + 1
    t = thetar(n);
    s = sin(t); c = cos(t);
    U = eye(3);
    U(i,i) =  c; U(i,j) = s;
    U(j,i) = -s; U(j,j) = c;
    Qr = U * Qr; n = n + 1;
  end
end
```

Now we will simulate the measured points. In order to give the reader repro-
ducible results we commented out the randomly generated data and replaced
them by fixed values. We construct an orthogonal matrix Qr by generating three
angles und using the MATLAB function ang2orth shown in Algorithm 23.3:

```
>> %thetar = rand(1,1:3)
>> thetar = [pi/4 pi/15 -pi/6];
>> Qr = ang2orth(thetar);
```

and also a random translation vector:

```
>> %tr = rand(3,1)*3;
>> tr = [1;3;2];
```

Now we can compute and plot the exactly transformed pyramid:

```
>> B = Qr * A + tr * ones(1, size(A,2));
>> plot3(B(1,v),B(2,v),B(3,v),':');
>> pause
```

We transform the nominal points and add some noise to simulate the measured points (again for reasons of reproducibility of results we give the points
explicitly):

```
>> % (randomly distorted) measured points:
>> % xi = (Qr*x+tr*ones(1,length(x))+randn(size(x))/10);
>>
>> xi = [ 0.8314 0.9821 1.0211 0.1425 0.2572 0.5229 1.7713
>>        3.0358 4.5232 3.8075 4.4826 5.0120 4.5364 3.3987
>>        1.9328 2.8703 1.0573 1.5803 3.1471 3.5394 1.9054];
>> plot3(xi(1,:),xi(2,:),xi(3,:),'o');
>> pause
```

FIGURE 23.1. *Point Fit.*

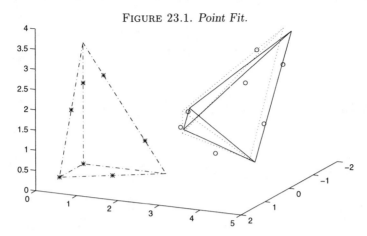

Using the function `pointfit` we now estimate Q and t from the measured points and we plot the fitted pyramid (see Figure 23.1):

```
>> [t,Q]= pointfit(xi,x);
>> % the fitted pyramid:
>> C = Q * A + t * ones(1, size(A,2));
>> plot3(C(1,v),C(2,v),C(3,v),'-');
>> hold off;
```

Finally we compute the rotation angles of the fitted pyramid and compare them with the original ones.

```
>> theta = rotangle(Q)

theta =

    0.8282    0.1772   -0.3964

>> thetar

thetar =

    0.7854    0.2094   -0.5236
```

As a final check we generate the orthogonal matrix S from the computed angles θ and compare it with the result Q of `pointfit`:

```
>> S = ang2orth(theta)>;
>> norm(Q - S)

ans =

    2.3497e-015
```

FIGURE 23.2.
Best Rotation: x_i nominal points, xi_i "measured points", c_i
best fit

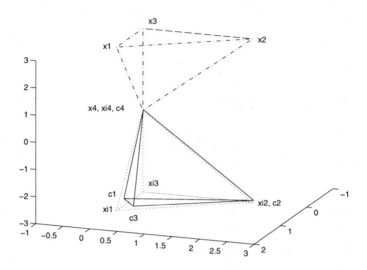

23.7.2   Second example

In this example we construct two sets of points which can be transformed exactly into one another by a reflection. We will compute the best solution using only rotations which cannot match the measured points since the orientation of the body would have to change. Consider again the vertices of a pyramid as the nominal points:

```
>> % Define nominal points.
>> A = [1 0 0 0;0 2 0 0;3 3 3 0];
>> v = [1 2 3 4 1 3 2 4];
>> plot3(A(1,v),A(2,v),A(3,v),'-.');
>> hold on;
>> view(110,20)
>> axis([-1 2 -1 3 -3 3])
```

As measured points we use the vertices of the pyramid which is reflected at the xy-plane:

```
>> B = [1 0 0 0;0 2 0 0;-3 -3 -3 0];
>> plot3(B(1,v),B(2,v),B(3,v),':')
```

We compute the best fit and plot the result in Figure 23.2:

```
>> [t,Q] = pointfit(B,A);
>> C = Q * A + t * ones(1, size(A,2))
>> plot3(C(1,v),C(2,v),C(3,v),'-')
```

```
>> text(1,-.25,3,'x1'); text(0,2.1,3,'x2')
>> text(0,.1,3.3,'x3'); text(0,-1,-0.1,'x4, xi4, c4')
>> text(1,-.25,-3,'xi1'); text(0,2.1,-3,'xi2, c2')
>> text(0,.1,-2.65,'xi3'); text(0.7,-0.3,-2.65,'c1')
>> text(1.75,.6,-2.65,'c3')
>> theta = 180/pi*rotangle(Q)
>> hold off;
```

Acknowledgments

I wish to thank Max Oertli for his help in finishing the last section.

References

[1] BUTLER, B. P., FORBES A. B. AND HARRIS, P. M., *Algorithms for Geometric Tolerance Assessment*. Technical Report DITC 228/94, National Physical Laboratory, Teddington, 1994.

[2] FORBES A. B., *Geometric Tolerance Assessment*. Technical Report DITC 210/92, National Physical Laboratory, Teddington, 1992.

[3] GANDER, W., *Algorithms for the Polar Decomposition*, SIAM J. on Sci. and Stat. Comp., Vol. 11, No. 6, 1990.

[4] HANSON, R. AND NORRIS, M. *Analysis of Measurements Based on the Singular Value Decomposition*, SIAM J. on Sci. and Stat. Comp., Vol. 2, No. 3, 1981.

[5] GOLUB, GENE H. AND VAN LOAN, CHARLES F., *Matrix Computations*. 2nd ed. Baltimore Johns Hopkins University Press, c1989.

Chapter 24. Modeling Social Processes

J. Hřebíček and T. Pitner

24.1 Introduction

Certain social processes can be studied using models introduced by Weidlich and Haag, see [3], [5]. They have attempted to develop a general background to quantitative sociology. We will apply their approach in this chapter using MAPLE in the solution of two models.

The general approach to modeling the evolution of a social system is based on an *aspect space*, which is defined as a linear vector space. Its vectors characterize members of the social system with respect to their behavior in society or incorporation into a social group, see [6]. These aspects could be occupation, bias towards a certain political party, standard of living, level of income, etc.

Vectors of the aspect space characterize the incorporation of a member of the system into a certain group i with respect to the given aspect α. Let m_i^α denote the number of those members. The evolution of a social system is determined by the changing opinion of members of its groups with respect to the aspect α.

Let us define the transient probability $w_{ij}^\alpha(t)$ as a probability (per unit time) of the change in opinion of members represented by the aspect α which implies change of the incorporation of a member from the group j to the group i. Then the increment of m_i^α is given by the *master equation*

$$\frac{dm_i^\alpha(t)}{dt} = \sum_{\substack{j=1 \\ j \neq i}}^{s} m_j^\alpha(t)w_{ij}^\alpha(t) - m_i^\alpha(t)w_{ji}^\alpha(t), \tag{24.1}$$

which is the fundamental equation describing one class of social systems, see [5]. The master equation can model processes with spatial structure (e.g. population migration) and time oscillation (e.g. Schumpeter's clock [4]).

24.2 Modeling Population Migration

An interesting application of Equation (24.1) is modeling a relationship between two groups of inhabitants ($s = 2$) living in two city quarters ($\alpha = 2$) and their migration. These groups are denoted by m and n. It is based on a migration model white and black inhabitants in Atlanta (Georgia, USA), see [6].

Let us assume that the number of inhabitants in the groups is constant and equal to $2m_0$ and $2n_0$, respectively. These groups will spread in time t into two city quarters. Their numbers in each quarter are $m_1(t)$, $m_2(t)$ and $n_1(t)$, $n_2(t)$, respectively. Then $m_1(t) + m_2(t) = 2m_0$ and $n_1(t) + n_2(t) = 2n_0$.

By introducing the new variables $m(t) = m_1(t) - m_0 = m_0 - m_2(t)$ and $n(t) = n_1(t) - n_0 = n_0 - n_2(t)$ we can reduce the master equations (24.1) to one equation per group. This gives us the following system (which for the moment is not coupled)

$$\begin{aligned} \frac{dm(t)}{dt} &= (m_0 - m(t))w_{12}^m(t) - (m_0 + m(t))w_{21}^m(t) , \\ \frac{dn(t)}{dt} &= (n_0 - n(t))w_{12}^n(t) - (n_0 + n(t))w_{21}^n(t) . \end{aligned} \tag{24.2}$$

Now we need to derive a model of transient probability functions $w_{ij}^\alpha(t)$. Three constants for each group m, n will be used:

1. *natural preferential parameters* a_m, a_n describe the natural effort of the inhabitants of the group m, n to live in the same city quarter,

2. *interior inclination parameters* b_m, b_n describing the effort of inhabitants of the group m, n to live together with the same group in common city quarter and

3. *exterior inclination parameter* c_m, c_n describing the effort of inhabitants of the group m, n to live together with the second group of inhabitants in the same city quarter.

Then the *transient probability functions* can be defined with these parameters by exponential functions, see [6]:

$$\begin{aligned} w_{12}^m(t) &= Ae^{a_m+b_m m(t)+c_m n(t)} \\ w_{21}^m(t) &= Ae^{-(a_m+b_m m(t)+c_m n(t))} \\ w_{12}^n(t) &= Ae^{a_n+b_n m(t)+c_n n(t)} \\ w_{21}^n(t) &= Ae^{-(a_n+b_n m(t)+c_n n(t))} \end{aligned} \tag{24.3}$$

where A is a time scaling factor.

Introducing new variables $x(t) = m(t)/m_0, y(t) = n(t)/n_0$ and denoting

$$\begin{aligned} u(t) &= a_m + b_m m_0 x(t) + c_m n_0 y(t) , \\ v(t) &= a_n + b_n m_0 x(t) + c_n n_0 y(t) , \end{aligned}$$

we obtain from (24.2) and (24.3):

$$\begin{aligned} \frac{dx(t)}{dt} &= 2A(\sinh u(t) - x(t)\cosh u(t)) , \\ \frac{dy(t)}{dt} &= 2A(\sinh v(t) - y(t)\cosh v(t)) , \end{aligned} \tag{24.4}$$

where $\sinh u = (e^u - e^{-u})/2, \cosh u = (e^u + e^{-u})/2$.

We try to solve (24.4) on the interval $\langle 0, T \rangle$ with initial conditions $x(0) = X$, $y(0) = Y$ by MAPLE:

```
> u  := am + bm*m0*x(t) + cm*n0*y(t):
> v  := an + bn*m0*x(t) + cn*n0*y(t):
> eq1 := diff(x(t),t) = 2*A*(sinh(u) - x(t)*cosh(u)):
> eq2 := diff(y(t),t) = 2*A*(sinh(v) - y(t)*cosh(v)):
> init := x(0) = X, y(0) = Y:
> sol := dsolve({eq1, eq2, init}, {x(t), y(t)});
```

$$sol :=$$

There seems not to be an analytical solution of the above initial value problem for general parameters a_m, a_n, b_m, b_n, c_m, c_n. However, we can compute with MAPLE a numerical solution.

24.2.1 Cyclic Migration without Regulation

The solution of the migration model (24.3), (24.4) strongly depends on the value of the parameters. There are three types of solution, see [1]. Two of the solution types converge to a constant steady state (one where in each quarter we will have m_0 of the first and n_0 of the second group, second the groups will be separated occupying one quarter by one single group). More interesting is a third solution converging to a limit cycle, which will be presented.

Parameters A, a_m, a_n, b_m, b_n, c_m, c_n of the transient probabilities have been investigated by Květoň, see [2]. Květoň studied migration between Czechs and Gipsies living in Most (Northern Bohemia, Czech Republic). Květoň found based on statistical investigations the following parameter values:

```
> param := A = 1/2, am = 0, bm = 1.2e-4, cm = 0.5e-3,
>            m0 = 10000, an = 0, bn = -1e-4, cn = 1.2e-3,
>            n0 = 1000:
> save(u, v, param, 'system.sav');
```

We solve System (24.4) numerically with initial conditions where both groups of inhabitants are split nearly equally in both quarters.

```
> init := x(0) = 0, y(0) = 0.01:
> eq1 := subs(param, eq1):  eq2 := subs(param, eq2):
> sol := dsolve({eq1, eq2, init}, {x(t), y(t)}, numeric):
> with(plots):
> odeplot(sol, [x(t), y(t)], 0..60, numpoints = 300,
>    view=[-1..1, -1..1]);
> odeplot(sol, [t,x(t), y(t)], 0..60, numpoints = 300,
>    orientation = [70,55], colour = black, axes = normal);
```

The phase plot of the solution $x(t)$ with respect to $y(t)$ (see Figure 24.1 – left) enables one to understand how the solution converges to the oscillation form of the periodical solution (see Figure 24.1 – right).

Květoň also found that different initial distributions of inhabitants from both groups does not have an influence on their cyclic migration. It is illustrated in the following example:

```
> init := x(0) = -1,   y(0) = 1:
> sold := dsolve({eq1, eq2, init}, {x(t), y(t)},numeric):
> odeplot(sold, [x(t), y(t)], 0..60, numpoints = 300,
>              view = [-1..1, -1..1], color = black);
```

Figure 24.2 shows that the phase plot of $x(t)$ with respect to $y(t)$ has the same envelope of functions as the phase plot on Figure 24.1 – left:

FIGURE 24.1.
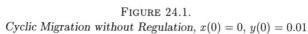
Cyclic Migration without Regulation, $x(0) = 0$, $y(0) = 0.01$

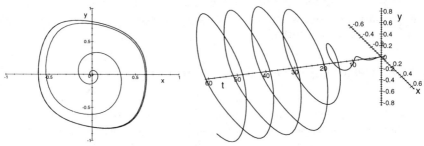

FIGURE 24.2.
Cyclic Migration without Regulation, $x(0) = -1$, $y(0) = 1$

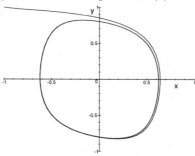

This solution shows a periodic migration which appears when inhabitants from one group don't like to live together with the second group forever in the same quarter.

24.2.2   Cyclic Migration with Regulation

Weidlich and Haag have developed a modified model (24.3), (24.4) for a directive transfer of inhabitants, see [6]:

$$\frac{dx(t)}{dt} = 2A(\sinh u(t) - x(t)\cosh u(t)) + r ,$$

$$\frac{dy(t)}{dt} = 2A(\sinh v(t) - y(t)\cosh v(t)) + s$$

where r and s are certain regulatory parameters (e.g., they represent the direct intervention of a government office to support the migration of a given group of inhabitants into a given quarter).

The role of regulatory parameters r and s can be shown using the same data (see [2]) as in the previous section:

```
> restart;
> with(plots):
> read('system.sav'):
> eq1r := diff(x(t), t) = 2*A*(sinh(u) - x(t)*cosh(u)) + r:
> eq2s := diff(y(t), t) = 2*A*(sinh(v) - y(t)*cosh(v)) + s:
> eq1r := subs(param, eq1r):  eq2s := subs(param, eq2s):
> initrs := x(0) = 0, y(0) = 0.01:
> r := -0.2:  s := 0.1:
> solrs := dsolve({eq1r, eq2s, initrs}, {x(t), y(t)}, numeric):
> odeplot(solrs, [x(t), y(t)], -30..50, numpoints = 200,
>          view = [-1..1, -1..1], color = black);
```

The phase plot on Figure 24.3 has a similar envelope of oscillating solution as the phase plot on Figure 24.2.

Increasing the regulatory parameters up to $r = -0.3$ and $s = 0.3$, the solution converges to a stable value, see Figure 24.4.

```
> r := -0.3:  s := 0.3:
> solrs := dsolve({eq1r, eq2s, initrs}, {x(t), y(t)}, numeric):
> odeplot(solrs, [t, x(t), y(t)], 0..100, numpoints = 300,
>          orientation = [-50, 45], axes = normal, colour = black);
```

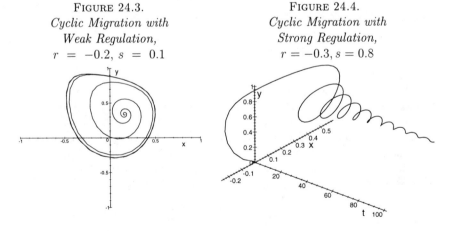

FIGURE 24.3.	FIGURE 24.4.
Cyclic Migration with	*Cyclic Migration with*
Weak Regulation,	*Strong Regulation,*
$r = -0.2, s = 0.1$	$r = -0.3, s = 0.8$

We can see that with the proper choice of regulatory parameters, it is possible to obtain a prescribed settlement structure.

24.3   Modeling Strategic Investment

We will consider periodical oscillation in the economy (depression and prosperity periods) known as *Schumpeter's clock*, see [4]. The base quantity in the model of an industry cycle is strategic investment $I(t) = E(t) + R(t)$, where $E(t)$ is expansive investment and $R(t)$ is rationalizing investment. Let us define the index of investment structure $Z(t) = (E(t) - R(t))/I(t)$ which we split in a constant Z_0 and a variable term $z(t)$ representing the ticking of Schumpeter's clock, see [1],

$$Z(t) = Z_0 + z(t) \ .$$

The model of Schumpeter's clock was constructed in this way because the investment structure depends on the distribution of investors. There are two kinds of investors, see [5]. An investor of type E prefers progressive or expansive investment and invests more than the average investment. An investor of type R prefers rational investment and invests less than the average investment.

Let $n_1(t)$ be the number of investors of type E, $n_2(t)$ the number of investors of type R and $2N$ the constant number of all investors, i.e. $2N = n_1(t) + n_2(t)$. Using the *investor configuration* $n(t) = (n_1(t) - n_2(t))/2$ we can write (similar as in the former section) $n_1(t) = N + n(t)$ and $n_2(t) = N - n(t)$.

The transient probability $w_{ij}^\alpha(t)$ was introduced in the previous section. In this section, the transient probability $w_{ij}^\alpha(t)$ means the probability (per unit time) of change of the subsystem α investor opinion which implies a change of his incorporation from group j (R-investors) to group i (E-investors). Using Equation (24.1) for the above defined numbers of investors $n(t), n_1(t)$ and $n_2(t)$, the following balance equation of the investors distribution can be obtained

$$\frac{dn(t)}{dt} = n_2(t)w_{21}(t) - n_1(t)w_{12}(t) \ . \tag{24.5}$$

When the new normalized variable $x(t) = n(t)/N$ is substituted into Equation (24.5), the following equation is obtained

$$\frac{dx(t)}{dt} = w_{21}(t)(1 - x(t)) - w_{12}(t)(1 + x(t)) \ . \tag{24.6}$$

The main problem is again to find an adequate formulation for the transient probability $w_{12}(t)$ and $w_{21}(t)$, see [3]. Let the *alternator* $p(t)$ represent changes of the investor's opinion from one type to the second. A positive value of $p(t)$ expresses the change from investment of the type R to type E. In the opposite case, the parameter $p(t)$ will be negative. Let the *coordinator* q represents the intensity of synchronization of the investor with effort of other investors. Then the transient probability functions can be defined as follows:

$$w_{12}(t) = Ae^{p(t)+qx(t)}$$

$$w_{21}(t) = Ae^{-(p(t)+qx(t))}$$

where A is a time scaling factor.

These transient probability functions are substituted into the Equation (24.6) and we obtain in a similar way as in the previous section:

$$\frac{dx(t)}{dt} = 2A[\sinh(p(t) + qx(t)) - x(t)\cosh(p(t) + qx(t))] . \qquad (24.7)$$

The investment oscillations are caused by time variability of the alternator $p(t)$. Mensch, in [3], derived the following equation for the alternator

$$\frac{dp(t)}{dt} = -2B(p_0 \sinh(ax(t)) + p(t)\cosh(ax(t))) , \qquad (24.8)$$

where $B > 0$, $a > 0$ and $p_0 > 0$ are the characteristic constants.

Equations (24.7) and (24.8) describe a evolution model of strategic investment. They have been used for modeling the economic growth in Germany from the fifties to the seventies (see [3]) with a good correspondence between reality and the calculated results.

Parameter values $B = 0.8$, $a = 4$, $p_0 = 0.5$, $A = 4B$ and $q = 1.6$ from [1] and the initial conditions $x(0) = 0.01$, $p(0) = 0$ are used for the solution and visualization Schumpeter's clock.

```
> restart:
> eq1 := diff(x(t), t) = 2*A*(sinh(p(t) + q*x(t)) -
>          x(t)*cosh(p(t) + q*x(t))):
> eq2 := diff(p(t),t) = -2*B*(p0*sinh(a*x(t)) +
>          p(t)*cosh(a*x(t))):
> B := 0.8:   A := 4*B:   a := 4:   p0 := 0.5:   q := 1.6:
> init := x(0) = 0.01, p(0) = 0:
> sol := dsolve({eq1, eq2, init}, {x(t), p(t)}, numeric):
> with(plots):
> odeplot(sol, [x(t), p(t)], 0..10, numpoints = 200,
>          color = black);
> odeplot(sol, [t, x(t), p(t)], 0..10, numpoints = 200,
>          orientation = [-100, 50], axes = normal, colour = black);
```

Figure 24.5 - left shows the solution and the limit cycle. This plot should be interpreted this way:

1. Quadrant $(p(t) > 0, x(t) > 0)$ illustrates the situation when the E-investors are in a majority and in addition the economic climate is good for expansive investment. We are here in a prosperity period.

2. Quadrant $(p(t) < 0, x(t) > 0)$ illustrates the state when E-investors are still in the majority but the economic climate supports a transition to rationalizing investment. The value of alternator $p(t)$ is already negative— that means the number of E-investors will decrease. This is a transition state from prosperity to depression.

3. Quadrant $(p(t) < 0, x(t) < 0)$ illustrates the state when there is a majority of R-investors and the economic climate is also good for rationalizing investment. We are in a state of economical depression.

4. Quadrant ($p(t) > 0$, $x(t) < 0$) illustrates that even if there is still a majority of R-investors, there a trend towards expansive investment (alternator $p(t) > 0$) has been started. This state is a transition from depression to prosperity.

FIGURE 24.5. *Schumpeter's clock*

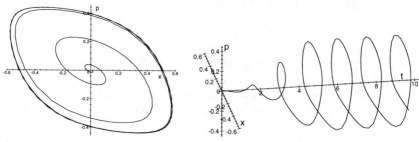

References

[1] J. KREMPASKÝ ET AL: *Synergetics*, Veda, Bratislava, 1987 (in Slovak)

[2] J. KREMPASKÝ, R. KVĚTOŇ: Acta Phys. Slov., Vol. 33, 1983, p. 115

[3] G. MENSCH, G. HAAG, W. WEIDLICH: Econometrica, Vol. 50, 1982, p. 15

[4] J. A. SCHUMPETER: *Konjunkturzyklen, eine theoretische, historische und statistische Analyse des kapitalistischen Prozesses. Vol. 2*, Vandenhoeck and Ruprecht, Göttingen, 1961

[5] W. WEIDLICH, G. HAAG: *Concepts and models of quantitative sociology. The dynamics of interacting populations.*, Springer Verlag, Berlin-Heidelberg-New York, 1983

[6] W. WEIDLICH, G. HAAG: *Dynamic of interacting groups in society.* in H. Haken: Dynamics of synergetics systems, Springer Verlag Berlin-Heidelberg-New York, 1980

Chapter 25. Contour Plots of Analytic Functions

W. Gautschi and J. Waldvogel

25.1 Introduction

There are two easy ways in MATLAB to construct contour plots of analytic functions, i.e., lines of constant modulus and constant phase. One is to use the MATLAB `contour` command for functions of two variables, another to solve the differential equations satisfied by the contour lines. This is illustrated here for the function $f(z) = e_n(z)$, where

$$e_n(z) = 1 + z + \frac{z^2}{2!} + \cdots + \frac{z^n}{n!} \qquad (25.1)$$

is the nth partial sum of the exponential series. The lines of constant modulus 1 of e_n are of interest in the numerical solution of ordinary differential equations, where they delineate regions of absolute stability for the Taylor expansion method of order n and also for any n-stage explicit Runge-Kutta method of order n, $1 \le n \le 4$ (cf. [4, §9.3.2]).

25.2 Contour Plots by the `contour` Command

Let f be analytic and $f(z) = re^{i\varphi}$. We may consider the modulus r as a function of two variables x, y, where $z = x + iy$; similarly for the phase φ, $-\pi < \varphi \le \pi$. Hence, we can apply the MATLAB command `contour` to r and φ to obtain the lines of constant modulus and phase.

In the MATLAB program below, the set of all x- and y-values is collected (in true MATLAB spirit) in a matrix a, which is operated upon to compute the desired values of r and φ for $f = e_n$ as input matrices to the routine `contour`.

The program begins with the definitions of the mesh h and the number nmax of contour plots to be generated. The vector bounds contains common lower and upper bounds for the x- and y-coordinates applicable for all plots. The bounds used here have been chosen to accommodate contour plots of the first four exponential sums. Then the contour levels vabs0 and vang0 for the modulus and phase of $f(z)$ are defined. The last preparatory step is generating the vectors x and y containing the discrete x- and y-values to be used in the matrix a of grid points. In the loop over n the values f of e_n on the entire grid are generated by almost the same statements that would evaluate e_n at a single point, where t stands for an individual term of the series (25.1). The only difference is the statement t=t.*a/n, in which the operation symbol .* invokes the element-by-element product of the matrices t and a. The last line of the program (here turned off by the comment sign %) generates the encapsulated postscript file fign.eps of figure(n), ready to be printed or incorporated into a text file.

```
% Contour plots of the first nmax exponential sums (Figure 1)
%
>> h = 1/64;  nmax = 4;  bounds = [-3.25  .75  -3.375  3.375];
>> vabs0 = [0:.1:1];  vang0 = [-.875:.125:1]*pi;
>> x = bounds(1):h:bounds(2);  y=bounds(3):h:bounds(4);
>> a = ones(size(y'))*x + i*y'*ones(size(x));
% Next line: a shorter way of generating a (more memory!)
>> % [xx,yy]=meshgrid(x,y); a=xx+i*yy;
>> t = ones(size(a));  f = t;
>> for n = 1: nmax
>>    if n <= 2, vabs = vabs0; vang = vang0;
>>      elseif n == 3, vabs = [vabs0 .47140452]; vang = vang0;
>>      else vabs = [vabs0 .58882535 .27039477];
>>          vang = [vang0 1.48185376 -1.48185376];
>>    end;
>>    t = t.*a/n;  f = f + t;
>>    figure(n);  clf;  hold on;
>>    axis(bounds);  axis image;
>>    contour(x, y, abs(f), vabs);
>>    contour(x, y, angle(f), vang);
>> end;
>> % figure(n); print -deps fign;
```

The results for $n = 1 : 1 : 4$[1] are shown in the plots below. Clearly visible are the n zeros of e_n from which emanate the lines of constant phase. Near these zeros, the lines of constant modulus become circle-like with radii tending to 0 as the zeros are approached. The contour lines are for $r = .1 : .1 : 1$ and $\varphi = -\frac{7}{8}\pi : \frac{1}{8}\pi : \pi$.

At points z_0 where $e'_n(z_0) = e_{n-1}(z_0) = 0$, $n \geq 2$, two lines of constant modulus intersect (cf. §3.1 below). The respective r-values are $r = |e_n(z_0)|$, or $r = |z_0|^n/n!$, since

$$e_n(z) = e_{n-1}(z) + \frac{z^n}{n!}. \tag{25.2}$$

These critical lines are also included in the plots (see the if statement of the program). When $n = 2$, they go through $z_0 = -1$, where $r = \frac{1}{2}$, while for $n = 3$ and $n = 4$, one has to 8 decimal digits: $z_0 = -1 \pm i$, $r = \sqrt{2}/3 = .47140452$ and $z_0 = -.70196418 \pm 1.80733949i$, $r = (1.93887332)^4/24 = .58882535$, $z_0 = -1.59607164$, $r = .27039477$, respectively.

---

[1]This MATLAB notation stands for $n = 1, 2, 3, 4$.

FIGURE 25.1.
Contour Plots of the First 4 Exponential Sums

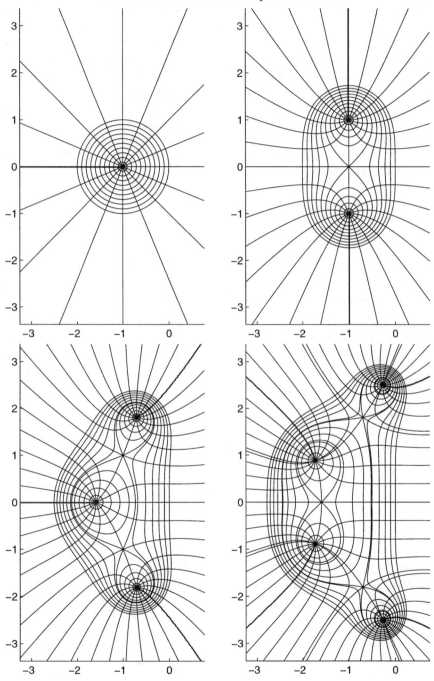

What's good for the r-lines is good for the φ-lines! The singular points for them are also the zeros z_0 of e_n' (cf. §3.2), to which there correspond φ-values defined by $e_n(z_0)/|e_n(z_0)| = (z_0/|z_0|)^n = e^{i\varphi}$, i.e., $\varphi = n \arg z_0$. Thus, for $n = 2$, we have $\varphi = 0$ (mod 2π), whereas for $n = 3$ we get $\varphi = \pm\frac{\pi}{4}$ corresponding to $z_0 = -1 \pm i$, respectively. All three of these φ-values are included among the values already listed above. For $n = 4$, the two complex values of z_0 shown in the previous paragraph yield $\varphi = 4 \arg z_0 = \pm 1.48185376$ (mod 2π), and the real value of z_0 yields $\varphi = 0$. These critical φ-lines are also shown in the plots in Figure 25.1

The figure was generated by means of the step size h=1/64 in order to obtain a good resolution, even for the "branch cuts" corresponding to |angle(f)|=pi. The choice h=1/32 is a good compromise, whereas h=1/16 is very fast while still producing satisfactory plots.

25.3   Differential Equations

For an analytic function f, let

$$w = f(z), \quad w = re^{i\varphi}, \quad z = x + iy. \tag{25.3}$$

25.3.1   Contour Lines $r = $ const.

To describe the lines $r = $ const, it is natural to take φ as independent variable. Differentiating

$$f(z(\varphi)) = re^{i\varphi}, \quad r = \text{const}, \tag{25.4}$$

with respect to φ then gives $f'(z)\frac{dz}{d\varphi} = ire^{i\varphi} = if(z)$, that is,

$$\frac{dz}{d\varphi} = i\, q(z) \,, \quad \text{where} \quad q(z) = \frac{f(z)}{f'(z)}. \tag{25.5}$$

With s the arc length, one has

$$\frac{ds}{d\varphi} = \sqrt{\left(\frac{dx}{d\varphi}\right)^2 + \left(\frac{dy}{d\varphi}\right)^2} = \left|\frac{dz}{d\varphi}\right| = |q(z)|,$$

so that

$$\frac{dz}{ds} = \frac{dz}{d\varphi}\frac{d\varphi}{ds} = i\,\frac{q(z)}{|q(z)|}. \tag{25.6}$$

Written as a system of differential equations, this is

$$\frac{dx}{ds} = -\text{Im}\left\{\frac{q(z)}{|q(z)|}\right\},$$
$$\hspace{3cm} z = x + iy. \tag{25.7}$$
$$\frac{dy}{ds} = \text{Re}\left\{\frac{q(z)}{|q(z)|}\right\},$$

If we are interested in a contour line crossing the real axis, we must find an initial point $x(0) = x_r$, $y(0) = 0$ for (25.7) with real x_r such that $f(x_r) = r$ (assuming $f(x)$ real for real x). In the case $f(x) = e_n(x)$, this is easy if $r \geq 1$, since $e_n(0) = 1$ and $e_n(x)$ monotonically increases for $x \geq 0$. There is thus a unique $x_r \geq 0$ such that $e_n(x_r) = r$. If $0 < r < 1$, this is still possible when n is odd. Then, $e'_n(x) = e_{n-1}(x) > 0$, since all zeros of e_m, when m is even, are known to be complex [3] (cf. also [1]). Thus, e_n monotonically increases from $-\infty$ to $+\infty$ as x increases from $-\infty$ to $+\infty$, and there is a unique $x_r < 0$ such that $e_n(x_r) = r$. When n is even, we have $e_n(x) > 0$ for all real x, and $e'_n = e_{n-1}$ vanishes at exactly one point $x_0 < 0$, where e_n has a minimum (cf. [2]). Owing to (25.2) and $e_{n-1}(x_0) = 0$, we have $e_n(x_0) = x_0^n/n!$, and there is a solution $x_r < 0$ of $e_n(x_r) = r$ if and only if $r \geq x_0^n/n!$. For smaller positive values of r, one must find a complex initial point $x(0)$, $y(0) > 0$ near one of the complex zeros of e_n.

The point z_0 where $f'(z_0) = 0$ is a singular point of (25.7), a point where two r-lines intersect at a right angle. This requires special care to get the integration of (25.7) started in all four directions. The initial point, of course, is z_0, that is, $x(0) = \text{Re } z_0$, $y(0) = \text{Im } z_0$. What needs some analysis is the value of the right-hand side of (25.7) at z_0. Let $h(z) = (z - z_0) q(z)$; then h is smooth near z_0 and has the Taylor expansion

$$h(z) = \frac{f_0 + \frac{1}{2}(z - z_0)^2 f_0'' + \cdots}{f_0'' + \frac{1}{2}(z - z_0) f_0''' + \cdots}, \quad h(z_0) = \frac{f_0}{f_0''},$$

where $f_0 = f(z_0)$, etc. (we assume $f_0 \neq 0$ and $f_0'' \neq 0$). Letting

$$\frac{z - z_0}{|z - z_0|} = e^{i\theta}, \quad -\frac{1}{2}\pi < \theta \leq \frac{1}{2}\pi, \quad h(z_0)/|h(z_0)| = e^{i\omega}, \quad -\pi < \omega \leq \pi,$$

(being mindful that to each θ there is a $\theta + \pi$ corresponding to the backward continuation of the line), we then have

$$\frac{q(z)}{|q(z)|} = \frac{|z - z_0|}{z - z_0} \frac{h(z)}{|h(z)|} \to e^{i(\omega - \theta_0)} \quad \text{as } z \to z_0,$$

where $\theta_0 = \lim_{z \to z_0} \theta$. It remains to determine θ_0.
Along an r-line through z_0, we have

$$r^2 = |f(z)|^2 = |f_0 + \frac{1}{2}(z - z_0)^2 f_0'' + \cdots|^2$$
$$= |f_0|^2 + \text{Re}[(z - z_0)^2 f_0'' \overline{f_0}] + O(|z - z_0|^3).$$

Since $|f_0|^2 = r^2$, this gives

$$\text{Re}\left\{\left(\frac{z - z_0}{|z - z_0|}\right)^2 f_0'' \overline{f_0}\right\} = O(|z - z_0|),$$

hence, as $z \to z_0$,

$$\operatorname{Re}\left(e^{2i\theta_0} f_0'' \overline{f_0}\right) = 0.$$

Therefore,

$$\tan 2\theta_0 = \frac{\operatorname{Re}(f_0'' \overline{f_0})}{\operatorname{Im}(f_0'' \overline{f_0})}. \tag{25.8}$$

There are exactly two solutions in $-\frac{1}{2}\pi < \theta_0 \le \frac{1}{2}\pi$, which differ by $\frac{1}{2}\pi$, confirming the orthogonality of the two r-lines through z_0.

Note that in the case $f(z) = e_n(z)$, we have $f'(z) = e_{n-1}(z)$, so that z_0 is a zero of e_{n-1}. This is clearly visible in the plots of §2. Furthermore, $f_0 = e_n(z_0)$, $f_0'' = e_{n-2}(z_0)$ if $n \ge 2$, so that (25.2) with $z = z_0$, once applied as is, and once with n replaced by $n-1$, gives $f_0 = z_0^n/n!$, $f_0'' = -z_0^{n-1}/(n-1)!$, and the equation for θ_0 reduces to

$$\tan 2\theta_0 = -\frac{\operatorname{Re} z_0}{\operatorname{Im} z_0} \quad (f = e_n).$$

25.3.2   Contour Lines $\varphi = \text{const.}$

For the lines $\varphi = \text{const}$, we take r as the independent variable and, by differentiating

$$f(z(r)) = re^{i\varphi}, \quad \varphi = \text{const},$$

with respect to r, obtain

$$\frac{dz}{dr} = \frac{e^{i\varphi}}{f'(z)}.$$

In terms of the arc length s, we now have

$$\frac{ds}{dr} = \left|\frac{dz}{dr}\right| = \frac{1}{|f'(z)|},$$

so that

$$\frac{dz}{ds} = \frac{dz}{dr}\frac{dr}{ds} = e^{i\varphi}\frac{|f'(z)|}{f'(z)},$$

or, written as a system of differential equations,

$$\frac{dx}{ds} = \operatorname{Re}\left\{e^{i\varphi}\frac{|f'(z)|}{f'(z)}\right\},$$
$$\hspace{4cm} z = x + iy. \tag{25.9}$$
$$\frac{dy}{ds} = \operatorname{Im}\left\{e^{i\varphi}\frac{|f'(z)|}{f'(z)}\right\},$$

The singular point of (25.9) is again z_0, a zero of f'. At this point,

$$\frac{f_0}{|f_0|} = e^{i\varphi}, \quad -\pi < \varphi \le \pi,$$

which determines φ. The limit of $|f'(z)|/f'(z)$ as $z \to z_0$ may be determined by a procedure similar to the one in §3.1. Instead, we directly use the Taylor series of f in z_0 in order to study the φ-lines (and the r-lines as well) near z_0 with $f'(z_0) = 0$. Let $z = z_0 + \zeta$, where ζ is a complex increment, and let

$$f_k := f^{(k)}(z_0), \quad k \geq 0, \quad f_0 \neq 0, \quad f_1 = 0, \quad f_2 \neq 0, \tag{25.10}$$

be the derivatives of f at z_0. Then the Taylor series is

$$f(z_0 + \zeta) = f_0 + f_2 \frac{\zeta^2}{2!} + f_3 \frac{\zeta^3}{3!} + \cdots . \tag{25.11}$$

Next, we observe that by defining $w = f_0 e^u$ in (25.3), i.e., by putting

$$f(z_0 + \zeta) = f(z_0) e^u, \tag{25.12}$$

the r-lines through z_0 are given by the values of ζ corresponding to purely imaginary values $u = it$, whereas the φ-lines through z_0 are given by $u \in R$. The point z_0 itself corresponds to $\zeta = u = 0$. We therefore need to solve Equ. (25.12), with $f(z_0 + \zeta)$ substituted from (25.11), for ζ, which is a typical task for MAPLE.

In the program below[2] the series (25.11) and the equation (25.12) are denoted by s and eq, respectively. The **solve** command automatically expands e^u in a Taylor series and solves the equation by means of a series progressing in appropriate powers of u (here half-integer powers). As expected, two solutions corresponding to the two possible values of the square root are found. Only the first solution zet0[1] is processed further: first by substituting the abbreviations fk defined in Equ. (25.10), then by introducing the variable v according to

$$u = \frac{v^2 f_2}{2 f_0} \quad \text{or} \quad v = \left(\frac{2 u f_0}{f_2} \right)^{\frac{1}{2}} . \tag{25.13}$$

The symbols D(f) and (D@@k)(f) stand for the derivative of f and the kth derivative of f, respectively. The call to the function map causes the operation defined by its first argument, here the simplification of the radicals, to be applied to each term of the expression defined by the second argument. Finally, the call to series causes the O-term to be simplified.

```
> N := 5:  Order := N:
> s := series(f(z0 + dz), dz):
> s0 := subs(D(f)(z0) = 0, s);
```

$$s0 := \mathrm{f}(z0) + \frac{1}{2} (D^{(2)})(f)(z0)\, dz^2 + \frac{1}{6} (D^{(3)})(f)(z0)\, dz^3 + \frac{1}{24} (D^{(4)})(f)(z0)$$
$$dz^4 + O(dz^5)$$

---

[2]The authors are indebted to Dominik Gruntz for this program.

```
> eq := s0 = f(z0)*exp(u):
> zet0 := solve(eq, dz);
```

$$zet0 := \frac{\mathrm{f}(z0)\,\sqrt{2}\,\sqrt{u}}{\sqrt{\%1}} - \frac{1}{3}\frac{(D^{(3)})(f)(z0)\,\mathrm{f}(z0)\,u}{(D^{(2)})(f)(z0)^2} + \frac{1}{288}($$
$$40\,\mathrm{f}(z0)^3\,(D^{(3)})(f)(z0)^2 - 24\,(D^{(2)})(f)(z0)\,\mathrm{f}(z0)^3\,(D^{(4)})(f)(z0)$$
$$+ 72\,(D^{(2)})(f)(z0)^3\,\mathrm{f}(z0)^2)\sqrt{2}\,u^{3/2} \Big/ ((D^{(2)})(f)(z0)^2\,\%1^{3/2}) + O(u^2),$$

$$-\frac{\mathrm{f}(z0)\,\sqrt{2}\,\sqrt{u}}{\sqrt{\%1}} - \frac{1}{3}\frac{(D^{(3)})(f)(z0)\,\mathrm{f}(z0)\,u}{(D^{(2)})(f)(z0)^2} - \frac{1}{288}(40\,\mathrm{f}(z0)^3\,(D^{(3)})(f)(z0)^2$$
$$- 24\,(D^{(2)})(f)(z0)\,\mathrm{f}(z0)^3\,(D^{(4)})(f)(z0) + 72\,(D^{(2)})(f)(z0)^3\,\mathrm{f}(z0)^2)\sqrt{2}\,u^{3/2}$$
$$\Big/ ((D^{(2)})(f)(z0)^2\,\%1^{3/2}) + O(u^2)$$
$$\%1 := \mathrm{f}(z0)\,(D^{(2)})(f)(z0)$$

```
> zet1 := subs(seq( (D@@k)(f)(z0) = f.k, k=0..N-1), zet0[1]):
> zet2 := map(radsimp, subs(u = v^2*f2/2/f0, zet1)):
> zeta := series(zet2, v);
```

$$\zeta := v - \frac{1}{6}\frac{f3}{f2}\,v^2 - \frac{1}{72}\frac{-5\,f0\,f3^2 + 3\,f2\,f0\,f4 - 9\,f2^3}{f0\,f2^2}\,v^3 + O(v^4)$$

The MAPLE program works for any $N \geq 3$, producing $N-2$ terms of the above series. However, it is fairly slow, since no "intelligence", such as information on the form of the resulting series, is built in. To be able to find this series, nevertheless, is a good accomplishment of a general-purpose symbolic manipulator. It can be seen that ζ may be written as a formal power series in the variable v defined in (25.13). If the original series (25.11) converges in a neighborhood of z_0, the resulting series converges in a neighborhood of $v = 0$.

The directions θ_0 of the r-lines at z_0 are now given by the values of ζ corresponding to $u = it$ in the limit $t \to 0$. The above series and Equ. (25.13) immediately yield

$$\theta_0 = \arg v = \frac{1}{2}\left(\arg f_0 - \arg f_2 \pm \frac{\pi}{2}\right).$$

Hence there are two r-lines through z_0 intersecting at a right angle, in perfect agreement with Equ. (25.8).

The directions of the φ-lines through z_0, on the other hand, are given by (25.13) for real values of u. We obtain the two directions $\theta_0 \pm \frac{\pi}{4}$, i.e., the tangents of the two φ-lines through z_0 are the bisectors of the tangents of the r-lines.

25.4   The Contour Lines $r = 1$ of $f = e_n$

As indicated in §3.1, we need to solve (25.7) with initial values $x = y = 0$. Let S_n be the point of intersection of the 1-line of e_n with the negative real axis. By symmetry, only the portion of each 1-line lying in the upper half of the complex plane needs to be computed.

We will discuss two implementations of this process: first a naive approach only using termination of the numerical integration at a precomputed value s_f of the independent variable (as available in the previous version MATLAB 4). At the end of this section we will present a simplified algorithm taking advantage of the Events capability of the current version MATLAB 5. For both implementations a good upper bound s_f for the arc length on the 1-line between the origin and S_n is needed.

Such an upper bound s_f may be obtained as follows. We first observe that the region $|e_n(z)| \leq 1$ approaches a semidisk of radius $\rho(n)$ as $n \to \infty$. An asymptotic analysis shows that

$$\rho(n) = \exp(-1) \cdot (n + \log \sqrt{2\pi n} + O(1)).$$

A good empirical choice of $O(1)$ is 3; then we obtain

$$s_f = (1 + \frac{\pi}{2}) \exp(-1) \cdot (n + \log \sqrt{2\pi n} + 3) \qquad (25.14)$$

as a close upper bound for the arc length up to the point S_n for $n \geq 1$.

If the Events capability is not used, we first integrate the differential equations up to the final value s_f. Then, only the points satisfying the condition $y \geq 0$ need to be plotted. The point S_n can be approximated by linear interpolation between the two points on the 1-line closest to S_n.

In the MATLAB program below this is done by using the find command with the parameter $y \geq 0$ in order to find the subset of all points satisfying the condition $y \geq 0$. Their indices are collected in the vector indices. The indices of the points used in linear interpolation are then l=length(indices) and ll=l+1. Finally, w is the normalized row vector containing the two interpolation weights, and the actual interpolation is carried out by the product w*z(1:ll,:).

```
% Level curves r = 1 for the first 21 exponential sums (Figure 2)
>> global n
>> nmin = 1;   nmax = 21;   tol = 3.e-8;
>> axis equal;   hold on; % arc = [];
>> options = odeset('RelTol',tol,'AbsTol',tol);
>> for n = nmin:nmax,
>>    sf = 0.94574*(n +.5*log(2*pi*n) + 3);
% 0.94574=(1+pi/2)*exp(-1)
>>    [s,z] = ode45('level4', [0 sf], [0; 0], options);
>>    indices = find(z(:, 2) >= 0);
>>    l = length(indices);   ll = l + 1;
>>    w = [-z(ll, 2), z(l, 2)]; w = w/sum(w);
>>    z(ll,:) = w*z(l:ll,:);
>>    plot(z(1:ll, 1), z(1:ll, 2));
% approximate arclengths and bounds sf
>> %   arc = [arc; w*s(l:ll), sf];
>> end;
```

ALGORITHM 25.1. *level4.m*

```
function zdot = level4(s, z)
%LEVEL4 generates the right-hand side of
%         the differential equation for the 1-lines of
%         f(z) = 1 + z + z^2/2! + ... + z^n/n!

global n

zc = z(1) + i*z(2);  t = 1;  f0 = 0;  f = t;
for k = 1: n, t = t*zc/k;  f0 = f;  f = f + t;  end;
q = f/f0;  zdot = [-imag(q); real(q)]/abs(q);
```

The MATLAB program begins with the definitions of the parameters nmin, nmax and the error tolerance tol to be used in the definition of the integrator options structure, options, by means of odeset. The 1-lines in the range nmin $\leq n \leq$ nmax are generated and plotted in the accuracy given by tol. Rerunning the program with new values of nmin and nmax adds new curves to the figure. The statements turned off by the comment marks % generate a table arc containing the actual arc lengths and the upper bounds s_f computed from (25.14).

The actual integration is done in the call to the integrator ode45. The input parameters of this procedure are: the name 'level4' (in string quotes) of the M-file defined in Alg. 25.1 according to Equ. (25.7), a vector containing the initial value 0 and the final value sf of the independent variable, the column vector [0;0] of the initial values, and the options structure, options. The choice tol = 3.0e-8 yields a high-resolution plot, whereas the default RelTol = 1.0e-3, AbsTol = 1.0e-6 (when the parameter options is omitted in the call) still yields a satisfactory plot. The values of the independent and dependent variables generated by the integrator are stored as the vectors s and z, respectively, ready to be plotted.

The results for $n = 1 : 1 : 21$ are shown in Figure 25.2 below. The features near the imaginary axis at the transition to the circular part seem to show a periodicity in n of a little over 5. For example, the curves corresponding to $n = 5, 10, 15, 21$ all show a particularly large protrusion into the right half-plane.

An investigation of this phenomenon is interesting, but exceeds the scope of this article. We limit ourselves to reporting that as $n \to \infty$, the period tends to

$$\frac{2\pi}{\frac{\pi}{2} - \exp(-1)} = 5.22329130.$$

This result was obtained by considering the function $e_\nu(z)$ for real values of ν (which leads to the incomplete gamma function) and requiring the 1-line of $e_\nu(z)$ to contain a saddle point with $e'_\nu(z) = 0$.

The Events capability of MATLAB 5 allows to stop a numerical integration at an "event", i.e. if a so-called "event function" passes through zero. The

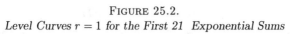

FIGURE 25.2.
Level Curves $r = 1$ for the First 21 Exponential Sums

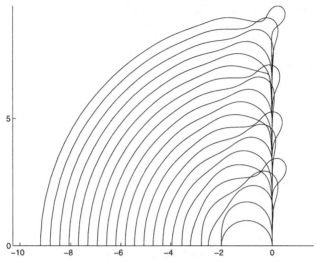

event function, in our case the imaginary part `imag(z)` of the complex dependent variable, has to be defined in the function `level` (Alg. 25.2), which also defines the right-hand sides `zdot` of the differential equations. This function needs to be coded with the additional input parameter `flag` and the additional output parameters `isterminal` and `direction` in such a way that `zdot` is the vector of the right-hand sides of the differential equations if `flag` is missing or undefined. If `flag` has the value `'events'`, however, the vector `zdot` must be defined as the event function, and the parameters `isterminal` (the indices of the relevant components of `zdot`) and `direction` (the direction of the zero passage) must be appropriately defined. Since `ode45` of MATLAB 5 allows to integrate complex-valued dependent variables, this simplification is taken advantage of in the program below.

```
% Level curves r = 1 for the first 21 exponential sums (Fig. 2)
% using the 'Events' capability and integration of complex
% dependent variables
>> global n
>> nmin = 1;   nmax = 21;   tol = 1e-6;
>> axis equal;   hold on; % arc = [];
>> options= odeset('RelTol',tol,'AbsTol',tol,'Events','on');
>> for n = nmin:nmax,
>>     sf = 0.94574*(n +.5*log(2*pi*n) + 3);
>>     [s,z] = ode45('level', [0 sf],0,options);
>>     plot(z);
% arclengths and bounds sf
>>     % arc = [arc; s(length(s)), sf];
>> end;
```

ALGORITHM 25.2. *level.m*

```
function [zdot, isterminal, direction] = level(s, z, flag)
%LEVEL generates the right-hand side of
%       the differential equation for the 1-lines of
%       f(z) = 1 + z + z^2/2! + ... + z^n/n!

global n

if nargin<3 | isempty(flag),
    t = 1;  f0 = 0; f = t;
    for k = 1:n, t = t*z/k; f0 = f;  f = f + t; end;
    q = f/f0; zdot = i*q/abs(q);
else
    switch(flag)
    case 'events'
        zdot= imag(z);
        isterminal= 1;
        direction= -1;
    otherwise
        error(['Unknown flag: ', flag]);
    end;
end
```

25.5   The Contour Lines $\varphi = \text{const}$ of $f = e_n$

Below is a MATLAB program that implements the method of §3.2 for any fixed $n > 0$, where the differential equations (25.9) must be implemented in the function phase and stored in the M-file phase.m (Alg. 25.3).

```
% Lines of constant phase for the 10th exponential sum (Fig 3)
%
>> global n phi
>> n = 10; tol = 1.e-5; sf = 1.5;
>> clf; axis([-6 6 -1 8]); hold on;
>> r = roots(1./gamma(n + 1:-1:1));
>> indices = find(imag(r) >= 0); zero = r(indices)
>> options= odeset('RelTol',tol,'AbsTol',tol);
>> for k = 1: length(zero),
>>     z0 = zero(k);
>>     for phi = -7/8*pi:pi/8:pi,
>>         [s,z] = ode45('phase', [0 sf], z0, options);
>>         plot(z);
>>     end;
>> end;
```

The program begins with the definitions of n, the error tolerance tol, and the desired arc length sf of the curve segments emanating from the zeros. Then, the vector r of the zeros of e_n is computed by means of the function roots, where the coefficients of e_n are generated by means of the gamma function. On the next line the subset of the zeros in the upper half-plane is formed by means

ALGORITHM 25.3. *phase.m*

```
function zdot = phase(s,z)

global n phi

eiphi = exp(i*phi);
t = 1; f = t;
for k=1:n-1, t = t*z/k; f = f+t; end;
zdot = eiphi*abs(f)/f;
```

of the `find` command with the argument `imag(r)>=0`. The statement used in the program stores all the indices defining the subset in the vector `indices`; then `r(indices)` is the vector of the zeros of e_n in the upper half-plane (which is printed for convenience).

The input parameters in the call to the integrator `ode45` are: the name `'phase'` (in string quotes) of the differential equations defined in Alg. 25.3 according to Equ. (25.9), a vector containing the initial value 0 and the final value `sf` of the independent variable, the complex initial value `z0`, and the options structure, `options` (defaults $10^{-3}, 10^{-6}$ when omitted). The values of the independent and dependent variables generated by the integrator are stored as the vectors `s` and `z`, respectively, ready to be plotted. The result for $n = 10$ is shown in Figure 25.3[3].

FIGURE 25.3.

Lines of Constant Phase for the 10th Exponential Sum

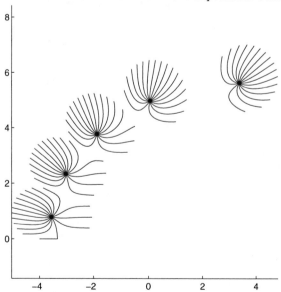

---

[3]We wrote and ran the script on August 1, 1996, while fireworks went off in celebration of the Swiss national holiday.

None of the complex singular points is in evidence in Figure 25.3 since the φ-values chosen do not correspond to level lines passing through a complex singular point. The real singular point at $z_0 = -3.333551485$, however, is clearly visible by an abrupt right-angled turn of the line $\varphi = 0$ (near the bottom of the figure). It is curious to note how the `ode45` integrator was able to integrate right through the singularity, or so it seems.

Acknowledgments

The first author was supported, in part, by the US National Science Foundation under grant DMS-9305430.

References

[1] A.J. CARPENTER, R.S. VARGA, J. WALDVOGEL, *Asymptotics for the zeros of the partial sums of e^z. I*, Rocky Mountain J. Math., 21, 1991, pp. 99–120.

[2] K.E. IVERSON, *The zeros of the partial sums of e^z*, Math. Tables and Other Aids to Computation, 7, 1953, pp. 165–168.

[3] G. PÓLYA, G. SZEGÖ, *Problems and Theorems in Analysis*, Vol. II, Part V, Exercise 74, Springer-Verlag, New York, 1976.

[4] H.R. SCHWARZ, *Numerical Analysis. A Comprehensive Introduction*, John Wiley & Sons, Chichester, 1989.

Chapter 26. Non Linear Least Squares: Finding the most accurate location of an aircraft

G. Gonnet

26.1   Introduction

Figure 26.1 illustrates a simplified typical situation of navigation with modern aircraft. The airplane is in an unknown position and receives signals from various beacons. The main purpose of this chapter is to develop a method for computing the most likely position of the aircraft based on the information available.

FIGURE 26.1. *Example of an aircraft and 4 beacons.*

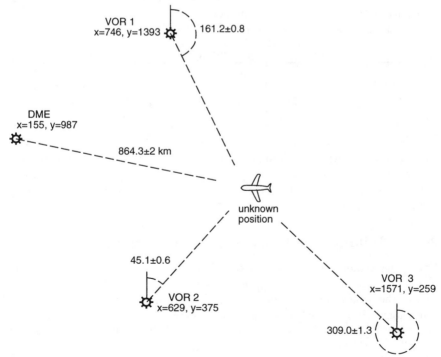

Three of the beacons are of the VOR (Very high frequency OmniRange) type. VOR beacons allow the airplane to read the angle from which the signal is coming. In other words, θ_1, θ_2 and θ_3 are known to the airplane. The DME (Distance Measuring Equipment) beacon allows, through a signal that is sent and bounced back, to measure the distance from the airplane to the beacon. In this case the distance is 864.3 ± 2 km.

Each of the measures is given with an estimate of its error. The standard notation for measures is $m \pm n$. This means that the true value being measured lies between $m - n$ and $m + n$. Different disciplines and have different interpretations for the statement *lies between*. It may mean an absolute statement, i.e. the true value is always between the two bounds, or a statistical statement, i.e. the true value lies within the two bounds $z\%$ of the time. It is common to assume that the error has a normal distribution, with average m and standard deviation n. For our analysis, it does not matter which definition of the error range is used, provided that all the measures use the same one.

We will simplify the problem by considering it in two dimensions only. That is, we will not consider the altitude, which could be read from other instruments and would unnecessarily complicate this example.

We will denote by x and y the unknown coordinates of the aircraft. It is easy to see, that unless we are in a pathological situation, any pair of two VOR/DME readouts will give enough information to compute x and y. With the four pieces of data, the problem is overdetermined. Since the measures are not exact, we want to compute x and y using all the information available, and hopefully, obtain a more accurate answer.

The input data is summarized in the following table.

	x coordinate	y coordinate	value	error
VOR 1	$x_1 = 746$	$y_1 = 1393$	$\theta_1 = 161.2$	$\sigma_1 = 0.8$
VOR 2	$x_2 = 629$	$y_2 = 375$	$\theta_2 = 45.10$	$\sigma_2 = 0.6$
VOR 3	$x_3 = 1571$	$y_3 = 259$	$\theta_3 = 309.0$	$\sigma_3 = 1.3$
DME	$x_4 = 155$	$y_4 = 987$	$d_4 = 864.3$	$\sigma_4 = 2.0$
aircraft	x	y		

This chapter is organized as follows. The first part will analyze how to pose the problem. The second section will do the necessary computations to find the optimal solution. The last section will analyze the confidence that we may derive from the results.

26.2 Building the Least Squares Equations

Under the assumption that the errors obey a normal distribution, it is completely appropriate to solve the problem of locating x and y by minimizing the sum of the squares of the errors. On the other hand, if we do not know anything about the distribution of the individual errors, it is still a good and stable procedure to minimize the sum of their squares. So, without further discussion, we will pose the problem as a *least squares* problem.

For any of the VOR beacons (1, 2 or 3), the equation to be satisfied is

$$\tan \tilde{\theta}_i \approx \frac{x - x_i}{y - y_i}$$

where $\tilde{\theta}_i = 2\pi\theta_i/360$ and for the DME (located at x_4, y_4), the equation is

$$\sqrt{(x - x_4)^2 + (y - y_4)^2} \approx d_4 .$$

(The standard measure of angles in aviation is clockwise from North and in degrees. This is different from trigonometry, which uses counterclockwise from East and in radians. Hence care has to be taken with the conversion from degrees to radians and the choosing of the right quadrant.)

The error is the difference between the left hand side and the right hand side of each equation. Although this would be correct for exactly determined systems, it is inadequate for overdetermined systems.

We can easily see that there is ambiguity in how we express the error, for example,

$$\tan \tilde{\theta}_i - \frac{x - x_i}{y - y_i} \tag{26.1}$$

$$\tan^{-1} \frac{x - x_i}{y - y_i} - \tilde{\theta}_i \tag{26.2}$$

$$(y - y_i) \tan \tilde{\theta}_i - (x - x_i)$$

are all possible expressions of the error for the VOR beacons. The DME equation shows the same ambiguity:

$$\sqrt{(x - x_4)^2 + (y - y_4)^2} - d_4$$

$$(x - x_4)^2 + (y - y_4)^2 - d_4^2 .$$

Which one should we use? Before we answer this question, we should direct our attention to an important technicality. Although Equation (26.1) is always correct, Equation (26.2) may not be always correct. This happens because $\tan^{-1}$ will return a principal value, that is a value between $-\pi/2$ and $\pi/2$. This brings two problems, one of them trivial, the second one more subtle. The trivial one is to convert aviation angles, which after normalization will be in the range from 0 to 2π, to the range $-\pi$ to π. This is done by subtracting 2π from the angle if it exceeds π. The second problem is that $\tan^{-1}$ returns values between $-\pi/2$ and $\pi/2$. This means that opposite directions, like for example the angles 135° and 315°, are indistinguishable. This may result in an equation that cannot be satisfied, or if the angles are reduced to be in the $\tan^{-1}$ range, then multiple, spurious, solutions are possible. To correct this problem we need to analyze the signs of $x - x_i$ and $y - y_i$, which is sometimes called quadrant analysis. This is a well known and common problem, and the function arctan with two arguments in MAPLE (atan2 in other languages) does the quadrant analysis and returns a value between $-\pi$ and π resolving the problem of opposite directions. Equation (26.2) should be written as:

$$\arctan(x - x_i, y - y_i) - \tilde{\theta}_i \ .$$

To compute the error we need to choose the most appropriate formula. Unless additional knowledge of the measuring equipment is known, the best and safest equation to use is the one which expresses the error in the same units as the magnitude being measured. For example, the VORs measure angles, hence the corresponding error should be an angle

$$\varepsilon_i = \arctan(x - x_i, y - y_i) - \tilde{\theta}_i \ .$$

The DME measures distances, and hence the error for this measure should be in kilometers (or any other measure of distance),

$$\varepsilon_4 = \sqrt{(x - x_4)^2 + (y - y_4)^2} - d_4 \ .$$

Before computing the sum of the squares of the errors, we need to make all these measures uniform, i.e. all to the same scale. This is easily obtained by dividing the ε_i by the standard deviation (i.e. by the value indicating the error of the measure).

In general, for a measure $z = m \pm n$, where z satisfies an equation $F(z, a, b, ...) = 0$, let $z = f(a, b, ...)$ be the inverse of F in its first argument. Then

$$\delta = \frac{m - f(a, b, ...)}{n}$$

is a normalized measure of the error. For unbiased errors following a normal distribution, then δ is $N(0, 1)$, that is normally distributed with average 0 and variance 1. Minimizing the sum of squares of variables, i.e. $\sum \delta_i^2$, belonging to the same distribution properly weighs each error.

The sum of squares that we want to minimize for our example is

$$S = \sum_{i=1}^{3} \left(\frac{\arctan(x - x_i, y - y_i) - \tilde{\theta}_i}{\sigma_i} \right)^2 + \left(\frac{\sqrt{(x - x_4)^2 + (y - y_4)^2} - d_4}{\sigma_4} \right)^2 \ .$$

It is rather obvious that this problem is non-linear in its unknowns x and y. It is also true, that no other representation of the errors would lead to a linear problem.

26.3　Solving the Non-linear System

To solve this problem we will use MAPLE, since we need to do some symbolic as well as numerical computations. First we define the input data. We use the vectors X and Y to store the beacon coordinates and x and y for the unknown coordinates of the airplane.

```
> theta := array( [ 161.2, 45.10, 309.0 ] );
> sigma := array( [ 0.8, 0.6, 1.3, 2.0 ] );
> X     := array( [ 746, 629, 1571, 155 ] );
> Y     := array( [ 1393, 375, 259, 987 ] );
> d4    := 864.3;
```

$$\theta := [161.2, \ 45.10, \ 309.0]$$

$$\sigma := [.8, \ .6, \ 1.3, \ 2.0]$$

$$X := [746, \ 629, \ 1571, \ 155]$$

$$Y := [1393, \ 375, \ 259, \ 987]$$

$$d4 := 864.3$$

The angles and the standard deviation of angles have to be converted to radians, as described earlier.

```
> for i to 3 do
>     theta[i] := evalf( 2*Pi*theta[i] / 360 );
>     if theta[i] > evalf(Pi) then
>         theta[i] := theta[i] - evalf(2*Pi) fi;
>     sigma[i] := evalf( 2*Pi*sigma[i] / 360 );
> od:
> i := 'i':
> print( theta );  print( sigma );
```

$$[2.813470755, \ .7871434929, \ -.890117918]$$

$$[.01396263402, \ .01047197552, \ .02268928028, \ 2.0]$$

We are now ready to construct the sum of squares.

```
> S := sum( ( (arctan( x-X[i], y-Y[i] ) - theta[i] ) /
            sigma[i] ) ^ 2, i = 1..3 ) +
          ( ( ( (x-X[4])^2 + (y-Y[4])^2 ) ^ (1/2) - d4 ) /
            sigma[4] ) ^ 2;
```

$$
\begin{aligned}
S := \ & 5129.384919 \, (\arctan(x - 746, \ y - 1393) - 2.813470755)^2 \\
& + 9118.906513 \, (\arctan(x - 629, \ y - 375) - .7871434929)^2 \\
& + 1942.488964 \, (\arctan(x - 1571, \ y - 259) + .890117918)^2 \\
& + .2500000000 \, (\sqrt{(x - 155)^2 + (y - 987)^2} - 864.3)^2
\end{aligned}
$$

Next we solve numerically for the derivatives equated to zero.

```
> sol := fsolve( { diff(S,x)=0, diff(S,y)=0 }, {x,y} );
```

$$sol := \{y = 723.9837773,\ x = 978.3070298\}$$

A solution has been found and it is definitely in the region that we expect it to be.

The first measure of success or failure of the approximation is to find what are the residues of the least squares approximation. Under the assumption that the errors are normally distributed, this will be the sum of the squares of four $N(0,1)$ variables. The expected value of such a sum is 4.

```
> S0 := evalf( subs( sol, S ));
```

$$S0 := .6684712637$$

This value is smaller than 4, and hence it indicates that either we are lucky, or the estimates for the error were too pessimistic. In either case, this is good news for the quality of the approximation. This together with the eigenvalue analysis from the next section give enough guarantees that we have found the right solution.

26.4 Confidence/Sensitivity Analysis

The procedure we will use to study the sensitivity of the solution is to expand the sum of the squares of the errors as a Taylor series around the minimum. Let $S(x,y) = S(\vec{p})$ be the sum of squares, which we will define as a function of the position vector $\vec{p} = [x,y]^T$. Let $\vec{p_0} = [978.30..., 723.98...]^T$ be the solution of the least squares problem. Then the 3-term Taylor series around $\vec{p_0}$ is

$$S(\vec{p}) = S(\vec{p_0}) + S'(\vec{p_0})(\vec{p} - \vec{p_0}) + (\vec{p} - \vec{p_0})^T S''(\vec{p_0})(\vec{p} - \vec{p_0}) + O(|\vec{p} - \vec{p_0}|^3) .$$

Since the gradient of S, $S'(\vec{p}) = [S'_x, S'_y]^T$, is zero at the minimum, (a numerical check shows that the gradient is within rounding error of $[0,0]$)

```
> S1 := evalf( subs( sol, linalg[grad]( S, [x,y] )));
```

$$S1 := [.36\,10^{-7}, .4\,10^{-8}]$$

The expression for S ignoring higher order terms is

$$S(\vec{p}) = S(\vec{p_0}) + (\vec{p} - \vec{p_0})^T S''(\vec{p_0})(\vec{p} - \vec{p_0}) .$$

In MAPLE the computation of the approximation to S in terms of the unknowns x and y is done by first computing the Hessian (the matrix with the second derivatives of S), evaluating it at the minimum and then computing the quadratic form.

```
> S2 := evalf( subs( sol, linalg[hessian]( S, [x,y] ) ));
> pmp0 := [ x-subs(sol,x), y-subs(sol,y) ];
> Sapprox := S0 + evalm( transpose(pmp0) &* S2 &* pmp0 );
```

$$S2 := \begin{bmatrix} .5118424680 & -.1726069278 \\ -.1726069278 & .09026159226 \end{bmatrix}$$

$$pmp0 := [x - 978.3070298, \; y - 723.9837773]$$

$$Sapprox := 292680.5513 + .5118424680\,x^2 - 751.5489380\,x$$
$$- .3452138556\,y\,x + 207.0292848\,y + .09026159226\,y^2$$

From here on we will work under the assumption that the errors of the measures have a normal distribution. The distribution of S, being a sum of 4 squares of variables from a distribution $N(0,1)$ is a chi-square distribution with 4 degrees of freedom. Knowing its distribution allows us to define a confidence interval for $S(\vec{p})$. Suppose that we are interested in a 95% confidence interval, then $S(\vec{p}) < 9.4877...$, where this value is obtained from the inverse of the cumulative (`icdf`) of a chi-square distribution,

```
> stats[ statevalf, icdf, chisquare[4] ] ( 0.95 );
```

$$9.487729037$$

The inequality $S(\vec{p}) < 9.4877...$ defines an ellipsis, an area which contains the true value of x and y with probability 95%. We can draw three ellipses for 3 different confidence intervals, e.g. 50%, 95% and 99.9%, all of which are reference values in statistical computations. Notice that the larger the confidence, the larger the ellipsis (cf. Figure 26.2).

The major axis of the ellipses is the direction for which the uncertainty is largest. The minor axis is the one for which we have the least uncertainty. The exact direction of both axes is given by the eigenvalue/eigenvector decomposition of S''.

```
> ev := linalg[eigenvects]( S2 );
```

$$ev := [.5734960798, \; 1, \; \{[-.9417276012, .3363764637]\}],$$
$$[.0286079803, \; 1, \; \{[-.3363764637, -.9417276012]\}]$$

The largest eigenvector, 0.5734... gives the direction of the steepest climb of $S(\vec{p})$, or the direction of the most confidence, in this case $[-0.941..., 0.336...]$.

FIGURE 26.2.

Ellipses of the expected locations for probabilities 50%,
95% and 99.9%.

```
> ellips := { seq( stats[ statevalf, icdf, chisquare[4]](c) =
              Sapprox, c = [0.5, 0.95, 0.999] ) }:
> plots[implicitplot]( ellips, x = 950..1000, y = 700..750,
              grid=[50,50], view=[950..1000, 700..750] );
```

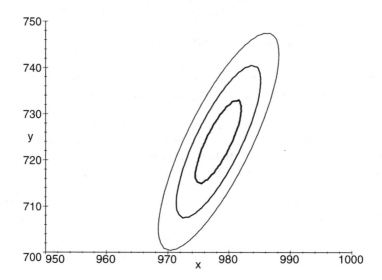

The smallest eigenvector, 0.0286... gives the direction of the least confidence, which is obviously perpendicular to the previous one. For this particular example, we see that the DME, together with the VOR 2 probably contribute the most information (the DME beacon had the smallest relative error), and hence the shape and orientation of the ellipses.

Computing the eigenvalues has one additional advantage. When we solve for the derivatives equal to zero, we could find a minimum, or a maximum or a saddle point. The proper way of checking that we obtained a minimum is to verify that the Hessian matrix is positive definite. Inspecting the eigenvalues gives more information. If all the eigenvalues are positive, then we have a minimum, if all are negative then we have a maximum, and if we have mixed signs, we are at a saddle point. Since both eigenvalues are positive, we can confirm that we have found a minimum.

The procedure described in this chapter can be trivially extended to any number of dimensions (unknown parameters).

Chapter 27. Computing Plane Sundials

M. Oettli and H. Schilt

27.1 Introduction

Many ancient sundials are found on the facades of old churches and other old houses. Beautiful examples are found in [3, 4]. Silent and often hardly noticed, they tell of a past age when man measured time using sundials. Even clocks and watches were set by the sun until the end of the 19th century. Today, sundials reappear in gardens or on the facades of houses mainly as a decorative element. Often, these sundials are not very accurate. From time to time, a look at one's watch confirms not insignificant differences. On the other hand, there are sundials whose accuracy surprise us.

There are many different types of sundials. Virtually anything casting a shadow can be made into a sundial. The aim of this chapter is to convey the mathematics which is necessary to design accurate plane sundials. These are sundials where the shadow cast by the tip of a pointer onto a plane surface marked with hour lines (the *dial*) indicates the time. The pointer is called the *gnomon*. Such sundials can not only show *local real* time but also the *mean* time, which we use in daily life, and even other time marks.

It is important to note, that to be accurate, a sundial must be specially designed for the spot it is to be used in and must also be pointed in the right direction. The algorithms written in MATLAB allow the reader to perform these calculations for his own sundial.

The chapter is based on [1] and its outline is as follows. First, the necessary astronomical fundamentals are introduced and some useful coordinate systems are defined for later use. The gnomonic projection conveys the basic understanding for all further calculations. In Section 27.3, various time marks are introduced for horizontal sundials. The restriction to horizontal dials is removed in Section 27.4. The Chapter concludes with a real example. To complete the picture we mention the related article [2].

27.2 Astronomical Fundamentals

We all know the astronomical fundamentals which are necessary to design a sundial. The Earth orbits on an elliptic path around the sun once a year (revolution) satisfying Kepler's three laws. On this path, the Earth rotates steadily around its axis (rotation), which is tilted with respect to its orbital plane around the sun. The orbital plane is called *ecliptic*. The change of day and night is due to the rotation, while the revolution gives us the seasons. Prominent points of the day are sunrise and sunset as well as the culmination, i.e., the time when the sun is in its highest position (noon).

FIGURE 27.1. *Earth's Elliptic Path*

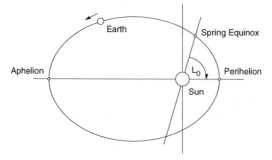

The sun is located at one focus of the Earth's elliptic path, see Figure 27.1. As the Earth travels away from the sun it slows down until it reaches the aphelion (farthest vertex). At this point the Earth is moving the slowest and is positioned at the summer solstice. The gravitational pull of the sun begins to pull the Earth back toward the sun and increases its speed. It continues to accelerate until it reaches the opposite vertex, the perihelion. At the spring equinox, the day and night become equal in length.

The astronomical constants we need, are the numerical eccentricity of the Earth's elliptic path. It has the value $e = 0.01672$. Further, we need the angle $\epsilon = 23.44°$ between ecliptic and the equatorial plane and finally the ecliptic longitude of the perihelion $L_0 = -77.11°$ counted counterclockwise starting from the spring equinox. Actually these constants are not exactly constant. For instance, the perihelion rotates around the sun once in 21'000 years. This is neglected in what follows.

27.2.1   Coordinate Systems

Various coordinate systems from the astronomers are used. All have a base of orthogonal vectors $(\mathbf{x}_a, \mathbf{y}_a, \mathbf{z}_a)$. If $\mathbf{s}$ is a vector in space then $\mathbf{s}_a$ denotes its vector representation regarding to the base $(\mathbf{x}_a, \mathbf{y}_a, \mathbf{z}_a)$.

The *horizontal coordinate system* is shown in Figure 27.2. The observer is located in the origin of a horizontal plane. The point located on the vertical above the observer is the *zenith*, the opposite on the sphere is the *nadir*. The elevation h gives the angle of a star above the horizon. Negative values refer to positions below the horizon. The azimuth a gives the angle between the south point S and the foot point of a vertical circle going through the star and the zenith.

FIGURE 27.2. *Horizontal System*

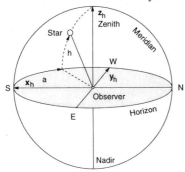

FIGURE 27.3. *Equatorial System*

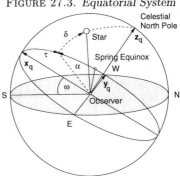

The horizontal coordinate system has the drawback that it refers to the zenith, which is not a fixed point in the cosmos. For astronomical observations it is more convenient to use the *local equatorial system*, which uses the north pole and the local equator as reference, see Figure 27.3. The hour angle τ of a star gives the angle between the meridian and the star measured on the equator in the direction of the daily motion (the meridian is the great circle from the north to the south pole). The *declination* δ of a star is its angular distance from the equator. It is positive if the star lies to the north of the equator, negative otherwise. Instead of the hour angle τ, the *right ascension* α, measured on the equator starting from the spring equinox in the positive direction, is often used.

Two additional celestial coordinate systems are convenient to describe the Earth's yearly motion around the sun. Both systems are geocentric, fade out the Earth's rotation and serve to specify the sun's position relative to the Earth. In contrast to the coordinate systems above they are independent of a specific location on the Earth.

The equinoxes are the points where the ecliptic intersects the celestial equator. The spring equinox is denoted by γ in Figure 27.4. On one hand we have the *system of the celestial equator* in which the sun's position is given by the two

FIGURE 27.4. *The Celestial Coordinate Systems*

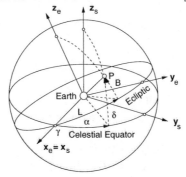

angles δ and α. The angular distance from the plane of celestial equator to the sun is the declination δ. The angle between the planes spring equinox – Earth's axis and sun – Earth's axis is the right ascension α. In the *ecliptic coordinate system* the position of the sun is given by its angle in the ecliptic counting from the spring equinox. This angle is called the ecliptic longitude L. A general point P is further specified by its ecliptic latitude B (positive in the direction of the north pole).

27.2.2   The Gnomonic Projection

Consider the shadow of a gnomon point cast onto a horizontal plane, see Figure 27.5. The direction of the shadow of the gnomon point can be expressed as

FIGURE 27.5. *Coordinate System of Dial*

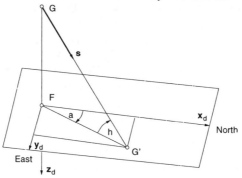

a vector $\mathbf{s}$ in the following coordinate system. Its origin is at the foot F of the perpendicular from the gnomon point G onto the plane. The $\mathbf{x}_d$-axis points to north, the $\mathbf{y}_d$-axis to east and the $\mathbf{z}_d$-axis in direction of the perpendicular. In this coordinate system, vector $\mathbf{s}$ is given as

$$\mathbf{s}_d = \begin{pmatrix} \cos(h)\cos(a) \\ \cos(h)\sin(a) \\ \sin(h) \end{pmatrix},$$

with azimuth a and altitude h. Further, the shadow G' of the gnomon point on the plane has the coordinates

$$g_1' = \frac{s_1}{s_3}g, \quad g_2' = \frac{s_2}{s_3}g, \quad g_3' = 0,$$

where g denotes the length of the gnomon, i.e., the distance GF.

Using the horizontal coordinate system of Section 27.2.1, the vector $\mathbf{s}$ can be expressed as

$$\mathbf{s}_h = \begin{pmatrix} \cos(h)\cos(a) \\ \cos(h)\sin(a) \\ \sin(h) \end{pmatrix} = -\mathbf{s}_d.$$

Consider now the task to determine the shadow of the gnomon point on the dial if the sun's position is given in the equatorial coordinate system. The sun's position $\mathbf{v}$ depending on the declination δ and the hour angle t is

$$\mathbf{v}_q = \begin{pmatrix} \cos(\delta)\cos(t) \\ \cos(\delta)\sin(t) \\ \sin(\delta) \end{pmatrix}.$$

A plane rotation is necessary to transform equatorial coordinates into horizontal coordinates, see Fig. 27.3. We have

$$\mathbf{v}_h = \begin{pmatrix} \cos(w) & 0 & -\sin(w) \\ 0 & 1 & 0 \\ \sin(w) & 0 & \cos(w) \end{pmatrix} \mathbf{v}_q,$$

where $w = 90° - \phi$ and ϕ is the observer's latitude. Thus, given the hour angle t and declination δ of the sun, the local latitude ϕ and the gnomon length g, we can calculate the shadow as given in the MATLAB procedure of Algorithm 27.1.

ALGORITHM 27.1. *Function* project

```
function project(t,d,phi,g,cmd)

X = [cos(d).*cos(t);               % rays in equatorial sys.
     cos(d).*sin(t);
     sin(d).*ones(size(t))];
w = pi/2 - phi;
R = [cos(w)  0   -sin(w);
        0    1    0;
     sin(w)  0    cos(w)];
X = R*X;                           % equatorial -> horizontal
ix = (X(3,:) > 0);                 % only rays from above
if any(ix),
   X = g*X(1:2,ix)./(ones(2,1)*X(3,ix));
                                   % shadow points
   plot(X(2,:),X(1,:),cmd);
end
```

The calculation of shadow points G' represents a point-wise mapping of the upper hemisphere onto the horizontal plane. This mapping is called the *gnomonical projection*. Great circles are mapped into straight lines, while other circles are mapped into conic sections. In particular, the equator is mapped into a straight line. For each nonzero declination, the sun's daily arcs are mapped into conic sections called the *declination lines* (the declination of the sun is practically constant along a given daily arc). The type of conic section depends on the latitude ϕ and the declination δ.

27.3   Time Marks

The previous section presented the fundamentals that allow one to determine the shadow of a gnomon point on a horizontal plane, knowing the declination δ and the hour angle t. Now various interesting time marks are presented. The hours are counted from 0 to 24 starting from midnight. Note that the hours on ancient sundials might be counted starting from the solar culmination, i.e., at noon.

27.3.1   Local Real Time

The time which is determined directly by the motion of the sun is called *local real time* or local apparent time. When the sun is in its highest point of the day, it is noon in local real time. But no two points share the same time unless they lie on the same meridian. This time is thus localized to a particular meridian. In addition, the hours of local real time are not equal in length during the year. Because of the Earth's elliptical orbit around the sun and the inclination of the Earth's axis to the ecliptic, the sun's apparent motion across the sky is not uniform throughout the year. A sundial showing local real time may run about 16 minutes fast or slow when compared with a watch, see Sec. 27.3.2.

The hour lines for local real time are especially simple. The hour angle t_r is independent of the declination and is given by

$$t_r = 15°(k - 12).$$

The hour lines are gnomonic projections of great circles and therefore are straight lines on the dial as shown in Figure 27.6. In addition, the dial in Figure 27.6 shows seven declination lines which mark the beginning of a new sign of zodiac. This is a division of the ecliptic into twelve equal sections of 30° according to an old Babylonic tradition. Table 27.1 gives the declination δ and ecliptic longitude L for the beginning of the twelve signs of zodiac. These values are easily derived using the relation between the ecliptic coordinate system and the system of the celestial equator, see Sec. 27.2.1.

TABLE 27.1. *The Signs of the Zodiac*

L	sign	δ	L	sign	δ
0°	Aries	0.0°	180°	Libra	0.0°
30°	Taurus	11.47°	210°	Scorpio	−11.47°
60°	Gemini	20.15°	240°	Sagittarius	−20.15°
90°	Cancer	23.44°	270°	Capricorn	−23.44°
120°	Leo	20.15°	300°	Aquarius	−20.15°
150°	Virgo	11.47°	330°	Pisces	−11.47°

FIGURE 27.6. *Horizontal Sundial for Real Time and Latitude +47°*

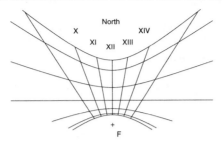

The program excerpt below uses the MATLAB procedure `project` to plot a dial for local real time given the latitude ϕ.

```
>> g = 1;  eps = 23.44;  p = pi/180;
>> clg;  axis([-5 5 -2 5]);  hold on;
>> phi = 47*p;                    % latitude
>> k = [7:17];
>> t = 15*p*(k-12);
>> d = p*[23.44 20.15 11.47 0 -11.47 -20.15 -23.44];
>> for i=1:length(d),             % declination lines
>>   project(t,d(i),phi,g,'-');
>> end
>> d = eps*p*[-1 1];
>> for k=8:16,                     % hour lines
>>   t = 15*p*(k-12);
>>   project(t,d,phi,g,'-');
>> end
```

27.3.2 Mean Time

The time we use in daily life is a *mean time*, where all hours have the same length regardless of the season. This mean time allows the rational use of clocks. The local mean time is defined by the motion of an imaginary sun progressing at a constant speed, which is just the average speed with which the real sun moves in the ecliptic, around the celestial equator instead of around the ecliptic. It leaves the spring equinox γ at the same instant as the true sun and meets it again after the lapse of a tropical year.

In order to indicate the mean time on a sundial a correction has to be applied to the real time. This correction, defined as the difference between local real time and local mean time, is called the *equation of time*. Before going into designing hour lines for the mean time the equation of time is introduced in the following two sections.

Kepler's Equation

Kepler's laws of planetary motion provide the fundamentals to introduce the mean time. The Earth's orbit is an ellipse with the sun at one focus, according to the first law, see Figure 27.7.

FIGURE 27.7. *The Elliptic Path of a Planet*

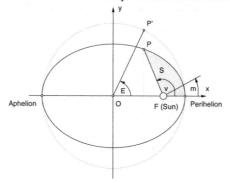

A point P on the ellipse has the coordinates

$$\mathbf{x} = (a\cos(E), \quad b\sin(E)),$$

where a and b are the larger and smaller half axis of the ellipse respectively. The angle E is called the *eccentric anomaly*. The distance of the focus F to the zero point O is ae, where e is the *numerical eccentricity*. Parameter e determines the shape of the ellipse. The Earth's orbit has nowadays the numerical eccentricity $e = 0.01672$. Further, the angle v is called the *true anomaly*. It is the polar angle for a coordinate system with F as origin. Given the coordinates of the points F and P we can calculate the length r of the vector FP. It is

$$r = a(1 - e\cos(E)), \qquad r\cos(v) = a(\cos(E) - e).$$

The first equation is obtained using the fact that the ellipse is an affine mapping of the dotted circle, and the latter is a simple comparison of distances on the x-axis. A tedious manipulation of these two equations yields a relation between v and E:

$$\tan(E/2) = \tan(v/2)\tan(\arccos(e)/2). \qquad (27.1)$$

The motion of a planet is such that the quantities v and E change irregularly. However, the radius vector FP sweeps out equal areas in equal intervals of time, according to Kepler's second law. Therefore, the hatched area S in Fig. 27.7 increases regularly. With the help of an affine mapping one finds

$$S = \frac{ab}{2}(E - e\sin(E)),$$

where E is given in radians. The quantity in parenthesis is called *mean anomaly* and is proportional to time. It is abbreviated as m. This leads us to *Kepler's equation*

$$m = E - e\sin(E),\tag{27.2}$$

which links the eccentric anomaly E with the mean anomaly m. The mean anomaly is the angular distance of a fictive planet from its perihelion as seen from the sun, where the fictive planet orbits around the sun with constant speed and the same orbiting time as the true planet.

The Equation of Time

Due to the Earth's varying speed in its orbit around the sun, which is described by Kepler's second law and the tilt of the Earth's axis relative to the ecliptic, the sun's apparent motion across the sky is not uniform throughout the year. Therefore, a correction has to be applied to the real time in order to indicate the mean time. This correction is called the equation of time.

The contribution z_k to the equation of time due to the Earth's varying speed is the difference between true and mean anomaly, i.e.,

$$z_k = m - v.$$

This quantity is best computed with respect to the ecliptic longitude L of the true sun. Remember that L is counted from spring equinox, see Sec. 27.2.1. On the other hand, the true anomaly v is counted from the Perihelion, which currently has the longitude $L_0 = -77.11°$, see Fig. 27.1. Therefore, the true anomaly v is given as

$$v = L - L_0.$$

Then, the eccentric anomaly E and the mean anomaly m are calculated using equations (27.1) and (27.2).

The second contribution to the equation of time is caused by the tilt of the Earth's axis with respect to the ecliptic. This tilt has the angle $\epsilon = 23.44°$ and points always in the same direction. If the Earth's axis was perpendicular to the ecliptic, this correction would not be necessary. Therefore, the fictive mean sun, which defines the mean time, moves around the celestial equator. Consequently, the contribution z_t to the equation of time is the difference between longitude L and right ascension α of the position of the true sun, i.e.,

$$z_t = L - \alpha.$$

The right ascension α is determined from L using the relation

$$\begin{pmatrix} 1 & 0 & 0 \\ 0 & \cos(\epsilon) & -\sin(\epsilon) \\ 0 & \sin(\epsilon) & \cos(\epsilon) \end{pmatrix} \begin{pmatrix} \cos(L) \\ \sin(L) \\ 0 \end{pmatrix} = \begin{pmatrix} \cos(\delta)\cos(\alpha) \\ \cos(\delta)\sin(\alpha) \\ \sin(\delta) \end{pmatrix}$$

for the true sun's position in the two celestial coordinate systems.

The equation of time z_g is the sum of z_k and z_t. As α makes a jump at $L=180°$ and E makes a jump at the aphelion ($v=180°$), the continuity of z_g must be forced by setting $z_g = \arctan(\tan(z_g))$. Thus, the observed value for the equation of time is

$$z_g = \arctan(\tan(z_k + z_t)).$$

Figure 27.8 shows the equation of time z_g as well as the two fractions z_k and z_t plotted against the signs of the zodiac (z_g depending on the day of the year varies slightly from year to year because of adjustments due to the leap years). While z_k is zero at the aphelion and perihelion of the Earth's orbit, z_t is in step

FIGURE 27.8. *The Equation of Time as a function of the zodiac*

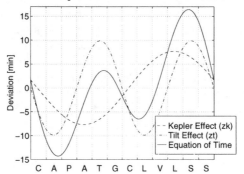

with the seasons, being zero at the equinoxes and the solstices (June 21 and December 21). Figure 27.9 shows the declination of the sun depending on the equation of time.

FIGURE 27.9. *The Analemma*

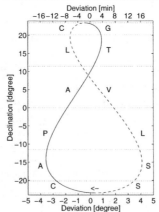

The graph looks like a figure-8 and is called the *analemma*. If one was to photograph the sun's position at the same time each day throughout the year,

the result would be an analemma. The solid part of the curve belongs to the
time from December 21 to June 21, while the dashed part belongs to the other
half of the year. The dotted declination lines mark the beginning of a new sign
of zodiac.

Local Mean Time

The time defined by the hour angle of the fictive mean sun introduced in the
previous section, is called the *local mean time*. That is, the local mean time T_m
is the local real time T_r subtracted by the equation of time:

$$T_m = T_r - z_g.$$

Consequently the hour angle t_m of the kth hour local mean time is

$$t_m = 15°(k - 12) + z_g.$$

By marking each hour of mean time by an analemma, the mean time can be
read directly from the dial without the need to apply an additional correction.
The following program excerpt first computes the declination of the sun and
the equation of time during the course of a year. With this information the
analemma for the kth hour is easily plotted.

```
>> L0 = -77.11;  e = 0.01672;  eps = 23.44;  p  = pi/180;
>> L  = 3*p*[-30:90];                % Sun's True Longitude
>> v  = L - L0*p;                     % True Anomaly
>> c  = sqrt((1-e)/(1+e));            % c=tan(arccos(e)/2)
>> E  = 2*atan(c*tan(v/2));           % Eccentric Anomaly
>> zk = E-e*sin(E)-v;
>> x  = [cos(L);                      % Sun's Coordinates in System
>>        sin(L)*cos(eps*p);          % of Celestial Equator
>>        sin(L)*sin(eps*p)];
>> r  = sqrt(x(1,:).^2+x(2,:).^2);
>> al = atan2(x(2,:),x(1,:));         % Right Ascension
>> d  = atan2(x(3,:),r);              % Declination
>> zt = L-al;
>> zg = atan(tan(zk+zt));             % Equation of Time
>> for k = 9:15,                      % Individual Hour Lines
>>    t  = 15*p*(k-12)+zg;
>>    project(t,d,phi,g,'-y');
>> end;
```

Zone Time

The local times discussed so far are astronomically correct for one meridian. A
sundial located farther west is behind a local one. But a system of timekeeping,
which is so narrowly localized, is impractical in daily life. Therefore, *time zones*
or standard times were introduced at the end of the 19th century. The Earth is

divided into 24 time zones, each approximately $15°$ of longitude apart beginning at the prime meridian. Time increases one hour for each zone east of the prime meridian where time is called Greenwich Mean Time (GMT) and decreases for each zone west of GMT.

The zone time differs from the local mean time by the difference in longitude between the local meridian λ and the meridian of the time zone λ_0. Thus, the hour angle for the kth hour zone time is

$$t_z = 15°(k - 12) + z_g + (\lambda_0 - \lambda).$$

27.3.3 Babylonic and Italic Hours

The people of Babylon, to whom we owe the division of the day into 24 hours, used to count the hours starting from sunrise. In contrast, the Italians of the Middle Ages started a new day at sunset. Both of these systems of timekeeping have been in use for a very long time. Countless old European dials are to be found even now, which refer to Babylonic or Italic hours. Let us study dials with these hour marks, although they are not used today anymore.
Let

$$\mathbf{u}_q = \begin{pmatrix} \cos(\delta)\cos(\tau) \\ \cos(\delta)\sin(\tau) \\ \sin(\delta) \end{pmatrix}$$

be the position of the sun at sunset. Vector $\mathbf{u}$ is in the horizontal plane. Thus

$$\mathbf{u}_h = \begin{pmatrix} \cos(w) & 0 & -\sin(w) \\ 0 & 1 & 0 \\ \sin(w) & 0 & \cos(w) \end{pmatrix} \mathbf{u}_q,$$

with $w = 90° - \phi$ and latitude ϕ, must have a zero z-component. This leads to the equation

$$\cos(\tau) = -\tan(\delta)\tan(\phi)$$

for the hour angle at sunset τ. The negative solution, i.e., $t = -\tau$ is the hour angle for the sunrise. Note that τ also is half the length of the light day. The hour angles for the bth Babylonic and ith Italic hours are

$$t_b = 15°b - \tau \qquad \text{and} \qquad t_i = 15°i + \tau$$

respectively. The hour angle t depends not only on b (or i respectively) but also on the declination of the sun. Remember the gnomonic projection: the Babylonic as well as the Italic hour lines are projections of great circles and therefore are straight lines. Further, the intersection of the bth Babylonic and the ith Italic hour line lies on the hour line $t_r = (b + i)/2$ real time.

For latitudes $\phi > |90° - \epsilon|$, Babylonic and Italic hours are only defined for declinations for which there is neither polar day nor polar night. That means, they are only defined for $\delta \in [-\delta_0, \delta_0]$ with

$$\delta_0 = \begin{cases} \epsilon & : \ |\phi| + \epsilon \leq 90° \\ 90° - \phi & : \ \text{otherwise.} \end{cases}$$

The program excerpt below plots Babylonic hour lines. Three points are calculated per hour line which must lie on a straight line.

```
>> % Babylonic hours
>> if (abs(phi)+eps*p > pi/2),    % declination restricted above
>>    d0 = pi/2-phi;              %  polar circle
>> else
>>    d0 = eps*p;
>> end
>> s0 = acos(-tan(d0)*tan(phi));
>> k3 = fix(s0/7.5/p);            % number of hours for longest day
>> d = d0*[-1:1];
>> s = acos(-tan(d)*tan(phi));
>> for k=0:k3,
>>    t = 15*k*p-s;
>>    project(t,d,phi,g,'-.r');
>> end
```

A dial with Babylonic and Italic hour lines is given in Figure 27.11.

27.4   Sundials on General Planes

So far, we discussed sundials on horizontal planes. It is not difficult however to generalize them to sundials on planes with any orientation. The orientation of a general plane is defined for instance by the azimuth a and elevation h of the perpendicular $\mathbf{n}$ onto the plane, see Figure 27.10.

FIGURE 27.10. *Horizontal and General Plane*

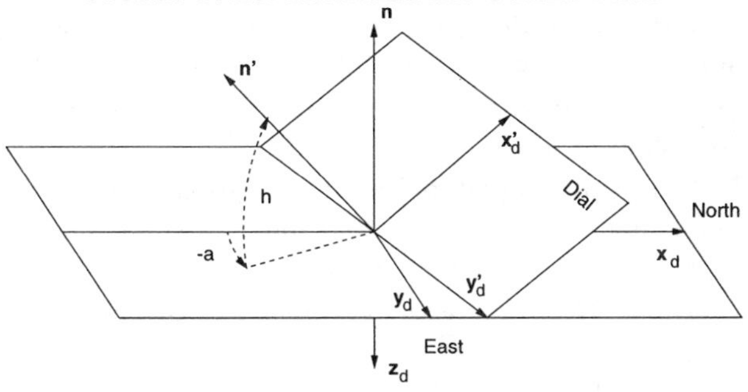

There are two different approaches to calculate the dial for this general plane. On the one hand, two rotations transform the coordinate system of the horizontal dial $(\mathbf{x}_d, \mathbf{y}_d, \mathbf{z}_d)$ into the coordinate system of the dial in the general plane $(\mathbf{x}'_d, \mathbf{y}'_d, \mathbf{z}'_d)$. On the other hand, we can determine a position on the Earth, for which the direction of the azimuth is parallel to the perpendicular of the desired plane. Then we can calculate a horizontal sundial for this new position. This approach requires some spherical trigonometry and modifications in several places of the existing programs would be necessary [1]. Let us take the first approach.

A sun-ray $\mathbf{v}_d$ in the local coordinate system of the horizontal dial is transformed to a sun-ray $\mathbf{v}'_d$ in the coordinate system of the general plane by the two rotations. In detail, vector $\mathbf{v}'_d$ is determined by

$$
\mathbf{v}'_d = \begin{pmatrix} \cos(b) & 0 & -\sin(b) \\ 0 & 1 & 0 \\ \sin(b) & 0 & \cos(b) \end{pmatrix} \begin{pmatrix} \cos(a) & \sin(a) & 0 \\ -\sin(a) & \cos(a) & 0 \\ 0 & 0 & 1 \end{pmatrix} \mathbf{v}_d,
$$

where a is the azimuth measured clockwise from the south point and $b = 90° - h$ with elevation h. The corresponding modification of the procedure `project` is straightforward and is left to the reader.

27.5   A Concluding Example

The code fragments given in this Chapter are easily combined into a MATLAB program, which calls the procedure `project` for plotting the dial. An appropriately magnified output can serve as a draft for a skilled painter to make an accurate sundial. As input the program requires the precise geographic position and orientation of the planed sundial as well as its desired size, what determines the length of the gnomon. The longitude of the time zone's meridian λ_0 is further needed if the dial shall also show zone time. The geographical parameters have to be determined with care to guarantee the accuracy. A final difficulty is the correct placement of the gnomon. Only the tip of the gnomon is used for reading the time. It must be placed in the predetermined distance above the zero point of the dial's coordinate system. The gnomon itself may have any orientation but should be parallel to the Earth's axis if the additional reading of the local real time is desired. More detailed instructions for setting up the sundial may be found in [1].

Figure 27.11 illustrates a dial for an almost vertical wall in Vingelz, in the western part of Switzerland. See Table 27.2 for the parameters of the sundial.

TABLE 27.2. *Geographical Position and Orientation*

Longitude:	−7°13′01″	Azimuth :	−48.95°
Latitude :	47°07′48″	Elevation:	5.71°

The dial indicates Central European Time (CET) as well as Babylonic and Italic hours. Even local real time can be read, as the gnomon is parallel to Earth's axis. The net of straight lines behind the analemmas shows the Babylonic and Italic hours. Only every second line is marked with an Arabic number. The lines 1, 3, 5, 7 indicate how many hours have past since sunrise (Babylonic hours) while the lines 8, 10, 12, 14 show how many hours are left until sunset (Italic hours).

Watch at the center of the filled circle to read the time. It represents the shadow of the tip of the gnomon.

FIGURE 27.11. *Sundial at Vingelz, Switzerland*

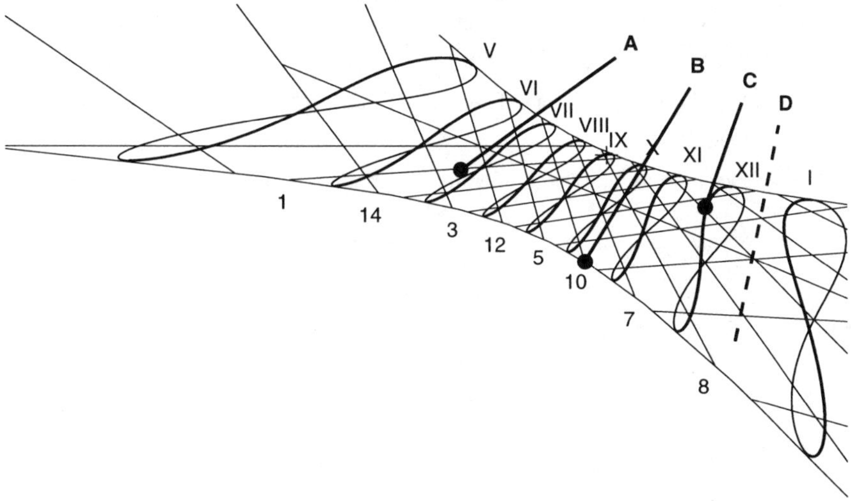

A: How many hours past since sunrise? Watch at the slightly inclined lines from lower left to upper right. Example A shows the shadow either on April 20 at 6 h 40 CET or on 23. August 6 h 45 CET. It is one hour after sunrise.

B: How long does it last until sunset? Watch at the steep lines from upper left to lower right. Example B shows the shadow on June 21 at 10 h 20 CET: ten hours are left until sunset.

C: What time is it? Watch at the analemmas with the Roman numbers. When the shadow casts onto the line of an analemma, it is exactly a complete hour CET. The thicker part of the analemma is good for the first half of the year (December 21 - June 21) and the thinner part for the second half of the year (June 21 - December 21). Example C shows the shadow on February 19 at 12 h CET.

D: Local real time. The intersection between Babylonic and Italic hour lines can be used to determine the local real time. Just watch at the shadow of the gnomon for this. Example D shows the shadow at noon 12 h local real time. The sun is at its highest point of the day for Vingelz.

References

[1] H. Schilt: "Ebene Sonnenuhren: verstehen und planen, berechnen und bauen",
9. Auflage, Felix Solis Tempus, Basel, 1997.

[2] Ch. Blatter: "Von den Keplerschen Gesetzen zu einer minutengenauen Sonnenuhr", Elemente der Mathematik (49), 155–165, 1994.

[3] R. Rohr: "Sundials: History, Theory, and Practice", University of Toronto Press, Toronto, 1970.

[4] A. Zenkert: "Faszination Sonnenuhr", Verlag Harri Deutsch, 2. Auflage, Thun, 1995.

Index

Index of used MAPLE Commands

Index of used MATLAB Commands

Computation
in mathematics

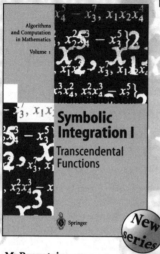

M. Bronstein
Symbolic Integration I
Transcendental Functions
(Algorithms and Computation in
Mathematics, Vol. 1)

1997. XIII, 299 pages, 3 figures.
Hardcover DM 78,–
ISBN 3-540-60521-5

This first volume in the series *Algorithms
and Computation in Mathematics*, is des-
tined to become the standard reference
work in the field. Professor Bronstein is
the number-one expert on this topic and
his book is the first to treat the subject
both comprehensively and in sufficient
detail - incorporating new results along
the way. The book addresses mathematicians and computer scientists interested
in symbolic computation, developers and
programmers of computer algebra systems as well as users of symbolic integration methods. Many algorithms are given
in pseudocode ready for immediate
implementation, making the book equally
suitable as a textbook for lecture courses
on symbolic integration.

F. Cucker, M. Shub, (Eds.)
Foundations of
Computational Mathematics
Selected papers of a conference held
at IMPA in Rio de Janeiro,
January 1997

1997. XV, 389 pages, 31 figures, 8 tabels
Softcover DM 128,–
ISBN 3-540-61647-0

FoCM brings together a novel constella-
tion of subjects in which the computational process itself and the foundational
mathematical underpinnings of algorithms are the objects of study. The conference was organized around nine work-
shops: systems of algebraic equations and
computational algebraic geometry, homo-
topy methods and real machines, information based complexity, numerical linear
algebra, approximation and PDE's, optimization, differential equations and
dynamical systems, relations to computer
science and vision and related computational tools.

Prices subject to change without notice.
In EU countries the local VAT is effective.

Please order by
Fax: +49 -30- 827 87- 301
e-mail: orders@springer.de
or through your bookseller

■ ■ ■ ■ ■ ■ ■ ■ ■ ■

Springer

Springer-Verlag, P. O. Box 31 13 40, D-10643 Berlin, Germany.

Springer
and the
environment

At Springer we firmly believe that an
international science publisher has a
special obligation to the environment,
and our corporate policies consistently
reflect this conviction.
We also expect our business partners –
paper mills, printers, packaging
manufacturers, etc. – to commit
themselves to using materials and
production processes that do not harm
the environment. The paper in this
book is made from low- or no-chlorine
pulp and is acid free, in conformance
with international standards for paper
permanency.